Nuclear Power: An Introduction

Table of Contents

Preface

One of the most path-breaking and controversial inventions of recent scientific history is nuclear power. It refers to the production of nuclear energy through nuclear fission and fusion. This energy is then used to produce electricity in nuclear power plants. This book is compiled in such a manner, that it will provide in-depth knowledge about the theory and practice of nuclear power. The topics included in this book are of utmost significance and bound to provide incredible insights to readers. This book is a complete source of knowledge on the present status of this important field. Those in search of information to further their knowledge will be greatly assisted by this textbook.

To facilitate a deeper understanding of the contents of this book a short introduction of every chapter is written below:

Chapter 1- Nuclear power is produced by nuclear reactions like nuclear fission, nuclear fusion and nuclear decay that create heat which is used by thermo-nuclear reactors to generate electricity. Nuclear power is clean power and uses very little fuel for producing large amounts of electricity. Most reactors are nuclear fission reactors. This introductory chapter provides exhaustive information on nuclear power and fusion power.

Chapter 2- A nuclear power plant uses energy produced by nuclear reactions to convert water into steam, which, in turn is used to run steam turbines to produce electricity. The radioactive fuel undergoes nuclear reaction in the reactor. There is a nuclear reactor coolant that is circulated to keep the temperature optimal and controlled. This chapter illustrates the various parts of a thermonuclear reactor and discusses the function and design of each.

Chapter 3- With the passage of time, nuclear reactors have evolved and have incorporated cutting-edge technology to improve thermal efficiency as well as reduce maintenance and optimize fuel technology. Based on their evolutionary improvements, the reactors have been arranged into generations. In this chapter, the reader is informed about Generations II, III and IV with differentiating characteristics and design of each generation.

Chapter 4- Nuclear reactions are those that involve the collision of two nuclei or the bombardment of a nucleus with subatomic particles like neutrons, protons or high energy electrons to produce one or more lighter nuclei. Nuclear reactions occur naturally when matter interacts with cosmic rays and can also be artificially induced in nuclear reactors. This chapter comprehensively studies nuclear reactions and the topics of energy conversion and nuclear binding energy.

Chapter 5- The radioactive isotopes used and produced in nuclear reactors make it imperative that there exist high standards of safety and security within the plant and in its operations. There have been several nuclear plant accidents that have led to stringent rules concerning the storage, usage and decommissioning of nuclear plant products and waste. This chapter highlights the efforts made in the direction of nuclear safety and security, nuclear reprocessing and nuclear decommissioning.

I owe the completion of this book to the never-ending support of my family, who supported me throughout the project.

Editor

Introduction to Nuclear Power

Nuclear power is produced by nuclear reactions like nuclear fission, nuclear fusion and nuclear decay that create heat which is used by thermo-nuclear reactors to generate electricity. Nuclear power is clean power and uses very little fuel for producing large amounts of electricity. Most reactors are nuclear fission reactors. This introductory chapter provides exhaustive information on nuclear power and fusion power.

The 1200 MWe, Leibstadt fission-electric power station in Switzerland. The boiling water reactor (BWR), located inside the dome capped cylindrical structure, is dwarfed in size by its cooling tower. The station produces a yearly average of 25 million kilowatt-hours per day, sufficient to power a city the size of Boston.

The Palo Verde Nuclear Generating Station, the largest in the US with 3 pressurized water reactors(PWRs), is situated in the Arizona desert. It uses sewage from cities as its cooling water in 9 squat mechanical draft cooling towers. Its total spent fuel/"waste" inventory produced since 1986, is contained in dry cask storage cylinders located between the artificial body of water and the electrical switchyard.

U.S. nuclear powered ships,(top to bottom) cruisers USS *Bainbridge*, the USS *Long Beach* and the *USS Enterprise*, the longest ever naval vessel, and the first nuclear-powered aircraft carrier. Picture taken in 1964 during a record setting voyage of 26,540 nmi (49,190 km) around the world in 65 days without refueling. Crew members are spelling out Einstein's mass-energy equivalence formula $E = mc^2$ on the flight deck.

Nuclear power is the use of nuclear reactions that release nuclear energy to generate heat, which most frequently is then used in steam turbines to produce electricity in a nuclear power plant. The term includes nuclear fission, nuclear decay and nuclear fusion. Presently, the nuclear fission of elements in the actinide series of the periodic table produce the vast majority of nuclear energy in the direct service of humankind, with nuclear decay processes, primarily in the form of geothermal energy, and radioisotope thermoelectric generators, in niche uses making up the rest.

Fission-electric power stations are one of the leading low carbon power generation methods of producing electricity, and in terms of total life-cycle greenhouse gas emissions per unit of energy generated, has emission values lower than "renewable energy" when the latter is taken as a single energy source. As all electricity supplying technologies use cement etc., during construction, emissions are yet to be brought to zero. A 2014 analysis of the carbon footprint literature by the Intergovernmental Panel on Climate Change (IPCC) reported that the embodied total life-cycle emission intensity of fission electricity has a median value of 12 g CO_2 eq/kWh which is the lowest out of all commercial baseload energy sources, and second lowest out of all commercial electricity technologies known, after wind power which is an Intermittent energy source with embodied greenhouse gas emissions, per unit of energy generated of 11 g CO_2eq/kWh. Each result is contrasted with coal & fossil gas at 820 and 490 g CO_2 eq/kWh. With this translating into, from the beginning of Fission-electric power station commercialization in the 1970s, having prevented the emission of about 64 billion tonnes of carbon dioxide equivalent, greenhouse gases that would have otherwise resulted from the burning of fossil fuels in thermal power stations.

There is a social debate about nuclear power. Proponents, such as the World Nuclear Association and Environmentalists for Nuclear Energy, contend that nuclear power is a safe, sustainable energy source that reduces carbon emissions. Opponents, such as Greenpeace International and NIRS, contend that nuclear power poses many threats to people and the environment.

Far-reaching fission power reactor accidents, or accidents that resulted in medium to long-lived fission product contamination of inhabited areas, have occurred in Generation I & II reactor designs, blueprinted between 1950 and 1980. These include the Chernobyl disaster which occurred in 1986, the Fukushima Daiichi nuclear disaster (2011), and the more contained Three Mile Island accident (1979). There have also been some nuclear submarine accidents. In terms of lives lost per unit of energy generated, analysis has determined that fission-electric reactors have caused fewer fatalities per unit of energy generated than the other major sources of energy generation. Energy production from coal, petroleum, natural gas and hydroelectricity has caused a greater number of fatalities per unit of energy generated due to air pollution and energy accident effects.

Four years after the Fukushima-Daiichi accident, there have been no fatalities due to exposure to radiation, and no discernible increased incidence of radiation-related health effects are expected among exposed members of the public and their descendants. The Japan Times estimated 1,600 deaths were the result of evacuation, due to physical and mental stress stemming from long stays at shelters, a lack of initial care as a result of hospitals being disabled by the tsunami, and suicides.

In 2015:

- Ten new reactors were connected to the grid.

- Seven reactors were permanently shut down.

- 441 reactors had a worldwide net capacity of 382,855 megawatts of electricity.

- 67 new nuclear reactors were under construction.

Most of the new activity is in China where there is an urgent need to control pollution from coal plants.

In October 2016, Watts Bar 2 became the first new United States reactor to enter commercial operation since 1996.

History

Origins

In 1932 physicist Ernest Rutherford discovered that when lithium atoms were "split" by protons from a proton accelerator, immense amounts of energy were released in accordance with the principle of mass–energy equivalence. However, he and other nuclear physics pioneers Niels Bohr and Albert Einstein believed harnessing the power of the atom for practical purposes anytime in the near future was unlikely, with Rutherford labeling such expectations "moonshine."

The same year, his doctoral student James Chadwick discovered the neutron, which was immediately recognized as a potential tool for nuclear experimentation because of its lack of an electric charge. Experimentation with bombardment of materials with neutrons led Frédéric and Irène Joliot-Curie to discover induced radioactivity in 1934, which allowed the creation of radium-like elements at much less the price of natural radium. Further work by Enrico Fermi in the 1930s focused on using slow neutrons to increase the effectiveness of induced radioactivity. Experiments bombarding uranium with neutrons led Fermi to believe he had created a new, transuranic element, which was dubbed hesperium.

December 2, 1942. A depiction of the scene when scientists observed the world's first man made nuclear reactor, the Chicago Pile-1, as it became self-sustaining/critical at the University of Chicago.

10Lise Meitner and Meitner's nephew, Otto Robert Frisch, conducted experiments with the products of neutron-bombarded uranium, as a means of further investigating Fermi's claims. They determined that the relatively tiny neutron split the nucleus of the massive uranium atoms into two roughly equal pieces, contradicting Fermi. This was an extremely surprising result: all other forms of nuclear decay involved only small changes to the mass of the nucleus, whereas this process—dubbed "fission" as a reference to biology—involved a complete rupture of the nucleus. Numerous scientists, including Leó Szilárd, who was one of the first, recognized that if fission reactions released additional neutrons, a self-sustaining nuclear chain reaction could result. Once this was experimentally confirmed and announced by Frédéric Joliot-Curie in 1939, scientists in many countries (including the United States, the United Kingdom, France, Germany, and the Soviet Union) petitioned their governments for support of nuclear fission research, just on the cusp of World War II, for the development of a nuclear weapon.

In the United States, where Fermi and Szilárd had both emigrated, this led to the creation of the first man-made reactor, known as Chicago Pile-1, which achieved criticality on December 2, 1942. This work became part of the Manhattan Project, which made enriched uranium and built large reactors to breed plutonium for use in the first nuclear weapons, which were used on the cities of Hiroshima and Nagasaki.

The first light bulbs ever lit by electricity generated by nuclear power at EBR-1 at Argonne National Laboratory-West, December 20, 1951.

In 1945, the pocketbook *The Atomic Age* heralded the untapped atomic power in everyday objects and depicted a future where fossil fuels would go unused. One science writer, David Dietz, wrote that instead of filling the gas tank of your car two or three times a week, you will travel for a year on a pellet of atomic energy the size of a vitamin pill. Glenn Seaborg, who chaired the Atomic Energy Commission, wrote "there will be nuclear powered earth-to-moon shuttles, nuclear powered artificial hearts, plutonium heated swimming pools for SCUBA divers, and much more". These overly optimistic predications remain unfulfilled.

United Kingdom, Canada, and USSR proceeded over the course of the late 1940s and early 1950s. Electricity was generated for the first time by a nuclear reactor on December 20, 1951, at the EBR-I experimental station near Arco, Idaho, which initially produced about 100 kW. Work was also strongly researched in the US on nuclear marine propulsion, with a test reactor being developed by 1953 (eventually, the USS Nautilus, the first nuclear-powered submarine, would launch in 1955). In 1953, US President Dwight Eisenhower gave his "Atoms for Peace" speech at the United Nations, emphasizing the need to develop "peaceful" uses of nuclear power quickly. This was followed by the 1954 Amendments to the Atomic Energy Act which allowed rapid declassification of U.S. reactor technology and encouraged development by the private sector.

Early Years

On June 27, 1954, the USSR's Obninsk Nuclear Power Plant became the world's first nuclear power plant to generate electricity for a power grid, and produced around 5 megawatts of electric power.

Later in 1954, Lewis Strauss, then chairman of the United States Atomic Energy Commission (U.S. AEC, forerunner of the U.S. Nuclear Regulatory Commission and the United States Department of Energy) spoke of electricity in the future being "too cheap to meter". Strauss was very likely referring to hydrogen fusion —which was secretly being developed as part of Project Sherwood at the time—but Strauss's statement was interpreted as a promise of very cheap energy from nuclear fission. The U.S. AEC itself had issued far more realistic testimony regarding nuclear fission to the U.S. Congress only months before, projecting that "costs can be brought down... [to]... about the same as the cost of electricity from conventional sources..."

In 1955 the United Nations' "First Geneva Conference", then the world's largest gathering of scientists and engineers, met to explore the technology. In 1957 EURATOM was launched alongside the European Economic Community (the latter is now the European Union). The same year also saw the launch of the International Atomic Energy Agency (IAEA).

Calder Hall, United Kingdom - The world's first commercial nuclear power station. First connected to the national power grid on 27 August 1956 and officially opened by Queen Elizabeth II on 17 October 1956

The Shippingport Atomic Power Station in Shippingport, Pennsylvania was the first commercial reactor in the USA and was opened in 1957.

The world's first commercial nuclear power station, Calder Hall at Windscale, England, was opened in 1956 with an initial capacity of 50 MW (later 200 MW). The first commercial nuclear generator to become operational in the United States was the Shippingport Reactor (Pennsylvania, December 1957).

One of the first organizations to develop nuclear power was the U.S. Navy, for the purpose of propelling submarines and aircraft carriers. The first nuclear-powered submarine, USS *Nautilus* (SSN-571), was put to sea in December 1954. As of 2016, the U.S. Navy submarine fleet is made up entirely of nuclear-powered vessels, with 75 submarines in service. Two U.S. nuclear submarines, USS *Scorpion* and USS *Thresher*, have been lost at sea. The Russian Navy is currently (2016) estimated to have 61 nuclear submarines in service; eight Soviet and Russian nuclear submarines have been lost at sea. This includes the Soviet submarine K-19 reactor accident in 1961 which resulted in 8 deaths and more than 30 other people were over-exposed to radiation. The Soviet submarine K-27 reactor accident in 1968 resulted in 9 fatalities and 83 other injuries. Moreover, Soviet submarine K-429 sank twice, but was raised after each incident. Several serious nuclear and radiation accidents have involved nuclear submarine mishaps.

The U.S. Army also had a nuclear power program, beginning in 1954. The SM-1 Nuclear Power Plant, at Fort Belvoir, Virginia, was the first power reactor in the U.S. to supply electrical energy to a commercial grid (VEPCO), in April 1957, before Shippingport. The SL-1 was a U.S. Army experimental nuclear power reactor at the National Reactor Testing Station in eastern Idaho. It underwent a steam explosion and meltdown in January 1961, which killed its three operators. In Soviet Union in The Mayak Production Association there were a number of accidents including an explosion that released 50-100 tonnes of high-level radioactive waste, contaminating a huge territory in the eastern Urals and causing numerous deaths and injuries. The Soviet regime kept this accident secret for about 30 years. The event was eventually rated at 6 on the seven-level INES scale (third in severity only to the disasters at Chernobyl and Fukushima).

Development

Installed nuclear capacity initially rose relatively quickly, rising from less than 1 gigawatt (GW) in 1960 to 100 GW in the late 1970s, and 300 GW in the late 1980s. Since the late 1980s worldwide capacity has risen much more slowly, reaching 366 GW in 2005. Between around 1970 and 1990, more than 50 GW of capacity was under construction (peaking at over 150 GW in the late 1970s and early 1980s) — in 2005, around 25 GW of new capacity was planned. More than two-thirds of all nuclear plants ordered after January 1970 were eventually cancelled. A total of 63 nuclear units were canceled in the USA between 1975 and 1980.

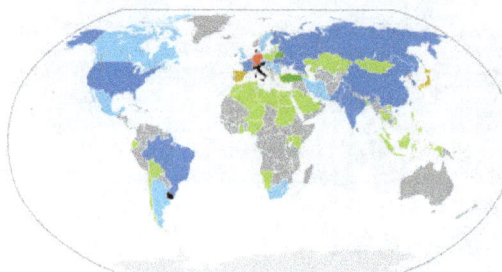

The status of nuclear power globally
(click image for legend)

During the 1970s and 1980s rising economic costs (related to extended construction times largely due to regulatory changes and pressure-group litigation) and falling fossil fuel prices made nuclear power plants then under construction less attractive. In the 1980s (U.S.) and 1990s (Europe), flat load growth and electricity liberalization also made the addition of large new baseload capacity unattractive.

Washington Public Power Supply System Nuclear Power Plants 3 and 5 were never completed.

The 1973 oil crisis had a significant effect on countries, such as France and Japan, which had relied more heavily on oil for electric generation (39% and 73% respectively) to invest in nuclear power.

Some local opposition to nuclear power emerged in the early 1960s, and in the late 1960s some members of the scientific community began to express their concerns. These concerns related to nuclear accidents, nuclear proliferation, high cost of nuclear power plants, nuclear terrorism and radioactive waste disposal. In the early 1970s, there were large protests about a proposed nuclear power plant in Wyhl, Germany. The project was cancelled in 1975 and anti-nuclear success at Wyhl inspired opposition to nuclear power in other parts of Europe and North America. By the mid-1970s anti-nuclear activism had moved beyond local protests and politics to gain a wider appeal and influence, and nuclear power became an issue of major public protest. Although it lacked a single co-ordinating organization, and did not have uniform goals, the movement's efforts gained a great deal of attention. In some countries, the nuclear power conflict "reached an intensity unprecedented in the history of technology controversies".

120,000 people attended an anti-nuclear protest in Bonn, Germany, on October 14, 1979, following the Three Mile Island accident.

In France, between 1975 and 1977, some 175,000 people protested against nuclear power in ten demonstrations. In West Germany, between February 1975 and April 1979, some 280,000 people were involved in seven demonstrations at nuclear sites. Several site occupations were also attempted. In the aftermath of the Three Mile Island accident in 1979, some 120,000 people attended a demonstration against nuclear power in Bonn. In May 1979, an estimated 70,000 people, including then governor of California Jerry Brown, attended a march and rally against nuclear power in Washington, D.C. Anti-nuclear power groups emerged in every country that has had a nuclear power programme.

Three Mile Island and Chernobyl

Health and safety concerns, the 1979 accident at Three Mile Island, and the 1986 Chernobyl disaster played a part in stopping new plant construction in many countries, although the public policy organization, the Brookings Institution states that new nuclear units, at the time of publishing in 2006, had not been built in the U.S. because of soft demand for electricity, and cost overruns on nuclear plants due to regulatory issues and construction delays. By the end of the 1970s it became clear that nuclear power would not grow nearly as dramatically as once believed. Eventually, more than 120 reactor orders in the U.S. were ultimately cancelled and the construction of new reactors ground to a halt. A cover story in the February 11, 1985, issue of *Forbes* magazine commented on the overall failure of the U.S. nuclear power program, saying it "ranks as the largest managerial disaster in business history".

The abandoned city of Pripyat with Chernobyl plant in the distance.

Unlike the Three Mile Island accident, the much more serious Chernobyl accident did not increase regulations affecting Western reactors since the Chernobyl reactors were of the problematic RBMK design only used in the Soviet Union, for example lacking "robust" containment buildings. Many of these RBMK reactors are still in use today. However, changes were made in both the reactors themselves (use of a safer enrichment of uranium) and in the control system (prevention of disabling safety systems), amongst other things, to reduce the possibility of a duplicate accident.

An international organization to promote safety awareness and professional development on operators in nuclear facilities was created: WANO; World Association of Nuclear Operators.

Opposition in Ireland and Poland prevented nuclear programs there, while Austria (1978), Sweden (1980) and Italy (1987) (influenced by Chernobyl) voted in referendums to oppose or phase out nuclear power. In July 2009, the Italian Parliament passed a law that cancelled the results of an earlier referendum and allowed the immediate start of the Italian nuclear program. After the

Fukushima Daiichi nuclear disaster a one-year moratorium was placed on nuclear power development, followed by a referendum in which over 94% of voters (turnout 57%) rejected plans for new nuclear power.

Nuclear Renaissance

Olkiluoto 3 under construction in 2009. It is the first EPR design, but problems with workmanship and supervision have created costly delays which led to an inquiry by the Finnish nuclear regulator STUK. In December 2012, Areva estimated that the full cost of building the reactor will be about €8.5 billion, or almost three times the original delivery price of €3 billion.

Since about 2001 the term *nuclear renaissance* has been used to refer to a possible nuclear power industry revival, driven by rising fossil fuel prices and new concerns about meeting greenhouse gas emission limits. In 2012, the World Nuclear Association reported that nuclear electricity generation was at its lowest level since 1999. As of January 2016, however, 65 new nuclear power reactors were under construction. Over 150 were planned, equivalent to nearly half of capacity at that time.

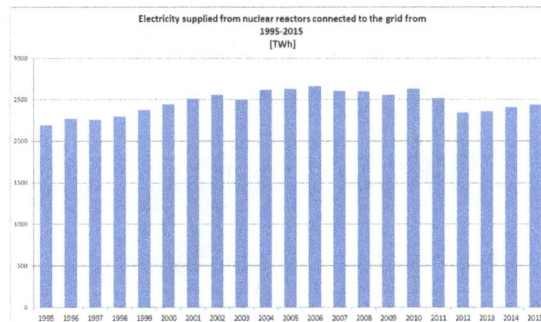

Production of nuclear power plants

Fukushima Daiichi Nuclear Disaster

Japan's 2011 Fukushima Daiichi nuclear accident, which occurred in a reactor design from the 1960s, prompted a re-examination of nuclear safety and nuclear energy policy in many countries. Germany plans to close all its reactors by 2022, and Italy has re-affirmed its ban on electric utilities generating, but not importing, fission derived electricity. In 2011 the International Energy Agency halved its prior estimate of new generating capacity to be built by 2035. In 2013 Japan signed a deal worth $22 billion, in which Mitsubishi Heavy Industries would build four modern *Atmea* reactors for Turkey. In August 2015, following 4 years of near zero fission-electricity generation, Japan began restarting its fission fleet, after safety upgrades were completed, beginning with Sendai fission-electric station.

In March 2011 the nuclear emergencies at Japan's Fukushima Daiichi Nuclear Power Plant and shutdowns at other nuclear facilities raised questions among some commentators over the future of the renaissance. China, Germany, Switzerland, Israel, Malaysia, Thailand, United Kingdom, Italy and the Philippines have reviewed their nuclear power programs. Indonesia and Vietnam still plan to build nuclear power plants.

The World Nuclear Association has said that "nuclear power generation suffered its biggest ever one-year fall through 2012 as the bulk of the Japanese fleet remained offline for a full calendar year". Data from the International Atomic Energy Agency showed that nuclear power plants globally produced 2346 TWh of electricity in 2012 – seven per cent less than in 2011. The figures illustrate the effects of a full year of 48 Japanese power reactors producing no power during the year. The permanent closure of eight reactor units in Germany was also a factor. Problems at Crystal River, Fort Calhoun and the two San Onofre units in the USA meant they produced no power for the full year, while in Belgium Doel 3 and Tihange 2 were out of action for six months. Compared to 2010, the nuclear industry produced 11% less electricity in 2012.

Post-fukushima Controversy

Eight of the seventeen operating reactors in Germany were permanently shut down as part of Germany's *Energiewende*.

The Fukushima Daiichi nuclear accident sparked controversy about the importance of the accident and its effect on nuclear's future. IAEA Director General Yukiya Amano said the Japanese nuclear accident "caused deep public anxiety throughout the world and damaged confidence in nuclear power", and the International Energy Agency halved its estimate of additional nuclear generating capacity to be built by 2035. But by 2015, the Agency's outlook had become more promising. "Nuclear power is a critical element in limiting greenhouse gas emissions," the agency noted, and "the prospects for nuclear energy remain positive in the medium to long term despite a negative impact in some countries in the aftermath of the [Fukushima-Daiichi] accident...it is still the second-largest source worldwide of low-carbon electricity. And the 72 reactors under construction at the start of last year were the most in 25 years." Though Platts reported in 2011 that "the crisis at Japan's

Fukushima nuclear plants has prompted leading energy-consuming countries to review the safety of their existing reactors and cast doubt on the speed and scale of planned expansions around the world", Progress Energy Chairman/CEO Bill Johnson made the observation that "Today there's an even more compelling case that greater use of nuclear power is a vital part of a balanced energy strategy". In 2011, *The Economist* opined that nuclear power "looks dangerous, unpopular, expensive and risky", and that "it is replaceable with relative ease and could be forgone with no huge structural shifts in the way the world works". Earth Institute Director Jeffrey Sachs disagreed, claiming combating climate change would require an expansion of nuclear power. "We won't meet the carbon targets if nuclear is taken off the table," he said. "We need to understand the scale of the challenge."

Investment banks were critical of nuclear soon after the accident. Many disputed their impartiality, however, due to significant investments in renewable energy, perceived by some as a valid alternative to nuclear. In early April 2011, analysts at Swiss-based investment bank UBS said: "At Fukushima, four reactors have been out of control for weeks, casting doubt on whether even an advanced economy can master nuclear safety...we believe the Fukushima accident was the most serious ever for the credibility of nuclear power". UBS has helped to raise more than $20 billion since 2006 and advised on more than a dozen deals for renewable energy and cleantech companies. Deutsche Bank advised that "the global impact of the Fukushima accident is a fundamental shift in public perception with regard to how a nation prioritizes and values its populations health, safety, security, and natural environment when determining its current and future energy pathways...renewable energy will be a clear long-term winner in most energy systems, a conclusion supported by many voter surveys conducted over the past few weeks. Deutsche Bank has over €1 billion in capital invested in renewables projects in Europe, North & South America, and Asia.

Manufacturers also recognized a profit opportunity in negative public perceptions about nuclear. In September 2011, German engineering giant Siemens announced it will withdraw entirely from the nuclear industry, as a response to the Fukushima nuclear accident in Japan, and said that it would no longer build nuclear power plants anywhere in the world. The company's chairman, Peter Löscher, said that "Siemens was ending plans to cooperate with Rosatom, the Russian state-controlled nuclear power company, in the construction of dozens of nuclear plants throughout Russia over the coming two decades". Renewable energy is a core component of Siemens's profit base. In February, 2016 the firm proposed a €10 billion renewable energy investment in Egypt.

In February 2012, the United States Nuclear Regulatory Commission approved the construction of two additional reactors at the Vogtle Electric Generating Plant, the first reactors to be approved in over 30 years since the Three Mile Island accident, but NRC Chairman Gregory Jaczko cast a dissenting vote citing safety concerns stemming from Japan's 2011 Fukushima nuclear disaster, and saying "I cannot support issuing this license as if Fukushima never happened". Jaczko resigned in April 2012. One week after Southern received the license to begin major construction on the two new reactors, a dozen environmental and anti-nuclear groups sued to stop the Plant Vogtle expansion project, saying "public safety and environmental problems since Japan's Fukushima Daiichi nuclear reactor accident have not been taken into account". In July 2012, the suit was rejected by the Washington, D.C. Circuit Court of Appeals.

Countries such as Australia, Austria, Denmark, Greece, Ireland, Italy, Latvia, Liechtenstein, Luxembourg, Malta, Portugal, Israel, Malaysia, New Zealand, and Norway have no nuclear power reactors and remain opposed to nuclear power. However, by contrast, some countries remain in favor, and financially support nuclear fusion research, including EU wide funding of the ITER project.

Industry

Companies

Capacity and Production

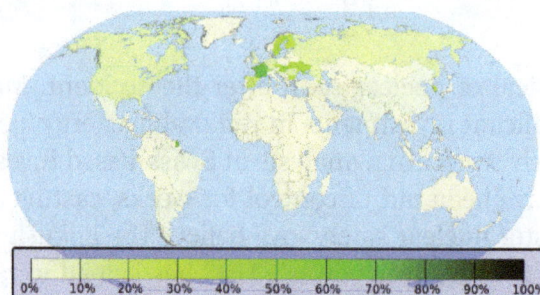

Percentage of a nations electricity, produced by fission-electric power stations.

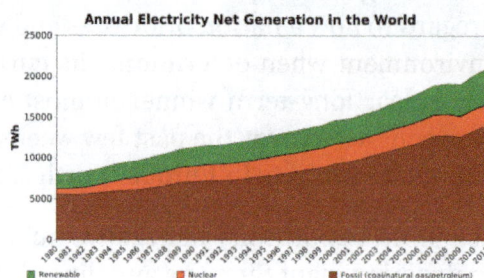

Net electrical generation by source and growth from 1980 to 2010. (Brown) - fossil fuels.(Red) - Fission.(Green)- "all renewables". In terms of energy generated between 1980 and 2010, the contribution from fission grew the fastest.

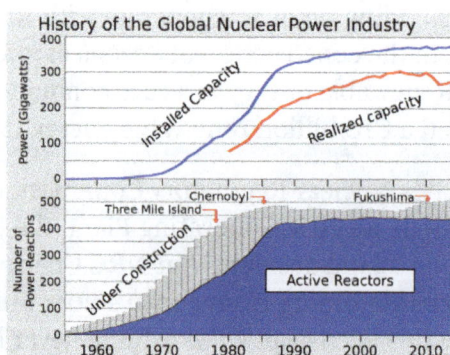

The rate of new construction builds for civilian fission-electric reactors essentially halted in the late 1980s, with the effects of accidents having a chilling effect. Increased capacity factor realizations in existing reactors was primarily responsible for the continuing increase in electrical energy produced during this period. The halting of new builds c. 1985, resulted in greater fossil fuel generation, see above graph.

Nuclear power capacity remained relatively stable between the mid 1980s until the accident at the Fukushima Daiichi reactor in March 2011. In June 2015, Platts reported global nuclear generation increased by 1% in 2014, the first annual increase since Fukushima.

The United States produces the most nuclear energy, with nuclear power providing 19% of the electricity it consumes, while France produces the highest percentage of its electrical energy from nuclear reactors—80% as of 2006. In the European Union as a whole, nuclear energy provides 30% of the electricity. Nuclear energy policy differs among European Union countries, and some, such as Austria, Estonia, Ireland and Italy, have no active nuclear power stations. In comparison, France has a large number of these plants, with 16 multi-unit stations in current use.

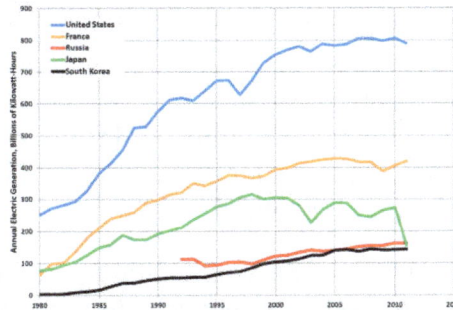

Electricity generation trends in the top five fission-energy producing countries (US EIA data)

Many military and some civilian (such as some icebreaker) ships use nuclear marine propulsion, a form of nuclear propulsion. A few space vehicles have been launched using full-fledged nuclear reactors: 33 reactors belong to the Soviet RORSAT series and one was the American SNAP-10A.

International research is continuing into safety improvements such as passively safe plants, the use of nuclear fusion, and additional uses of process heat such as hydrogen production (in support of a hydrogen economy), for desalinating sea water, and for use in district heating systems.

Nuclear (fission) power stations, excluding the contribution from naval nuclear fission reactors, provided 11% of the world's electricity in 2012, somewhat less than that generated by hydro-electric stations at 16%. Since electricity accounts for about 25% of humanity's energy usage with the majority of the rest coming from fossil fuel reliant sectors such as transport, manufacture and home heating, nuclear fission's contribution to the global final energy consumption is about 2.5%, a little more than the combined global electricity production from "new renewables"; wind, solar, biofuel and geothermal power, which together provided 2% of global final energy consumption in 2014.

Regional differences in the use of fission energy are large. Fission energy generation, with a 20% share of the U.S. electricity production, is the single largest deployed technology among current low-carbon power sources in the country. In addition, two-thirds of the European Union's twenty-seven nations' low-carbon energy is produced by fission. Some of these nations have banned its generation, such as Italy, which ended the use of fission-electric generation, which started in 1963, in 1990. France is the largest user of nuclear energy, deriving 75% of its electricity from fission.

In 2013, the IAEA reported that there were 437 operational civil fission-electric reactors in 31 countries, although not every reactor was producing electricity. In addition, there were approximately 140 naval vessels using nuclear propulsion in operation, powered by some 180 reactors. As of 2013, attaining a net energy gain from sustained nuclear fusion reactions, excluding natural fusion power sources such as the Sun, remains an ongoing area of international physics and engineering research. With commercial fusion power production remaining unlikely before 2050.

Since commercial nuclear energy began in the mid-1950s, 2008 was the first year that no new nuclear power plant was connected to the grid, although two were connected in 2009.

In 2015, the IAEA reported that worldwide there were 67 civil fission-electric power reactors under construction in 15 countries including Gulf states such as the United Arab Emirates (UAE). Over half of the 67 total being built were in Asia, with 28 in China. Eight new grid connections were completed by China in 2015 and the most recently completed reactor to be connected to the electrical grid, as of January 2016, was at the Kori Nuclear Power Plant in the Republic of Korea. In the US, four new Generation III reactors were under construction at Vogtle and Summer station, while a fifth was nearing completion at Watts Bar station, all five were expected to become operational before 2020. In 2013, four aging uncompetitive U.S reactors were closed. According to the World Nuclear Association, the global trend is for new nuclear power stations coming online to be balanced by the number of old plants being retired.

Analysis in 2015 by Professor and Chair of Environmental Sustainability Barry W. Brook and his colleagues on the topic of replacing fossil fuels entirely, from the electric grid of the world, has determined that at the historically modest and proven-rate at which nuclear energy was added to and replaced fossil fuels in France and Sweden during each nation's building programs in the 1980s, within 10 years nuclear energy could displace or remove fossil fuels from the electric grid completely, "allow[ing] the world to meet the most stringent greenhouse-gas mitigation targets.". In a similar analysis, Brook had earlier determined that 50% of all global energy, that is not solely electricity, but transportation synfuels etc. could be generated within approximately 30 years, if the global nuclear fission build rate was identical to each of these nation's already proven decadal rates(in units of installed nameplate capacity, GW per year, per unit of global GDP(GW/year/$).

This is in contrast to the completely conceptual paper-studies for a *100% renewable energy* world, which would require an orders of magnitude more costly global investment per year, which has no historical precedent, having never been attempted due to its prohibitive cost, along with far greater land that would need to be devoted to the wind, wave and solar projects, and the inherent assumption that humanity will use less, and not more, energy in the future. As Brook notes the "principal limitations on nuclear fission are not technical, economic or fuel-related, but are instead linked to complex issues of societal acceptance, fiscal and political inertia, and inadequate critical evaluation of the real-world constraints facing [the other] low-carbon alternatives."

Economics

George W. Bush signing the Energy Policy Act of 2005, which was designed to promote the US nuclear power industry, through incentives and subsidies, including cost-overrun support up to a total of $2 billion for six new nuclear plants. However, as of 2014 some electric utilities have rebuffed the loan package, including South Carolina Electric and Gas which operates Summer Station (the location of 2 new builds), noting instead that "it was easier to raise [loan] money commercially."

The Ikata Nuclear Power Plant, a pressurized water reactor that cools by utilizing a secondary coolant heat exchanger with a large body of water, an alternative cooling approach to large cooling towers.

Nuclear power plants typically have high capital costs for building the plant, but low fuel costs. Although nuclear power plants can vary their output the electricity is generally less favorably priced when doing so. Nuclear power plants are therefore typically run as much as possible to keep the cost of the generated electrical energy as low as possible, supplying mostly base-load electricity.

Internationally the price of nuclear plants rose 15% annually in 1970-1990. Yet, nuclear power has total costs in 2012 of about $96 per megawatt hour (MWh), most of which involves capital construction costs, compared with solar power at $130 per MWh, and natural gas at the low end at $64 per MWh.

In 2015, the *Bulletin of the Atomic Scientists* unveiled the Nuclear Fuel Cycle Cost Calculator, an online tool that estimates the full cost of electricity produced by three configurations of the nuclear fuel cycle. Two years in the making, this interactive calculator is the first generally accessible model to provide a nuanced look at the economic costs of nuclear power; it lets users test how sensitive the price of electricity is to a full range of components—more than 60 parameters that can be adjusted for the three configurations of the nuclear fuel cycle considered by this tool (once-through, limited-recycle, full-recycle). Users can select the fuel cycle they would like to examine, change cost estimates for each component of that cycle, and even choose uncertainty ranges for the cost of particular components. This approach allows users around the world to compare the cost of different nuclear power approaches in a sophisticated way, while taking account of prices relevant to their own countries or regions.

In recent years there has been a slowdown of electricity demand growth. In Eastern Europe, a number of long-established projects are struggling to find finance, notably Belene in Bulgaria and the additional reactors at Cernavoda in Romania, and some potential backers have pulled out. Where the electricity market is competitive, cheap natural gas is available, and its future supply relatively secure, this also poses a major problem for nuclear projects and existing plants.

Analysis of the economics of nuclear power must take into account who bears the risks of future uncertainties. To date all operating nuclear power plants were developed by state-owned or regulated utility monopolies where many of the risks associated with construction costs, operating performance, fuel price, accident liability and other factors were borne by consumers rather than suppliers. In addition, because the potential liability from a nuclear accident is so great, the full cost of liability insurance is generally limited/capped by the government, which the U.S. Nuclear Regulatory Commission concluded constituted a significant subsidy. Many countries have now liberalized the electricity market where these risks, and the risk of cheaper competitors emerging before capital costs are recovered, are borne by plant suppliers and operators rather than consumers, which leads to a significantly different evaluation of the economics of new nuclear power plants.

Following the 2011 Fukushima Daiichi nuclear disaster, costs are expected to increase for currently operating and new nuclear power plants, due to increased requirements for on-site spent fuel management and elevated design basis threats.

The economics of new nuclear power plants is a controversial subject, since there are diverging views on this topic, and multibillion-dollar investments ride on the choice of an energy source. Comparison with other power generation methods is strongly dependent on assumptions about construction timescales and capital financing for nuclear plants as well as the future costs of fossil fuels and renewables as well as for energy storage solutions for intermittent power sources. Cost estimates also need to take into account plant decommissioning and nuclear waste storage costs. On the other hand, measures to mitigate global warming, such as a carbon tax or carbon emissions trading, may favor the economics of nuclear power.

Nuclear Power Organizations

There are multiple organizations which have taken a position on nuclear power and the nuclear power industry– some are proponents, and some are opponents.

Proponents

The majority of pro-nuclear energy organizations and associations is either industry-supported or directly formed from industry members as advocacy groups or trade associations.

- Environmentalists for Nuclear Energy (International)
- Nuclear Industry Association (United Kingdom)
- World Nuclear Association, a confederation of companies connected with nuclear power production. (International)
- International Atomic Energy Agency (IAEA)
- Nuclear Energy Institute (United States)
- American Nuclear Society (United States)
- United Kingdom Atomic Energy Authority (United Kingdom)
- EURATOM (Europe)
- European Nuclear Education Network (Europe)
- Atomic Energy of Canada Limited (Canada)

- Nuclear Matters (United States)
- Breakthrough Institute (United States)
- Thorium Energy Alliance (United States)
- Californians for Green Nuclear Power (United States)
- Save Diablo Canyon (United States)
- Thorium Now (United States)
- Category:Nuclear industry organizations

Opponents

- Friends of the Earth International, a network of environmental organizations.
- Greenpeace International, a non-governmental organization
- Nuclear Information and Resource Service (International)
- World Information Service on Energy (International)
- Sortir du nucléaire (France)
- Pembina Institute (Canada)
- Institute for Energy and Environmental Research (United States)
- Sayonara Nuclear Power Plants (Japan)
- Category:Anti-nuclear organizations

Future of The Industry

The future of nuclear power varies greatly between countries, depending on government policies. Some countries, many of them in Europe, such as Germany, Belgium, and Lithuania, have adopted policies of nuclear power phase-out. At the same time, some Asian countries, such as China, South Korea, and India, have committed to rapid expansion of nuclear power. Many other countries, such as the United Kingdom and the United States, have policies in between. Japan was a major generator of nuclear power before the Fukushima accident, but as of August 2016, Japan has restarted only three of its nuclear plants, and the extent to which it will resume its nuclear program is uncertain.

Brunswick Nuclear Plant discharge canal

In 2015, the International Energy Agency reported that the Fukushima accident had a strongly negative effect on nuclear power, yet "the prospects for nuclear energy remain positive in the medium to long term despite a negative impact in some countries in the aftermath of the accident." The IEA noted that at the start of 2014, there were 72 nuclear reactors under construction worldwide, the largest number in 25 years, and that China planned to increase nuclear power capacity from 17 gigawatts (GW) in 2014, to 58 GW in 2020.

The Bruce Nuclear Generating Station, the largest nuclear power facility in the world

In 2016, the US Energy Information Administration projected for its "base case" that world nuclear power generation would increase from 2,344 billion kW-hr in 2012 to 4,501 billion kW-hr in 2040. Most of the predicted increase was expected to be in Asia.

The nuclear power industry in western nations has a history of construction delays, cost overruns, plant cancellations, and nuclear safety issues despite significant government subsidies and support. In December 2013, *Forbes* magazine cited a report which concluded that, in western countries, "reactors are not a viable source of new power". Even where they make economic sense, they are not feasible because nuclear's "enormous costs, political and popular opposition, and regulatory uncertainty". This view echoes the statement of former Exelon CEO John Rowe, who said in 2012 that new nuclear plants in the United States "don't make any sense right now" and won't be economically viable in the foreseeable future. John Quiggin, economics professor, also says the main problem with the nuclear option is that it is not economically-viable. Quiggin says that we need more efficient energy use and more renewable energy commercialization. Former NRC member Peter Bradford and Professor Ian Lowe made similar statements in 2011. However, some "nuclear cheerleaders" and lobbyists in the West continue to champion reactors, often with proposed new but largely untested designs, as a source of new power.

Much more new build activity is occurring in developing countries like South Korea, India and China. In March 2016, China had 30 reactors in operation, 24 under construction and plans to build more, However, according to a government research unit, China must not build "too many nuclear power reactors too quickly", in order to avoid a shortfall of fuel, equipment and qualified plant workers.

In the US, licenses of almost half its reactors have been extended to 60 years, Two new Generation III reactors are under construction at Vogtle, a dual construction project which marks the end of a 34-year period of stagnation in the US construction of civil nuclear power reactors. The station operator licenses of almost half the present 104 power reactors in the US, as of 2008, have been given extensions to 60 years. As of 2012, U.S. nuclear industry officials expect five new reactors to enter service by 2020, all at existing plants. In 2013, four aging, uncompetitive, reactors were permanently closed. Relevant state legislatures are trying to close Vermont Yankee and Indian Point Nuclear Power Plant.

The U.S. NRC and the U.S. Department of Energy have initiated research into Light water reactor sustainability which is hoped will lead to allowing extensions of reactor licenses beyond 60 years, provided that safety can be maintained, as the loss in non-CO_2-emitting generation capacity by retiring reactors "may serve to challenge U.S. energy security, potentially resulting in increased greenhouse gas emissions, and contributing to an imbalance between electric supply and demand."

There is a possible impediment to production of nuclear power plants as only a few companies worldwide have the capacity to forge single-piece reactor pressure vessels, which are necessary in the most common reactor designs. Utilities across the world are submitting orders years in advance of any actual need for these vessels. Other manufacturers are examining various options, including making the component themselves, or finding ways to make a similar item using alternate methods.

According to the World Nuclear Association, globally during the 1980s one new nuclear reactor started up every 17 days on average, and in the year 2015 it was estimated that this rate could in theory eventually increase to one every 5 days, although no plans exist for that. As of 2007, Watts Bar 1 in Tennessee, which came on-line on February 7, 1996, was the last U.S. commercial nuclear reactor to go on-line. This is often quoted as evidence of a successful worldwide campaign for nuclear power phase-out. Electricity shortages, fossil fuel price increases, global warming, and heavy metal emissions from fossil fuel use, new technology such as passively safe plants, and national energy security may renew the demand for nuclear power plants.

Nuclear Power Plant

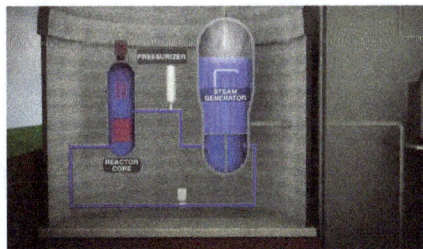

An animation of a Pressurized water reactor in operation.

Unlike fossil fuel power plants, the only substance leaving the cooling towers of nuclear power plants is water vapour and thus does not pollute the air or cause global warming.

Just as many conventional thermal power stations generate electricity by harnessing the thermal energy released from burning fossil fuels, nuclear power plants convert the energy released from the nucleus of an atom via nuclear fission that takes place in a nuclear reactor. The heat is removed from the reactor core by a cooling system that uses the heat to generate steam, which drives a steam turbine connected to a generator producing electricity.

Life Cycle of Nuclear Fuel

A nuclear reactor is only part of the life-cycle for nuclear power. The process starts with mining. Uranium mines are underground, open-pit, or in-situ leach mines. In any case, the uranium ore is extracted, usually converted into a stable and compact form such as yellowcake, and then transported to a processing facility. Here, the yellowcake is converted to uranium hexafluoride, which is then enriched using various techniques. At this point, the enriched uranium, containing more than the natural 0.7% U-235, is used to make rods of the proper composition and geometry for the particular reactor that the fuel is destined for. The fuel rods will spend about 3 operational cycles (typically 6 years total now) inside the reactor, generally until about 3% of their uranium has been fissioned, then they will be moved to a spent fuel pool where the short lived isotopes generated by fission can decay away. After about 5 years in a spent fuel pool the spent fuel is radioactively and thermally cool enough to handle, and it can be moved to dry storage casks or reprocessed.

The nuclear fuel cycle begins when uranium is mined, enriched, and manufactured into nuclear fuel, (1) which is delivered to a nuclear power plant. After usage in the power plant, the spent fuel is delivered to a reprocessing plant (2) or to a final repository (3) for geological disposition. In reprocessing 95% of spent fuel can potentially be recycled to be returned to usage in a power plant (4).

Conventional Fuel Resources

Uranium is a fairly common element in the Earth's crust. Uranium is approximately as common as tin or germanium in the Earth's crust, and is about 40 times more common than silver. Uranium is present in trace concentrations in most rocks, dirt, and ocean water, but can be economically extracted currently only only where it is present in high concentrations. Still, the world's present measured resources of uranium, economically recoverable at the arbitrary price ceiling of 130 USD/kg, are enough to last for between 70 and 100 years.

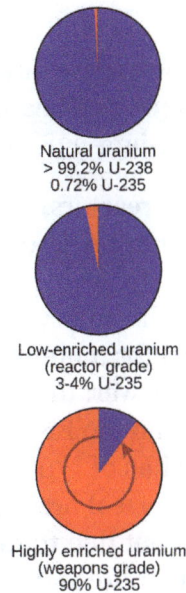

Proportions of the isotopes, uranium-238 (blue) and uranium-235 (red) found naturally, versus grades that are enriched. light water reactors require fuel enriched to (3-4%), while others such as the CANDU reactor uses natural uranium.

According to the OECD in 2006, there was an expected 85 years worth of uranium in already identified resources, when that uranium is used in present reactor technology, in the OECD's red book of 2011, due to increased exploration, known uranium resources have grown by 12.5% since 2008, with this increase translating into greater than a century of uranium available if the metals usage rate were to continue at the 2011 level. The OECD also estimate 670 years of economically recoverable uranium in total conventional resources and phosphate ores, while also using present reactor technology, a resource that is recoverable from between 60-100 US$/kg of Uranium. In a similar manner to every other natural metal resource, for every tenfold increase in the cost per kilogram of uranium, there is a three-hundredfold increase in available lower quality ores that would then become economical. As the OECD note:

Even if the nuclear industry expands significantly, sufficient fuel is available for centuries. If advanced breeder reactors could be designed in the future to efficiently utilize recycled or depleted uranium and all actinides, then the resource utilization efficiency would be further improved by an additional factor of eight.

For example, the OECD have determined that with a pure fast reactor fuel cycle with a burn up of, and recycling of, all the Uranium and actinides, actinides which presently make up the most hazardous substances in nuclear waste, there is 160,000 years worth of Uranium in total conventional resources and phosphate ore, at the price of 60-100 US$/kg of Uranium.

Current light water reactors make relatively inefficient use of nuclear fuel, mostly fissioning only the very rare uranium-235 isotope. Nuclear reprocessing can make this waste reusable, and more efficient reactor designs, such as the currently under construction Generation III reactors achieve a higher efficiency burn up of the available resources, than the current vintage generation II reactors, which make up the vast majority of reactors worldwide.

Breeding

As opposed to current light water reactors which use uranium-235 (0.7% of all natural uranium), fast breeder reactors use uranium-238 (99.3% of all natural uranium). It has been estimated that there is up to five billion years' worth of uranium-238 for use in these power plants.

Breeder technology has been used in several reactors, but the high cost of reprocessing fuel safely, at 2006 technological levels, requires uranium prices of more than 200 USD/kg before becoming justified economically. Breeder reactors are still however being pursued as they have the potential to burn up all of the actinides in the present inventory of nuclear waste while also producing power and creating additional quantities of fuel for more reactors via the breeding process. In 2005, there were two breeder reactors producing power: the Phénix in France, which has since powered down in 2009 after 36 years of operation, and the BN-600 reactor, a reactor constructed in 1980 Beloyarsk, Russia which is still operational as of 2013. The electricity output of BN-600 is 600 MW — Russia plans to expand the nation's use of breeder reactors with the BN-800 reactor, was scheduled to become operational in 2014, but due to delays is not scheduled to produce power until 2017. The technical design of a yet larger breeder, the BN-1200 reactor was originally scheduled to be finalized in 2013, with construction slated for 2015 but has also been delayed. Japan's Monju breeder reactor restarted (having been shut down in 1995) in 2010 for 3 months, but shut down again after equipment fell into the reactor during reactor checkups, it is planned to become re-operational in late 2013. Both China and India are building breeder reactors. With the Indian 500 MWe Prototype Fast Breeder Reactor scheduled to become operational in 2014, with plans to build five more by 2020. The China Experimental Fast Reactor began producing power in 2011.

Another alternative to fast breeders is thermal breeder reactors that use uranium-233 bred from thorium as fission fuel in the thorium fuel cycle. Thorium is about 3.5 times more common than uranium in the Earth's crust, and has different geographic characteristics. This would extend the total practical fissionable resource base by 450%. India's three-stage nuclear power programme features the use of a thorium fuel cycle in the third stage, as it has abundant thorium reserves but little uranium.

Solid Waste

The most important waste stream from nuclear power plants is spent nuclear fuel. It is primarily composed of unconverted uranium as well as significant quantities of transuranic actinides (plutonium and curium, mostly). In addition, about 3% of it is fission products from nuclear reactions. The actinides (uranium, plutonium, and curium) are responsible for the bulk of the long-term radioactivity, whereas the fission products are responsible for the bulk of the short-term radioactivity.

High-level Radioactive Waste

Following interim storage in a spent fuel pool, the bundles of used fuel assemblies of a typical nuclear power station are often stored on site in the likes of the eight dry cask storage vessels pictured above. At Yankee Rowe Nuclear Power Station, which generated 44 billion kilowatt hours of electricity over its lifetime, its complete spent fuel inventory is contained within sixteen casks.

High-level radioactive waste management concerns management and disposal of highly radioactive materials created during production of nuclear power. The technical issues in accomplishing this are daunting, due to the extremely long periods radioactive wastes remain deadly to living organisms. Of particular concern are two long-lived fission products, Technetium-99 (half-life 220,000 years) and Iodine-129 (half-life 15.7 million years), which dominate spent nuclear fuel radioactivity after a few thousand years. The most troublesome transuranic elements in spent fuel are Neptunium-237 (half-life two million years) and Plutonium-239 (half-life 24,000 years). Consequently, high-level radioactive waste requires sophisticated treatment and management to successfully isolate it from the biosphere. This usually necessitates treatment, followed by a long-term management strategy involving permanent storage, disposal or transformation of the waste into a non-toxic form.

A nuclear fuel rod assembly bundle being inspected before entering a reactor.

Governments around the world are considering a range of waste management and disposal options, usually involving deep-geologic placement, although there has been limited progress toward implementing long-term waste management solutions. This is partly because the timeframes in question when dealing with radioactive waste range from 10,000 to millions of years, according to studies based on the effect of estimated radiation doses.

Some proposed nuclear reactor designs however such as the American Integral Fast Reactor and the Molten salt reactor can use the nuclear waste from light water reactors as a fuel, transmutating it to isotopes that would be safe after hundreds, instead of tens of thousands of years. This offers a potentially more attractive alternative to deep geological disposal.

Another possibility is the use of thorium in a reactor especially designed for thorium (rather than mixing in thorium with uranium and plutonium (i.e. in existing reactors). Used thorium fuel remains only a few hundreds of years radioactive, instead of tens of thousands of years.

Since the fraction of a radioisotope's atoms decaying per unit of time is inversely proportional to its half-life, the relative radioactivity of a quantity of buried human radioactive waste would diminish over time compared to natural radioisotopes (such as the decay chains of 120 trillion tons of thorium and 40 trillion tons of uranium which are at relatively trace concentrations of parts per million each over the crust's $3 * 10^{19}$ ton mass). For instance, over a timeframe of thousands of years, after the most active short half-life radioisotopes decayed, burying U.S. nuclear waste would increase the radioactivity in the top 2000 feet of rock and soil in the United States (10 million km²) by ≈ 1 part in 10 million over the cumulative amount of natural radioisotopes in such a volume, although the vicinity of the site would have a far higher concentration of artificial radioisotopes underground than such an average.

Low-level Radioactive Waste

The nuclear industry also produces a large volume of low-level radioactive waste in the form of contaminated items like clothing, hand tools, water purifier resins, and (upon decommissioning) the materials of which the reactor itself is built. In the US, the Nuclear Regulatory Commission has repeatedly attempted to allow low-level materials to be handled as normal waste: landfilled, recycled into consumer items, etcetera.

Comparing Radioactive Waste to Industrial Toxic Waste

In countries with nuclear power, radioactive wastes comprise less than 1% of total industrial toxic wastes, much of which remains hazardous for long periods. Overall, nuclear power produces far less waste material by volume than fossil-fuel based power plants. Coal-burning plants are particularly noted for producing large amounts of toxic and mildly radioactive ash due to concentrating naturally occurring metals and mildly radioactive material from the coal. A 2008 report from Oak Ridge National Laboratory concluded that coal power actually results in more radioactivity being released into the environment than nuclear power operation, and that the population effective dose equivalent, or dose to the public from radiation from coal plants is 100 times as much as from the operation of nuclear plants. Indeed, coal ash is much less radioactive than spent nuclear fuel on a weight per weight basis, but coal ash is produced in much higher quantities per unit of energy generated, and this is released directly into the environment as fly ash, whereas nuclear plants use shielding to protect the environment from radioactive materials, for example, in dry cask storage vessels.

Waste Disposal

Disposal of nuclear waste is often said to be the Achilles' heel of the industry. Presently, waste is mainly stored at individual reactor sites and there are over 430 locations around the world where radioactive material continues to accumulate. Some experts suggest that centralized underground repositories which are well-managed, guarded, and monitored, would be a vast improvement. There is an "international consensus on the advisability of storing nuclear waste in deep geological repositories", with the lack of movement of nuclear waste in the 2 billion year old natural nuclear fission reactors in Oklo, Gabon being cited as "a source of essential information today."

There are no commercial scale purpose built underground repositories in operation. The Waste Isolation Pilot Plant (WIPP) in New Mexico has been taking nuclear waste since 1999 from production reactors, but as the name suggests is a research and development facility. A radiation leak at WIPP in 2014 brought renewed attention to the need for R&D on disposal or radioactive waste and spent fuel.

Reprocessing

Reprocessing can potentially recover up to 95% of the remaining uranium and plutonium in spent nuclear fuel, putting it into new mixed oxide fuel. This produces a reduction in long term radioactivity within the remaining waste, since this is largely short-lived fission products, and reduces its volume by over 90%. Reprocessing of civilian fuel from power reactors is currently done in Europe, Russia, Japan, and India. The full potential of reprocessing has not been achieved because it requires breeder reactors, which are not commercially available.

Nuclear reprocessing reduces the volume of high-level waste, but by itself does not reduce radioactivity or heat generation and therefore does not eliminate the need for a geological waste repository. Reprocessing has been politically controversial because of the potential to contribute to nuclear proliferation, the potential vulnerability to nuclear terrorism, the political challenges of repository siting (a problem that applies equally to direct disposal of spent fuel), and because of its high cost compared to the once-through fuel cycle. Several different methods for reprocessing been tried, but many have had safety and practicality problems which have led to their discontinuation.

In the United States, the Obama administration stepped back from President Bush's plans for commercial-scale reprocessing and reverted to a program focused on reprocessing-related scientific research. Reprocessing is not allowed in the U.S. In the U.S., spent nuclear fuel is currently all treated as waste. A major recommendation of the Blue Ribbon Commission on America's Nuclear Future was that "the United States should undertake an integrated nuclear waste management program that leads to the timely development of one or more permanent deep geological facilities for the safe disposal of spent fuel and high-level nuclear waste".

Depleted Uranium

Uranium enrichment produces many tons of depleted uranium (DU) which consists of U-238 with most of the easily fissile U-235 isotope removed. U-238 is a tough metal with several commercial uses—for example, aircraft production, radiation shielding, and armor—as it has a higher density than lead. Depleted uranium is also controversially used in munitions; DU penetrators (bullets or APFSDS tips) "self sharpen", due to uranium's tendency to fracture along shear bands.

Accidents, Attacks and Safety

Accidents

Some serious nuclear and radiation accidents have occurred. Benjamin K. Sovacool has reported that worldwide there have been 99 accidents at nuclear power plants. Fifty-seven accidents have occurred since the Chernobyl disaster, and 57% (56 out of 99) of all nuclear-related accidents have occurred in the USA.

Nuclear power plant accidents include the Chernobyl accident (1986) with approximately 60 deaths so far attributed to the accident and a predicted, eventual total death toll, of from 4000 to 25,000 latent cancers deaths. The Fukushima Daiichi nuclear disaster (2011), has not caused any radiation related deaths, with a predicted, eventual total death toll, of from 0 to 1000, and the Three Mile Island accident (1979), no causal deaths, cancer or otherwise, have been found in follow up studies of this accident. Nuclear-powered submarine mishaps include the K-19 reactor accident (1961), the K-27 reactor accident (1968), and the K-431 reactor accident (1985). International research is continuing into safety improvements such as passively safe plants, and the possible future use of nuclear fusion.

In terms of lives lost per unit of energy generated, nuclear power has caused fewer accidental deaths per unit of energy generated than all other major sources of energy generation. Energy produced by coal, petroleum, natural gas and hydropower has caused more deaths per unit of energy generated, from air pollution and energy accidents. This is found in the following comparisons, when the immediate nuclear related deaths from accidents are compared to the immediate deaths from these other energy sources, when the latent, or predicted, indirect cancer deaths from nuclear energy accidents are compared to the immediate deaths from the above energy sources, and when the combined immediate and indirect fatalities from nuclear power and all fossil fuels are compared, fatalities resulting from the mining of the necessary natural resources to power generation and to air pollution. With these data, the use of nuclear power has been calculated to have prevented in the region of 1.8 million deaths between 1971 and 2009, by reducing the proportion of energy that would otherwise have been generated by fossil fuels, and is projected to continue to do so.

Although according to Benjamin K. Sovacool, fission energy accidents ranked first in terms of their total economic cost, accounting for 41 percent of all property damage attributed to energy accidents. Analysis presented in the international Journal, *Human and Ecological Risk Assessment* found that coal, oil, Liquid petroleum gas and hydroelectric accidents(primarily due to the Banqiao dam burst) have resulted in greater economic impacts than nuclear power accidents.

Following the 2011 Japanese Fukushima nuclear disaster, authorities shut down the nation's 54 nuclear power plants, but it has been estimated that if Japan had never adopted nuclear power, accidents and pollution from coal or gas plants would have caused more lost years of life. As of 2013, the Fukushima site remains highly radioactive, with some 160,000 evacuees still living in temporary housing, and some land will be unfarmable for centuries. The difficult Fukushima disaster cleanup will take 40 or more years, and cost tens of billions of dollars.

Forced evacuation from a nuclear accident may lead to social isolation, anxiety, depression, psychosomatic medical problems, reckless behavior, even suicide. Such was the outcome of the 1986 Chernobyl nuclear disaster in Ukraine. A comprehensive 2005 study concluded that "the mental health impact of Chernobyl is the largest public health problem unleashed by the accident to date". Frank N. von Hippel, a U.S. scientist, commented on the 2011 Fukushima nuclear disaster, saying that "fear of ionizing radiation could have long-term psychological effects on a large portion of the population in the contaminated areas". A 2015 report in *Lancet* explained that serious impacts of nuclear accidents were often not directly attributable to radiation exposure, but rather social and psychological effects. Evacuation and long-term displacement of affected populations created problems for many people, especially the elderly and hospital patients. But long-term displacement is not a unique feature to nuclear accidents, with hydropower and lignite surface mining projects routinely displacing thousands during normal, non-accident, operations, e.g. Three Gorges Dam resp. Garzweiler surface mine.

Attacks and Sabotage

Terrorists could target nuclear power plants in an attempt to release radioactive contamination into the community. The United States 9/11 Commission has said that nuclear power plants were potential targets originally considered for the September 11, 2001 attacks. An attack on a reactor's spent fuel pool could also be serious, as these pools are less protected than the reactor core. The release of radioactivity could lead to thousands of near-term deaths and greater numbers of long-term fatalities.

If nuclear power use is to expand significantly, nuclear facilities will have to be made extremely safe from attacks that could release massive quantities of radioactivity into the community. New reactor designs have features of passive safety, such as the flooding of the reactor core without active intervention by reactor operators. But these safety measures have generally been developed and studied with respect to accidents, not to the deliberate reactor attack by a terrorist group. However, the US Nuclear Regulatory Commission does now also require new reactor license applications to consider security during the design stage. In the United States, the NRC carries out "Force on Force" (FOF) exercises at all Nuclear Power Plant (NPP) sites at least once every three years. In the U.S., plants are surrounded by a double row of tall fences which are electronically monitored. The plant grounds are patrolled by a sizeable force of armed guards.

Insider sabotage regularly occurs, because insiders can observe and work around security measures. Successful insider crimes depended on the perpetrators' observation and knowledge of security vulnerabilities. A fire caused 5–10 million dollars worth of damage to New York's Indian Point Energy Center in 1971. The arsonist turned out to be a plant maintenance worker. Sabotage by workers has been reported at many other reactors in the United States: at Zion Nuclear Power Station (1974), Quad Cities Nuclear Generating Station, Peach Bottom Nuclear Generating Station, Fort St. Vrain Generating Station, Trojan Nuclear Power Plant (1974), Browns Ferry Nuclear Power Plant (1980), and Beaver Valley Nuclear Generating Station (1981). Many reactors overseas have also reported sabotage by workers.

Nuclear Proliferation

Many technologies and materials associated with the creation of a nuclear power program have a dual-use capability, in that they can be used to make nuclear weapons if a country chooses to do so. When this happens a nuclear power program can become a route leading to a nuclear weapon or a public annex to a "secret" weapons program. The concern over Iran's nuclear activities is a case in point.

United States and USSR/Russian nuclear weapons stockpiles, 1945-2006.The Megatons to Megawatts Program was the main driving force behind the sharp reduction in the quantity of nuclear weapons worldwide since the cold war ended. However without an increase in nuclear reactors and greater demand for fissile fuel, the cost of dismantling has dissuaded Russia from continuing their disarmament.

A fundamental goal for American and global security is to minimize the nuclear proliferation risks associated with the expansion of nuclear power. If this development is "poorly managed or efforts to contain risks are unsuccessful, the nuclear future will be dangerous". The Global Nuclear Energy Partnership is one such international effort to create a distribution network in which developing countries in need of energy, would receive nuclear fuel at a discounted rate, in exchange for that nation agreeing to forgo their own indigenous develop of a uranium enrichment program. The France-based Eurodif/*European Gaseous Diffusion Uranium Enrichment Consortium* was/ is one such program that successfully implemented this concept, with Spain and other countries without enrichment facilities buying a share of the fuel produced at the French controlled enrichment facility, but without a transfer of technology. Iran was an early participant from 1974, and remains a shareholder of Eurodif via Sofidif.

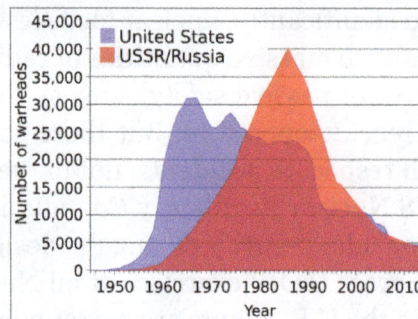

According to Benjamin K. Sovacool, a "number of high-ranking officials, even within the United Nations, have argued that they can do little to stop states using nuclear reactors to produce nuclear weapons". A 2009 United Nations report said that:

the revival of interest in nuclear power could result in the worldwide dissemination of uranium enrichment and spent fuel reprocessing technologies, which present obvious risks of proliferation as these technologies can produce fissile materials that are directly usable in nuclear weapons.

On the other hand, one factor influencing the support of power reactors is due to the appeal that these reactors have at reducing nuclear weapons arsenals through the Megatons to Megawatts Program, a program which eliminated 425 metric tons of highly enriched uranium(HEU), the equivalent of 17,000 nuclear warheads, by diluting it with natural uranium making it equivalent to low enriched uranium(LEU), and thus suitable as nuclear fuel for commercial fission reactors. This is the single most successful non-proliferation program to date.

The Megatons to Megawatts Program, the brainchild of Thomas Neff of MIT, was hailed as a major success by anti-nuclear weapon advocates as it has largely been the driving force behind the sharp reduction in the quantity of nuclear weapons worldwide since the cold war ended. However without an increase in nuclear reactors and greater demand for fissile fuel, the cost of dismantling and down blending has dissuaded Russia from continuing their disarmament.

Currently, according to Harvard professor Matthew Bunn: "The Russians are not remotely interested in extending the program beyond 2013. We've managed to set it up in a way that costs them more and profits them less than them just making new low-enriched uranium for reactors from scratch. But there are other ways to set it up that would be very profitable for them and would also serve some of their strategic interests in boosting their nuclear exports."

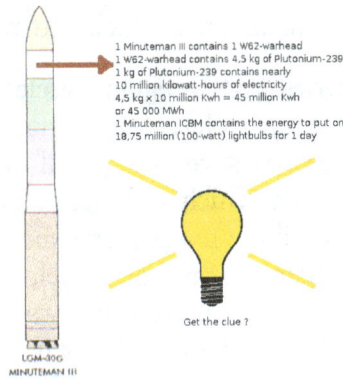

How much usable nuclear energy is placed in 1 (Minuteman) ICBM ?

1 Minuteman III contains 1 W62-warhead
1 W62-warhead contains 4,5 kg of Plutonium-239
1 kg of Plutonium-239 contains nearly
10 million kilowatt-hours of electricity
4,5 kg x 10 million Kwh = 45 million Kwh
or 45 000 MWh
1 Minuteman ICBM contains the energy to put on
18,75 million (100-watt) lightbulbs for 1 day

Get the clue ?

LGM-30G
MINUTEMAN III

Up to 2005, the Megatons to Megawatts Program had processed $8 billion of HEU/weapons grade uranium into LEU/reactor grade uranium, with that corresponding to the elimination of 10,000 nuclear weapons.

For approximately two decades, this material generated nearly 10 percent of all the electricity consumed in the United States (about half of all US nuclear electricity generated) with a total of around 7 trillion kilowatt-hours of electricity produced. Enough energy to energize the entire United States electric grid for about two years. In total it is estimated to have cost $17 billion, a "bargain for US ratepayers", with Russia profiting $12 billion from the deal. Much needed profit for the Russian nuclear oversight industry, which after the collapse of the Soviet economy, had difficulties paying for the maintenance and security of the Russian Federations highly enriched uranium and warheads.

In April 2012 there were thirty one countries that have civil nuclear power plants, of which nine have nuclear weapons, with the vast majority of these nuclear weapons states having first produced weapons, before commercial fission electricity stations. Moreover, the re-purposing of civilian nuclear industries for military purposes would be a breach of the Non-proliferation treaty, of which 190 countries adhere to.

Environmental Issues

Carbon emissions from nuclear power
Sovacool life cycle study survey, 2008

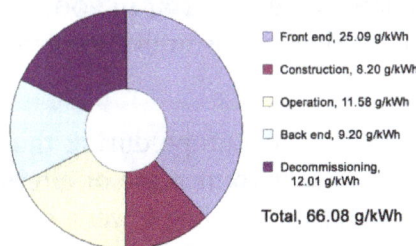

Front end, 25.09 g/kWh

Construction, 8.20 g/kWh

Operation, 11.58 g/kWh

Back end, 9.20 g/kWh

Decommissioning, 12.01 g/kWh

Total, 66.08 g/kWh

Mean value of carbon dioxide emissions from qualified life cycle studies among 103 surveyed. Includes results of 1997 Vattenfall study.

A 2008 synthesis of 103 studies, published by Benjamin K. Sovacool, estimated that the value of CO_2 emissions for nuclear power over the lifecycle of a plant was 66.08 g/kW·h. Comparative results for various renewable power sources were 9–32 g/kW·h. A 2012 study by Yale University arrived at a different value, with the mean value, depending on which Reactor design was analyzed, ranging from 11 to 25 g/kW·h of total life cycle nuclear power CO_2 emissions.

Life cycle analysis (LCA) of carbon dioxide emissions show nuclear power as comparable to renewable energy sources. Emissions from burning fossil fuels are many times higher.

According to the United Nations (UNSCEAR), regular nuclear power plant operation including the nuclear fuel cycle causes radioisotope releases into the environment amounting to 0.0002 millisieverts (mSv) per year of public exposure as a global average. (Such is small compared to variation in natural background radiation, which averages 2.4 mSv/a globally but frequently varies between 1 mSv/a and 13 mSv/a depending on a person's location as determined by UNSCEAR). As of a 2008 report, the remaining legacy of the worst nuclear power plant accident (Chernobyl) is 0.002 mSv/a in global average exposure (a figure which was 0.04 mSv per person averaged over the entire populace of the Northern Hemisphere in the year of the accident in 1986, although far higher among the most affected local populations and recovery workers).

Climate C1hange

Climate change causing weather extremes such as heat waves, reduced precipitation levels and droughts can have a significant impact on all thermal power station infrastructure, including large biomass-electric and fission-electric stations alike, if cooling in these power stations, namely in the steam condenser is provided by certain freshwater sources. While many thermal stations use indirect seawater cooling or cooling towers that in comparison use little to no freshwater, those that were designed to heat exchange with rivers and lakes, can run into economic problems.

This presently infrequent generic problem may become increasingly significant over time. This can force nuclear reactors to be shut down, as happened in France during the 2003 and 2006 heat waves. Nuclear power supply was severely diminished by low river flow rates and droughts, which meant rivers had reached the maximum temperatures for cooling reactors. During the heat waves, 17 reactors had to limit output or shut down. 77% of French electricity is produced by nuclear power and in 2009 a similar situation created a 8GW shortage and forced the French government to import electricity. Other cases have been reported from Germany, where extreme temperatures have reduced nuclear power production only 9 times due to high temperatures between 1979 and 2007. In particular:

- the Unterweser nuclear power plant reduced output by 90% between June and September 2003

- the Isar nuclear power plant cut production by 60% for 14 days due to excess river temperatures and low stream flow in the river Isar in 2006 However the more modern Isar II station did not have to cut production, as unlike its sister station Isar I, Isar II was built with a cooling tower.

Similar events have happened elsewhere in Europe during those same hot summers. If global warming continues, this disruption is likely to increase or alternatively, station operators could instead retro-fit other means of cooling, like cooling towers, despite these frequently being large structures and therefore sometimes unpopular with the public.

Comparison With Renewable Energy

As of 2013, the World Nuclear Association has said "There is unprecedented interest in renewable energy, particularly solar and wind energy, which provide electricity without giving rise to any carbon dioxide emission. Harnessing these for electricity depends on the cost and efficiency of the technology, which is constantly improving, thus reducing costs per peak kilowatt".

Renewable electricity production, from sources such as wind power and solar power, is frequently criticized for being intermittent or variable.

Like nuclear energy, renewable electricity supply, of primarily hydropower, in the 20-50+% range has already been implemented in several European systems, albeit in the context of an integrated European grid system. In 2012, the share of electricity generated by all types of renewable sources in Germany was 21.9%, compared to 16.0% for nuclear power after Germany shut down 7-8 of its 18 nuclear reactors in 2011. In the United Kingdom, the amount of energy produced from renewable energy is expected to exceed that from nuclear power by 2018, and Scotland plans to obtain all electricity from renewable energy by 2020. The majority of installed renewable energy across the world is in the form of hydro power.

The IPCC has said that if governments were supportive, and the full complement of renewable energy technologies were deployed, renewable energy supply could account for almost 80% of the world's energy use within forty years. Rajendra Pachauri, chairman of the IPCC, said the necessary investment in renewables would cost only about 1% of global GDP annually. This approach could contain greenhouse gas levels to less than 450 parts per million, the safe level beyond which climate change becomes catastrophic and irreversible.

In 2014, Brookings Institution published *The Net Benefits of Low and No-Carbon Electricity Technologies* which states, after performing an energy and emissions cost analysis, that "The net benefits of new nuclear, hydro, and natural gas combined cycle plants far outweigh the net benefits of new wind or solar plants", with the most cost effective low carbon power technology being determined to be nuclear power.

Similarly, analysis in 2015 by Professor and Chair of Environmental Sustainability Barry W. Brook and his colleagues on the topic of replacing fossil fuels entirely, from the electric grid of the world, has determined that at the historically modest and proven-rate at which nuclear energy was added to and replaced fossil fuels in France and Sweden during each nation's building programs in the 1980s, within 10 years nuclear energy could displace or remove fossil fuels from the electric grid completely, "allow[ing] the world to meet the most stringent greenhouse-gas mitigation targets.". In a similar analysis, Brook had earlier determined that 50% of all global energy, that is not solely electricity, but transportation synfuels etc. could be generated within approximately 30 years, if the global nuclear fission build rate was identical to each of these nation's already proven decadal rates(in units of installed nameplate capacity, GW per year, per unit of global GDP(GW/year/$).

This is in contrast to the completely conceptual paper-studies for a *100% renewable energy* world, which would require an orders of magnitude more costly global investment per year, which has no historical precedent, having never been attempted due to its prohibitive cost, along with far greater land that would have to be devoted to the wind, wave and solar projects, and the inherent assumption that humanity will use less, and not more, energy in the future. As Brook notes the "principal limitations on nuclear fission are not technical, economic or fuel-related, but are instead linked to complex issues of societal acceptance, fiscal and political inertia, and inadequate critical evaluation of the real-world constraints facing [the other] low-carbon alternatives."

While the cost of constructing established nuclear power reactor designs has followed an increasing trend due to regulations and court cases whereas the levelized cost of electricity is declining for wind power. In about 2011, wind power became as inexpensive as natural gas, and anti-nuclear groups have suggested that in 2010 solar power became cheaper than nuclear power. Data from the EIA in 2011 estimated that in 2016, solar will have a levelized cost of electricity almost twice that of nuclear (21¢/kWh for solar, 11.39¢/kWh for nuclear), and wind somewhat less (9.7¢/kWh). However, the US EIA has also cautioned that levelized costs of intermittent sources such as wind and solar are not directly comparable to costs of "dispatchable" sources (those that can be adjusted to meet demand), as intermittent sources need costly large-scale back-up power supplies for when the weather changes.

A 2010 study by the Global Subsidies Initiative compared global relative energy subsidies, or government financial aid to different energy sources, with this aid not solely funnelled into research and development but into bribing or "incentivizing" utilities to pursue renewable energy systems, over other options. Results show that fossil fuels receive about 1 US cents per kWh of energy they produce, nuclear energy receives 1.7 cents / kWh, renewable energy (excluding hydroelectricity) receives 5.0 cents / kWh and biofuels receive 5.1 cents / kWh in subsidies.

There is however no small volume of intensely radioactive spent fuel that needs to be stored or reprocessed with conventional renewable energy sources. A nuclear plant needs to be disassembled and removed. Much of the disassembled nuclear plant needs to be stored as low level nuclear waste for a few decades. However, from a safety stand point, nuclear power, in terms of lives lost per unit of electricity delivered, is comparable to and in some cases, lower than many renewable energy sources.

Nuclear Decommissioning

The financial costs of every nuclear power plant continues for some time after the facility has finished generating its last useful electricity. Once no longer economically viable, nuclear reactors and uranium enrichment facilities are generally decommissioned, returning the facility and its parts to a safe enough level to be entrusted for other uses, such as greenfield status. After a cooling-off period that may last decades, reactor core materials are dismantled and cut into small pieces to be packed in containers for interim storage or transmutation experiments. The consensus on how to approach the task is one that is relatively inexpensive, but it has the potential to be hazardous to the natural environment as it presents opportunities for human error, accidents or sabotage.

In the USA a Nuclear Waste Policy Act and Nuclear Decommissioning Trust Fund is legally required, with utilities banking 0.1 to 0.2 cents/kWh during operations to fund future decommissioning. They must report regularly to the NRC on the status of their decommissioning funds. About 70% of the total estimated cost of decommissioning all US nuclear power reactors has already been collected (on the basis of the average cost of $320 million per reactor-steam turbine unit).

In the U.S. in 2011, there are 13 reactors that had permanently shut down and are in some phase of decommissioning. With Connecticut Yankee Nuclear Power Plant and Yankee Rowe Nuclear Power Station having completed the process in 2006-2007, after ceasing commercial electricity production circa 1992. The majority of the 15 years, was used to allow the station to naturally cooldown on its own, which makes the manual disassembly process both safer and cheaper. Decommissioning at nuclear sites which have experienced a serious accident are the most expensive and time-consuming.

Working under an insurance framework that limits or structures accident liabilities in accordance with the Paris convention on nuclear third-party liability, the Brussels supplementary convention, and the Vienna convention on civil liability for nuclear damage and in the U.S. the Price-Anderson Act. It is often argued that this potential shortfall in liability represents an external cost not included in the cost of nuclear electricity; but the cost is small, amounting to about 0.1% of the levelized cost of electricity, according to a CBO study.

These beyond-regular-insurance costs for worst-case scenarios are not unique to nuclear power, as hydroelectric power plants are similarly not fully insured against a catastrophic event such as the Banqiao Dam disaster, where 11 million people lost their homes and from 30,000 to 200,000 people died, or large dam failures in general. As private insurers base dam insurance premiums on limited scenarios, major disaster insurance in this sector is likewise provided by the state.

Debate on Nuclear Power

The nuclear power debate concerns the controversy which has surrounded the deployment and use of nuclear fission reactors to generate electricity from nuclear fuel for civilian purposes. The debate about nuclear power peaked during the 1970s and 1980s, when it "reached an intensity unprecedented in the history of technology controversies", in some countries.

Proponents of nuclear energy contend that nuclear power is a sustainable energy source that reduces carbon emissions and increases energy security by decreasing dependence on imported energy sources. Proponents claim that nuclear power produces virtually no conventional air pollution, such as greenhouse gases and smog, in contrast to the chief viable alternative of fossil fuel. Nuclear power can produce base-load power unlike many renewables which are intermittent energy sources lacking large-scale and cheap ways of storing energy. M. King Hubbert saw oil as a resource that would run out, and proposed nuclear energy as a replacement energy source. Proponents claim that the risks of storing waste are small and can be further reduced by using the latest technology in newer reactors, and the operational safety record in the Western world is excellent when compared to the other major kinds of power plants.

Opponents believe that nuclear power poses many threats to people and the environment. These threats include the problems of processing, transport and storage of radioactive nuclear waste, the risk of nuclear weapons proliferation and terrorism, as well as health risks and environmental damage from uranium mining. They also contend that reactors themselves are enormously complex machines where many things can and do go wrong; and there have been serious nuclear accidents. Critics do not believe that the risks of using nuclear fission as a power source can be fully offset through the development of new technology. They also argue that when all the energy-intensive stages of the nuclear fuel chain are considered, from uranium mining to nuclear decommissioning, nuclear power is neither a low-carbon nor an economical electricity source.

Arguments of economics and safety are used by both sides of the debate.

Use in Space

Both fission and fusion appear promising for space propulsion applications, generating higher mission velocities with less reaction mass. This is due to the much higher energy density of nuclear reactions: some 7 orders of magnitude (10,000,000 times) more energetic than the chemical reactions which power the current generation of rockets.

Radioactive decay has been used on a relatively small scale (few kW), mostly to power space missions and experiments by using radioisotope thermoelectric generators such as those developed at Idaho National Laboratory.

Research

Advanced Concepts

Current fission reactors in operation around the world are second or third generation systems, with most of the first-generation systems having been retired some time ago. Research into advanced generation IV reactor types was officially started by the Generation IV International Forum (GIF) based on eight technology goals, including to improve nuclear safety, improve proliferation resistance, minimize waste, improve natural resource utilization, the ability to consume existing nuclear waste in the production of electricity, and decrease the cost to build and run such plants. Most of these reactors differ significantly from current operating light water reactors, and are generally not expected to be available for commercial construction before 2030.

The nuclear reactors to be built at Vogtle are new AP1000 third generation reactors, which are said to have safety improvements over older power reactors. However, John Ma, a senior structural engineer at the NRC, is concerned that some parts of the AP1000 steel skin are so brittle that the "impact energy" from a plane strike or storm driven projectile could shatter the wall. Edwin Lyman, a senior staff scientist at the Union of Concerned Scientists, is concerned about the strength of the steel containment vessel and the concrete shield building around the AP1000.

The Union of Concerned Scientists has referred to the EPR (nuclear reactor), currently under construction in China, Finland and France, as the only new reactor design under consideration in the United States that "...appears to have the potential to be significantly safer and more secure against attack than today's reactors."

One disadvantage of any new reactor technology is that safety risks may be greater initially as reactor operators have little experience with the new design. Nuclear engineer David Lochbaum has explained that almost all serious nuclear accidents have occurred with what was at the time the most recent technology. He argues that "the problem with new reactors and accidents is twofold: scenarios arise that are impossible to plan for in simulations; and humans make mistakes". As one director of a U.S. research laboratory put it, "fabrication, construction, operation, and maintenance of new reactors will face a steep learning curve: advanced technologies will have a heightened risk of accidents and mistakes. The technology may be proven, but people are not".

Hybrid Nuclear Fusion-fission

Hybrid nuclear power is a proposed means of generating power by use of a combination of nuclear fusion and fission processes. The concept dates to the 1950s, and was briefly advocated by Hans

Bethe during the 1970s, but largely remained unexplored until a revival of interest in 2009, due to delays in the realization of pure fusion. When a sustained nuclear fusion power plant is built, it has the potential to be capable of extracting all the fission energy that remains in spent fission fuel, reducing the volume of nuclear waste by orders of magnitude, and more importantly, eliminating all actinides present in the spent fuel, substances which cause security concerns.

Nuclear Fusion

Nuclear fusion reactions have the potential to be safer and generate less radioactive waste than fission. These reactions appear potentially viable, though technically quite difficult and have yet to be created on a scale that could be used in a functional power plant. Fusion power has been under theoretical and experimental investigation since the 1950s.

Construction of the ITER facility began in 2007, but the project has run into many delays and budget overruns. The facility is now not expected to begin operations until the year 2027 – 11 years after initially anticipated. A follow on commercial nuclear fusion power station, DEMO, has been proposed. There are also suggestions for a power plant based upon a different fusion approach, that of an inertial fusion power plant.

Fusion powered electricity generation was initially believed to be readily achievable, as fission-electric power had been. However, the extreme requirements for continuous reactions and plasma containment led to projections being extended by several decades. In 2010, more than 60 years after the first attempts, commercial power production was still believed to be unlikely before 2050.

References

- Bain, Alastair S.; et al. (1997). Canada enters the nuclear age: a technical history of Atomic Energy of Canada. Magill-Queen's University Press. p. ix. ISBN 0-7735-1601-8.

- Pfau, Richard (1984) No Sacrifice Too Great: The Life of Lewis L. Strauss University Press of Virginia, Charlottesville, Virginia, p. 187, ISBN 978-0-8139-1038-3

- David Bodansky (2004). Nuclear Energy: Principles, Practices, and Prospects. Springer. p. 32. ISBN 978-0-387-20778-0. Retrieved 2008-01-31.

- Kragh, Helge (1999). Quantum Generations: A History of Physics in the Twentieth Century. Princeton NJ: Princeton University Press. p. 286. ISBN 0-691-09552-3.

- McKeown, William (2003). Idaho Falls: The Untold Story of America's First Nuclear Accident. Toronto: ECW Press. ISBN 978-1-55022-562-4.

- Bernard L. Cohen (1990). The Nuclear Energy Option: An Alternative for the 90s. New York: Plenum Press. ISBN 978-0-306-43567-6.

- Rüdig, Wolfgang, ed. (1990). Anti-nuclear Movements: A World Survey of Opposition to Nuclear Energy. Detroit, MI: Longman Current Affairs. p. 1. ISBN 0-8103-9000-0.

- Falk, Jim (1982). Global Fission: The Battle Over Nuclear Power. Melbourne: Oxford University Press. pp. 95–96. ISBN 978-0-19-554315-5.

- Uranium 2007 – Resources, Production and Demand. Nuclear Energy Agency, Organisation for Economic Co-operation and Development. 2008-06-10. ISBN 978-92-64-04766-2.

- Ojovan, M. I.; Lee, W.E. (2005). An Introduction to Nuclear Waste Immobilisation. Amsterdam: Elsevier Science Publishers. p. 315. ISBN 0-08-044462-8.

- National Research Council (1995). Technical Bases for Yucca Mountain Standards. Washington, D.C.: National Academy Press. p. 91. ISBN 0-309-05289-0.

- "World doubles new build reactor capacity in 2015". London, UK: World Nuclear News. 4 January 2016. Retrieved 7 March 2016.

- "World's First APR-1400 Connected to Grid". Washington DC, USA: NEI (Nuclear Energy Institute). 21 January 2016. Retrieved 7 March 2016.

- Blue Ribbon Commission on America's Nuclear Future. "Disposal Subcommittee Report to the Full Commission" (PDF). Retrieved 1 January 2016.

- Blau, Max (2016-10-20). "First new US nuclear reactor in 20 years goes live". CNN.com. Cable News Network. Turner Broadcasting System, Inc. Retrieved 2016-10-20.

Nuclear Power Plant: An Overview

A nuclear power plant uses energy produced by nuclear reactions to convert water into steam, which, in turn is used to run steam turbines to produce electricity. The radioactive fuel undergoes nuclear reaction in the reactor. There is a nuclear reactor coolant that is circulated to keep the temperature optimal and controlled. This chapter illustrates the various parts of a thermonuclear reactor and discusses the function and design of each.

Nuclear Power Plant

A nuclear power plant or nuclear power station is a thermal power station in which the heat source is a nuclear reactor. As is typical in all conventional thermal power stations the heat is used to generate steam which drives a steam turbine connected to an electric generator which produces electricity. As of 23 April 2014, the IAEA report there are 435 nuclear power reactors in operation operating in 31 countries. Nuclear power stations are usually considered to be base load stations, since fuel is a small part of the cost of production. Their operations and maintenance (O&M) and fuel costs are, along with hydropower stations, at the low end of the spectrum and make them suitable as base-load power suppliers. The cost of spent fuel management, however, is somewhat uncertain.

A nuclear power station (Grafenrheinfeld Nuclear Power Plant, Grafenrheinfeld, Bavaria, Germany). The nuclear reactor is contained inside the spherical containment building in the center – left and right are cooling towers which are common cooling devices used in all thermal power stations, and likewise, emit water vapor from the non-radioactive steam turbine section of the power plant.

History

Electricity was generated by a nuclear reactor for the first time ever on September 3, 1948 at the X-10 Graphite Reactor in Oak Ridge, Tennessee in the United States, and was the first nuclear power station to power a light bulb. The second, larger experiment occurred on December 20, 1951 at the EBR-I experimental station near Arco, Idaho in the United States. On June 27, 1954, the world's first nuclear power station to generate electricity for a power grid started operations at the Soviet city of Obninsk. The world's first full scale power station, Calder Hall in England opened on October 17, 1956.

The control room at an American nuclear power station

Systems

The conversion to electrical energy takes place indirectly, as in conventional thermal power stations. The fission in a nuclear reactor heats the reactor coolant. The coolant may be water or gas or even liquid metal depending on the type of reactor. The reactor coolant then goes to a steam generator and heats water to produce steam. The pressurized steam is then usually fed to a multi-stage steam turbine. After the steam turbine has expanded and partially condensed the steam, the remaining vapor is condensed in a condenser. The condenser is a heat exchanger which is connected to a secondary side such as a river or a cooling tower. The water is then pumped back into the steam generator and the cycle begins again. The water-steam cycle corresponds to the Rankine cycle.

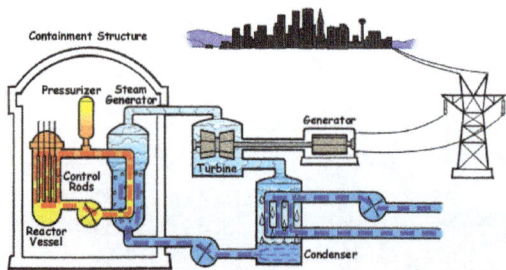

Pressurized water reactor

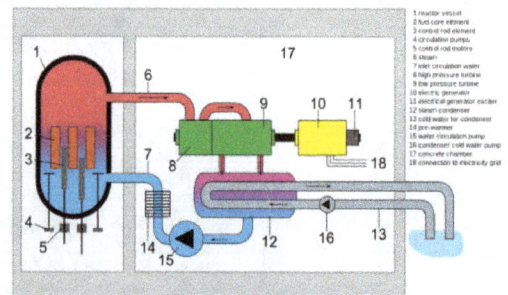

BWR schematic

Nuclear Reactors

The nuclear reactor is the heart of the station. In its central part, the reactor core's heat is generated by controlled nuclear fission. With this heat, a coolant is heated as it is pumped through the reactor and thereby removes the energy from the reactor. Heat from nuclear fission is used to raise steam, which runs through turbines, which in turn powers the electrical generators.

Nuclear reactors usually rely on uranium to fuel the chain reaction. Uranium is a very heavy metal that is abundant on Earth and is found in sea water as well as most rocks. Naturally occurring uranium is found in two different isotopes: uranium-238 (U-238), accounting for 99.3% and ura-nium-235 (U-235) accounting for about 0.7%. Isotopes are atoms of the same element with a different number of neutrons. Thus, U-238 has 146 neutrons and U-235 has 143 neutrons. Different isotopes have different behaviors. For instance, U-235 is fissile which means that it is easily split and gives off a lot of energy making it ideal for nuclear energy. On the other hand, U-238 does not have that property despite it being the same element. Different isotopes also have different half-lives. A half-life is the amount of time it takes for half of a sample of a radioactive element to decay. U-238 has a longer half-life than U-235, so it takes longer to decay over time. This also means that U-238 is less radioactive than U-235.

Since nuclear fission creates radioactivity, the reactor core is surrounded by a protective shield. This containment absorbs radiation and prevents radioactive material from being released into the environment. In addition, many reactors are equipped with a dome of concrete to protect the reactor against both internal casualties and external impacts.

Steam Turbine

The purpose of the steam turbine is to convert the heat contained in steam into mechanical energy. The engine house with the steam turbine is usually structurally separated from the main reactor building. It is so aligned to prevent debris from the destruction of a turbine in operation from flying towards the reactor.

In the case of a pressurized water reactor, the steam turbine is separated from the nuclear system. To detect a leak in the steam generator and thus the passage of radioactive water at an early stage, an activity meter is mounted to track the outlet steam of the steam generator. In contrast, boiling water reactors pass radioactive water through the steam turbine, so the turbine is kept as part of the radiologically controlled area of the nuclear power station.

Generator

The generator converts mechanical power supplied by the turbine into electrical power. Low-pole AC synchronous generators of high rated power are used.

Cooling System

A cooling system removes heat from the reactor core and transports it to another area of the station, where the thermal energy can be harnessed to produce electricity or to do other useful work. Typically the hot coolant is used as a heat source for a boiler, and the pressurized steam from that drives one or more steam turbine driven electrical generators.

Safety Valves

In the event of an emergency, safety valves can be used to prevent pipes from bursting or the reactor from exploding. The valves are designed so that they can derive all of the supplied flow rates with little increase in pressure. In the case of the BWR, the steam is directed into the suppression chamber and condenses there. The chambers on a heat exchanger are connected to the intermediate cooling circuit.

Feedwater Pump

The water level in the steam generator and nuclear reactor is controlled using the feedwater system. The feedwater pump has the task of taking the water from the condensate system, increasing the pressure and forcing it into either the steam generators (in the case of a pressurized water reactor) or directly into the reactor (for boiling water reactors).

Emergency Power Supply

Most nuclear stations require two distinct sources of offsite power feeding station service transformers that are sufficiently separated in the stations's switchyard and can receive power from multiple transmission lines. In addition in some nuclear stations the turbine generator can power the station's house loads while the station is online via station service transformers which tap power from the generator output bus bars before they reach the step-up transformer (these stations also have station service transformers that receive offsite power directly from the switch yard). Even with the redundancy of two power sources total loss of offsite power is still possible. Nuclear power stations are equipped with emergency power.

Workers in A Nuclear Power Station

- Nuclear engineers
- Reactor operators
- Health physicists
- Emergency response team personnel
- Nuclear Regulatory Commission Resident Inspectors

In the United States and Canada, workers except for management, professional (such as engineers) and security personnel are likely to be members of either the International Brotherhood of Electrical Workers (IBEW) or the Utility Workers Union of America (UWUA), or one of the various trades and labor unions representing Machinist, laborers, boilermakers, millwrights, iron workers etc.

Economics

The Bruce Nuclear Generating Station, the largest nuclear power facility in the world

The economics of new nuclear power stations is a controversial subject, and multibillion-dollar investments ride on the choice of an energy source. Nuclear power stations typically have high capital costs, but low direct fuel costs, with the costs of fuel extraction, processing, use and spent fuel storage internalized costs. Therefore, comparison with other power generation methods is strongly dependent on assumptions about construction timescales and capital financing for nuclear stations. Cost estimates take into account station decommissioning and nuclear waste storage or recycling costs in the United States due to the Price Anderson Act. With the prospect that all spent nuclear fuel/"nuclear waste" could potentially be recycled by using future reactors, generation IV reactors are being designed to completely close the nuclear fuel cycle. Presently, however there has not yet been any actual bulk recycling of waste from a NPP, and on-site temporary storage is still being used at almost all plant sites due to waste repository construction problems. Only Finland has stable repository plans, therefore from a world-wide perspective, long-term waste storage costs are uncertain.

Nuclear reactors in operation releasing hot steam as a side product

On the other hand, construction, or capital cost aside, measures to mitigate global warming such as a carbon tax or carbon emissions trading, increasingly favor the economics of nuclear power. Further efficiencies are hoped to be achieved through more advanced reactor designs, Generation III reactors promise to be at least 17% more fuel efficient, and have lower capital costs, while futuristic Generation IV reactors promise 10000-30000% greater fuel efficiency and the elimination of nuclear waste.

In Eastern Europe, a number of long-established projects are struggling to find finance, notably Belene in Bulgaria and the additional reactors at Cernavoda in Romania, and some potential backers have pulled out. Where cheap gas is available and its future supply relatively secure, this also poses a major problem for nuclear projects.

Analysis of the economics of nuclear power must take into account who bears the risks of future uncertainties. To date all operating nuclear power stations were developed by state-owned or regulated utility monopolies where many of the risks associated with construction costs, operating performance, fuel price, and other factors were borne by consumers rather than suppliers. Many countries have now liberalized the electricity market where these risks, and the risk of cheaper competitors emerging before capital costs are recovered, are borne by station suppliers and operators rather than consumers, which leads to a significantly different evaluation of the economics of new nuclear power stations.

Following the 2011 Fukushima I nuclear accidents, costs are likely to go up for currently operating and new nuclear power stations, due to increased requirements for on-site spent fuel management and elevated design basis threats. However many designs, such as the currently under construction AP1000, use passive nuclear safety cooling systems, unlike those of Fukushima I which required active cooling systems, this largely eliminates the necessity to spend more on redundant back up safety equipment.

Safety and Accidents

In his book, *Normal accidents*, Charles Perrow says that multiple and unexpected failures are built into society's complex and tightly-coupled nuclear reactor systems. Such accidents are unavoidable and cannot be designed around. An interdisciplinary team from MIT has estimated that given the expected growth of nuclear power from 2005 – 2055, at least four serious nuclear accidents would be expected in that period. However the MIT study does not take into account improvements in safety since 1970. To date, there have been five serious accidents (core damage) in the world since 1970 (one at Three Mile Island in 1979; one at Chernobyl in 1986; and three at Fukushima-Daiichi in 2011), corresponding to the beginning of the operation of generation II reactors. This leads to on average one serious accident happening every eight years worldwide.

Nuclear power plants cannot explode like a nuclear bomb because the fuel for uranium reactors is not enriched enough, and nuclear weapons require precision explosives to force fuel into a small enough volume to go supercritical. Most reactors require continuous temperature control to prevent a core meltdown, which has occurred on a few occasions through accident or natural disaster, releasing radiation and making the surrounding area uninhabitable. Modern nuclear reactor designs however, have had numerous safety improvements since the first generation nuclear reactors. Plants must be defended against theft of nuclear material (for example to make a dirty bomb) and attack by enemy military (which has occurred) planes or missiles, or planes hijacked by terrorists.

Controversy

The nuclear power debate is about the controversy which has surrounded the deployment and use of nuclear fission reactors to generate electricity from nuclear fuel for civilian purposes. The debate about nuclear power peaked during the 1970s and 1980s, when it "reached an intensity unprecedented in the history of technology controversies", in some countries.

The abandoned city of Prypiat, Ukraine, following the Chernobyl disaster. The Chernobyl nuclear power station is in the background.

Proponents argue that nuclear power is a sustainable energy source which reduces carbon emissions and can increase energy security if its use supplants a dependence on imported fuels. Proponents advance the notion that nuclear power produces virtually no air pollution, in contrast to the chief viable alternative of fossil fuel. Proponents also believe that nuclear power is the only viable course to achieve energy independence for most Western countries. They emphasize that the risks of storing waste are small and can be further reduced by using the latest technology in newer reactors, and the operational safety record in the Western world is excellent when compared to the other major kinds of power plants.

Opponents say that nuclear power poses many threats to people and the environment. These threats include health risks and environmental damage from uranium mining, processing and transport, the risk of nuclear weapons proliferation or sabotage, and the unsolved problem of radioactive nuclear waste. The environment issue is also regarding discharge of hot water into the sea. The hot water modifies the environmental conditions for the marine flora fauna. They also contend that reactors themselves are enormously complex machines where many things can and do go wrong, and there have been many serious nuclear accidents. Critics do not believe that these risks can be reduced through new technology. They argue that when all the energy-intensive stages of the nuclear fuel chain are considered, from uranium mining to nuclear decommissioning, nuclear power is not a low-carbon electricity source.

Reprocessing

Nuclear reprocessing technology was developed to chemically separate and recover fissionable plutonium from irradiated nuclear fuel. Reprocessing serves multiple purposes, whose relative importance has changed over time. Originally reprocessing was used solely to extract plutonium for producing nuclear weapons. With the commercialization of nuclear power, the reprocessed plutonium was recycled back into MOX nuclear fuel for thermal reactors. The reprocessed uranium, which constitutes the bulk of the spent fuel material, can in principle also be re-used as fuel, but that is only economic when uranium prices are high or disposal is expensive. Finally, the breeder reactor can employ not only the recycled plutonium and uranium in spent fuel, but all the actinides, closing the nuclear fuel cycle and potentially multiplying the energy extracted from natural uranium by more than 60 times.

Nuclear reprocessing reduces the volume of high-level waste, but by itself does not reduce radioactivity or heat generation and therefore does not eliminate the need for a geological waste repository. Reprocessing has been politically controversial because of the potential to contribute to nuclear proliferation, the potential vulnerability to nuclear terrorism, the political challenges of repository siting (a problem that applies equally to direct disposal of spent fuel), and because of its high cost compared to the once-through fuel cycle. In the United States, the Obama administration stepped back from President Bush's plans for commercial-scale reprocessing and reverted to a program focused on reprocessing-related scientific research.

Accident Indemnification

The Vienna Convention on Civil Liability for Nuclear Damage puts in place an international framework for nuclear liability. However states with a majority of the world's nuclear power stations, including the U.S., Russia, China and Japan, are not party to international nuclear liability conventions.

In the U.S., insurance for nuclear or radiological incidents is covered (for facilities licensed through 2025) by the Price-Anderson Nuclear Industries Indemnity Act.

Under the Energy policy of the United Kingdom through its Nuclear Installations Act of 1965, liability is governed for nuclear damage for which a UK nuclear licensee is responsible. The Act requires compensation to be paid for damage up to a limit of £150 million by the liable operator for ten years after the incident. Between ten and thirty years afterwards, the Government meets this obligation. The Government is also liable for additional limited cross-border liability (about £300 million) under international conventions (Paris Convention on Third Party Liability in the Field of Nuclear Energy and Brussels Convention supplementary to the Paris Convention).

Decommissioning

Nuclear decommissioning is the dismantling of a nuclear power station and decontamination of the site to a state no longer requiring protection from radiation for the general public. The main difference from the dismantling of other power stations is the presence of radioactive material that requires special precautions to remove and safely relocate to a waste repository.

Warranty period of operation of nuclear power stations is 30 years. One of the wear factors is the deterioration of the reactors shell under the action of neutron bombardment.

Generally speaking, nuclear stations were designed for a life of about 30 years. Newer stations are designed for a 40 to 60-year operating life.

Decommissioning involves many administrative and technical actions. It includes all clean-up of radioactivity and progressive demolition of the station. Once a facility is decommissioned, there should no longer be any danger of a radioactive accident or to any persons visiting it. After a facility has been completely decommissioned it is released from regulatory control, and the licensee of the station no longer has responsibility for its nuclear safety.

Historic Accidents

The nuclear industry says that new technology and oversight have made nuclear station much safer, but 57 small accidents have occurred since the Chernobyl disaster in 1986 until 2008. Two thirds of these mishaps occurred in the US. The French Atomic Energy Agency (CEA) has concluded that technical innovation cannot eliminate the risk of human errors in nuclear station operation.

According to Benjamin Sovacool, an interdisciplinary team from MIT in 2003 estimated that given the expected growth of nuclear power from 2005 – 2055, at least four serious nuclear accidents would be expected in that period. However the MIT study does not take into account improvements in safety since 1970.

Flexibility of Nuclear Power Stations

Nuclear stations are used primarily for base load because of economic considerations. The fuel cost of operations for a nuclear station is smaller than the fuel cost for operation of coal or gas plants. There is no cost saving if you run a nuclear station at less than full capacity.

However, nuclear stations are routinely used in load following mode on a large scale in France, although "it is generally accepted that this is not an ideal economic situation for nuclear stations." Unit A at the German Biblis Nuclear Power Plant is designed to in- and decrease its output 15% per minute between 40 and 100% of its nominal power. Boiling water reactors normally have load-following capability, implemented by varying the recirculation water flow.

Future Power Stations

A new generation of designs for nuclear power stations, known as the Generation IV reactors, are the subject of active research. Many of these new designs specifically attempt to make fission reactors cleaner, safer and/or less of a risk to the proliferation of nuclear weapons. Passively safe stations (such as the ESBWR) are available to be built and other reactors that are designed to be nearly fool-proof are being pursued. Fusion reactors, which are still in the early stages of development, diminish or eliminate some of the risks associated with nuclear fission.

Two 1600 MWe European Pressurized Reactors (EPRs) are being built in Europe, and two are being built in China. The reactors are a joint effort of French AREVA and German Siemens AG, and will be the largest reactors in the world. One EPR is in Olkiluoto, Finland, as part of the Olkiluoto Nuclear Power Plant. The reactor was originally scheduled to go online in 2009, but has been repeatedly delayed, and as of September 2014 has been pushed back to 2018. Preparatory work for the EPR at the Flamanville Nuclear Power Plant in Flamanville, Manche, France was started in 2006, with a scheduled completion date of 2012. The French reactor has also been delayed, and was projected, in 2013, to launch in 2016. The two Chinese EPRs are part of the Taishan Nuclear Power Plant in Taishan, Guangdong. The Taishan reactors were scheduled to go online in 2014 and 2015, but that has been delayed to 2017.

As of March 2007, there are seven nuclear power stations under construction in India, and five in China.

In November 2011 Gulf Power stated that by the end of 2012 it hopes to finish buying off 4000 acres of land north of Pensacola, Florida in order to build a possible nuclear power station.

In 2010 Russia launched a floating nuclear power station. The £100 million vessel, the Akademik Lomonosov, is the first of seven stations that will bring vital energy resources to remote Russian regions.

By 2025, Southeast Asia nations would have a total of 29 nuclear power stations, Indonesia will have 4 nuclear power stations, Malaysia 4, Thailand 5 and Vietnam 16 from nothing at all in 2011.

In 2013 China had 32 nuclear reactors under construction, the highest number in the world.

Expansion at two Nuclear Power Stations in the United States, Vogtle and V. C. Summer Nuclear Power Station, located in Georgia and South Carolina, respectively, are scheduled to be completed between 2016 and 2019. The two new Vogtle reactors, and the two new reactors at Virgil C. Summer Nuclear Station, represent the first nuclear power construction projects in the United States since the Three Mile Island nuclear accident in 1979.

Several countries have begun Thorium-based nuclear power programs. Thorium is four times more abundant within nature than uranium. Over 60% of thorium's ore monazite is found in five countries: Australia, the United States, India, Brazil, and Norway. These thorium resources are enough to power current energy needs for thousands of years. The thorium fuel cycle is able to generate nuclear energy with a lower output of radiotoxic waste than the uranium fuel cycle.

Nuclear Reactor

A nuclear reactor, formerly known as an atomic pile, is a device used to initiate and control a sustained nuclear chain reaction. Nuclear reactors are used at nuclear power plants for electricity generation and in propulsion of ships. Heat from nuclear fission is passed to a working fluid (water or gas), which runs through steam turbines. These either drive a ship's propellers or turn electrical generators. Nuclear generated steam in principle can be used for industrial process heat or for district heating. Some reactors are used to produce isotopes for medical and industrial use, or for production of weapons-grade plutonium. Some are run only for research. Today there are about 450 nuclear power reactors that are used to generate electricity in about 30 countries around the world.

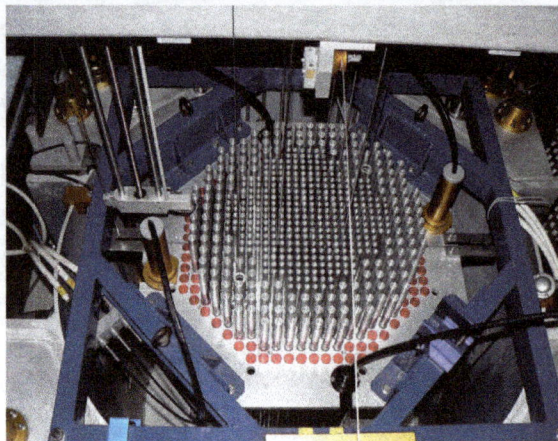

Core of CROCUS, a small nuclear reactor used for research at the EPFL in Switzerland

Mechanism

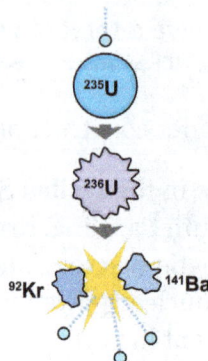

An induced nuclear fission event. A neutron is absorbed by the nucleus of a uranium-235 atom, which in turn splits into fast-moving lighter elements (fission products) and free neutrons. Though both reactors and nuclear weapons rely on nuclear chain-reactions, the rate of reactions in a reactor occurs much more slowly than in a bomb.

Just as conventional power-stations generate electricity by harnessing the thermal energy released from burning fossil fuels, nuclear reactors convert the energy released by controlled nuclear fission into thermal energy for further conversion to mechanical or electrical forms.

Fission

When a large fissile atomic nucleus such as uranium-235 or plutonium-239 absorbs a neutron, it may undergo nuclear fission. The heavy nucleus splits into two or more lighter nuclei, (the fission products), releasing kinetic energy, gamma radiation, and free neutrons. A portion of these neutrons may later be absorbed by other fissile atoms and trigger further fission events, which release more neutrons, and so on. This is known as a nuclear chain reaction.

To control such a nuclear chain reaction, neutron poisons and neutron moderators can change the portion of neutrons that will go on to cause more fission. Nuclear reactors generally have automatic and manual systems to shut the fission reaction down if monitoring detects unsafe conditions.

Commonly-used moderators include regular (light) water (in 74.8% of the world's reactors), solid graphite (20% of reactors) and heavy water (5% of reactors). Some experimental types of reactor have used beryllium, and hydrocarbons have been suggested as another possibility.

Heat Generation

The reactor core generates heat in a number of ways:

- The kinetic energy of fission products is converted to thermal energy when these nuclei collide with nearby atoms.

- The reactor absorbs some of the gamma rays produced during fission and converts their energy into heat.

- Heat is produced by the radioactive decay of fission products and materials that have been activated by neutron absorption. This decay heat-source will remain for some time even after the reactor is shut down.

A kilogram of uranium-235 (U-235) converted via nuclear processes releases approximately three million times more energy than a kilogram of coal burned conventionally (7.2×10^{13} joules per kilogram of uranium-235 versus 2.4×10^7 joules per kilogram of coal).

Cooling

A nuclear reactor coolant — usually water but sometimes a gas or a liquid metal (like liquid sodium) or molten salt — is circulated past the reactor core to absorb the heat that it generates. The heat is carried away from the reactor and is then used to generate steam. Most reactor systems employ a cooling system that is physically separated from the water that will be boiled to produce pressurized steam for the turbines, like the pressurized water reactor. However, in some reactors the water for the steam turbines is boiled directly by the reactor core; for example the boiling water reactor.

Reactivity Control

The power output of the reactor is adjusted by controlling how many neutrons are able to create more fissions.

Control rods that are made of a neutron poison are used to absorb neutrons. Absorbing more neutrons in a control rod means that there are fewer neutrons available to cause fission, so pushing the control rod deeper into the reactor will reduce its power output, and extracting the control rod will increase it.

At the first level of control in all nuclear reactors, a process of delayed neutron emission by a number of neutron-rich fission isotopes is an important physical process. These delayed neutrons account for about 0.65% of the total neutrons produced in fission, with the remainder (termed "prompt neutrons") released immediately upon fission. The fission products which produce delayed neutrons have half lives for their decay by neutron emission that range from milliseconds to as long as several minutes, and so considerable time is required to determine exactly when a reactor reaches the critical point. Keeping the reactor in the zone of chain-reactivity where delayed neutrons are *necessary* to achieve a critical mass state allows mechanical devices or human operators to control a chain reaction in "real time"; otherwise the time between achievement of criticality and nuclear meltdown as a result of an exponential power surge from the normal nuclear chain reaction, would be too short to allow for intervention. This last stage, where delayed neutrons are no longer required to maintain criticality, is known as the prompt critical point. There is a scale for describing criticality in numerical form, in which bare criticality is known as *zero dollars* and the prompt critical point is *one dollar*, and other points in the process interpolated in cents.

In some reactors, the coolant also acts as a neutron moderator. A moderator increases the power of the reactor by causing the fast neutrons that are released from fission to lose energy and become thermal neutrons. Thermal neutrons are more likely than fast neutrons to cause fission. If the coolant is a moderator, then temperature changes can affect the density of the coolant/moderator and therefore change power output. A higher temperature coolant would be less dense, and therefore a less effective moderator.

In other reactors the coolant acts as a poison by absorbing neutrons in the same way that the control rods do. In these reactors power output can be increased by heating the coolant, which makes it a less dense poison. Nuclear reactors generally have automatic and manual systems to scram the reactor in an emergency shut down. These systems insert large amounts of poison (often boron in the form of boric acid) into the reactor to shut the fission reaction down if unsafe conditions are detected or anticipated.

Most types of reactors are sensitive to a process variously known as xenon poisoning, or the iodine pit. The common fission product Xenon-135 produced in the fission process acts as a neutron poison that absorbs neutrons and therefore tends to shut the reactor down. Xenon-135 accumulation can be controlled by keeping power levels high enough to destroy it by neutron absorption as fast as it is produced. Fission also produces iodine-135, which in turn decays (with a half-life of 6.57 hours) to new xenon-135. When the reactor is shut down, iodine-135 continues to decay to xenon-135, making restarting the reactor more difficult for a day or two, as the xenon-135 decays into cesium-135, which is not nearly as poisonous as xenon-135, with a half-life of 9.2 hours. This temporary state is the "iodine pit." If the reactor has sufficient extra reactivity capacity, it can be re-started. As the extra xenon-135 is transmuted to xenon-136, which is much less a neutron poison, within a few hours the reactor experiences a "xenon burnoff (power) transient". Control rods must be further inserted to replace the neutron absorption of the lost xenon-135. Failure to properly follow such a procedure was a key step in the Chernobyl disaster.

Reactors used in nuclear marine propulsion (especially nuclear submarines) often cannot be run at continuous power around the clock in the same way that land-based power reactors are normally run, and in addition often need to have a very long core life without refueling. For this reason many designs use highly enriched uranium but incorporate burnable neutron poison in the fuel rods. This allows the reactor to be constructed with an excess of fissionable material, which is nevertheless made relatively safe early in the reactor's fuel burn-cycle by the presence of the neutron-absorbing material which is later replaced by normally produced long-lived neutron poisons (far longer-lived than xenon-135) which gradually accumulate over the fuel load's operating life.

Electrical Power Generation

The energy released in the fission process generates heat, some of which can be converted into usable energy. A common method of harnessing this thermal energy is to use it to boil water to produce pressurized steam which will then drive a steam turbine that turns an alternator and generates electricity.

Early Reactors

The neutron was discovered in 1932. The concept of a nuclear chain reaction brought about by nuclear reactions mediated by neutrons was first realized shortly thereafter, by Hungarian scientist Leó Szilárd, in 1933. He filed a patent for his idea of a simple reactor the following year while working at the Admiralty in London. However, Szilárd's idea did not incorporate the idea of nuclear fission as a neutron source, since that process was not yet discovered. Szilárd's ideas for nuclear reactors using neutron-mediated nuclear chain reactions in light elements proved unworkable.

The Chicago Pile, the first nuclear reactor, built in secrecy at the University of Chicago in 1942 during World War 2 as part of the US's Manhattan project.

Inspiration for a new type of reactor using uranium came from the discovery by Lise Meitner, Fritz Strassmann and Otto Hahn in 1938 that bombardment of uranium with neutrons (provided by an alpha-on-beryllium fusion reaction, a "neutron howitzer") produced a barium residue, which they reasoned was created by the fissioning of the uranium nuclei. Subsequent studies in early 1939 (one of them by Szilárd and Fermi) revealed that several neutrons were also released during the fissioning, making available the opportunity for the nuclear chain reaction that Szilárd had envisioned six years previously.

Lise Meitner and Otto Hahn in their laboratory.

On 2 August 1939 Albert Einstein signed a letter to President Franklin D. Roosevelt (written by Szilárd) suggesting that the discovery of uranium's fission could lead to the development of "extremely powerful bombs of a new type", giving impetus to the study of reactors and fission. Szilárd and Einstein knew each other well and had worked together years previously, but Einstein had never thought about this possibility for nuclear energy until Szilard reported it to him, at the beginning of his quest to produce the Einstein-Szilárd letter to alert the U.S. government.

Shortly after, Hitler's Germany invaded Poland in 1939, starting World War II in Europe. The U.S. was not yet officially at war, but in October, when the Einstein-Szilárd letter was delivered to him, Roosevelt commented that the purpose of doing the research was to make sure "the Nazis don't blow us up." The U.S. nuclear project followed, although with some delay as there remained skepticism (some of it from Fermi) and also little action from the small number of officials in the government who were initially charged with moving the project forward.

The following year the U.S. Government received the Frisch–Peierls memorandum from the UK, which stated that the amount of uranium needed for a chain reaction was far lower than had previously been thought. The memorandum was a product of the MAUD Committee, which was working on the UK atomic bomb project, known as Tube Alloys, later to be subsumed within the Manhattan Project.

The Chicago Pile Team, including Enrico Fermi and Leó Szilárd.

Eventually, the first artificial nuclear reactor, Chicago Pile-1, was constructed at the University of Chicago, by a team led by Enrico Fermi, in late 1942. By this time, the program had been pressured for a year by U.S. entry into the war. The Chicago Pile achieved criticality on 2 December 1942 at 3:25 PM. The reactor support structure was made of wood, which supported a pile (hence the name) of graphite blocks, embedded in which was natural uranium-oxide 'pseudospheres' or 'briquettes'.

Soon after the Chicago Pile, the U.S. military developed a number of nuclear reactors for the Manhattan Project starting in 1943. The primary purpose for the largest reactors (located at the Hanford Site in Washington state), was the mass production of plutonium for nuclear weapons. Fermi and Szilard applied for a patent on reactors on 19 December 1944. Its issuance was delayed for 10 years because of wartime secrecy.

"World's first nuclear power plant" is the claim made by signs at the site of the EBR-I, which is now a museum near Arco, Idaho. Originally called "Chicago Pile-4", it was carried out under the direction of Walter Zinn for Argonne National Laboratory. This experimental LMFBR operated by the U.S. Atomic Energy Commission produced 0.8 kW in a test on 20 December 1951 and 100 kW (electrical) the following day, having a design output of 200 kW (electrical).

Besides the military uses of nuclear reactors, there were political reasons to pursue civilian use of atomic energy. U.S. President Dwight Eisenhower made his famous Atoms for Peace speech to the UN General Assembly on 8 December 1953. This diplomacy led to the dissemination of reactor technology to U.S. institutions and worldwide.

The first nuclear power plant built for civil purposes was the AM-1 Obninsk Nuclear Power Plant, launched on 27 June 1954 in the Soviet Union. It produced around 5 MW (electrical).

After World War II, the U.S. military sought other uses for nuclear reactor technology. Research by the Army and the Air Force never came to fruition; however, the U.S. Navy succeeded when they steamed the USS *Nautilus* (SSN-571) on nuclear power 17 January 1955.

The first commercial nuclear power station, Calder Hall in Sellafield, England was opened in 1956 with an initial capacity of 50 MW (later 200 MW).

The first portable nuclear reactor "Alco PM-2A" used to generate electrical power (2 MW) for Camp Century from 1960.

Components

The control room of NC State's Pulstar Nuclear Reactor.

The key components common to most types of nuclear power plants are:

- Nuclear fuel
- Nuclear reactor core
- Neutron moderator
- Neutron poison
- Neutron howitzer (provides steady source of neutrons to re-initiate reaction following shutdown)
- Coolant (often the Neutron Moderator and the Coolant are the same, usually both purified water)
- Control rods
- Reactor vessel
- Boiler feedwater pump
- Steam generators (not in BWRs)
- Steam turbine
- Electrical generator
- Condenser
- Cooling tower (not always required)
- Radwaste System (a section of the plant handling radioactive waste)
- Refueling Floor
- Spent fuel pool
- Nuclear safety systems
  - Reactor Protective System (RPS)
  - Emergency Diesel Generators
  - Emergency Core Cooling Systems (ECCS)
  - Standby Liquid Control System (emergency boron injection, in BWRs only)
- Essential service water system (ESWS)
- Containment building
- Control room
- Emergency Operations Facility
- Nuclear training facility (usually contains a Control Room simulator)

Reactor Types

NC State's PULSTAR Reactor is a 1 MW pool-type research reactor with 4% enriched, pin-type fuel consisting of UO_2 pellets in zircaloy cladding.

Classifications

Nuclear Reactors are classified by several methods; a brief outline of these classification methods is provided.

Classification By Type of Nuclear Reaction

Nuclear Fission

All commercial power reactors are based on nuclear fission. They generally use uranium and its product plutonium as nuclear fuel, though a thorium fuel cycle is also possible. Fission reactors can be divided roughly into two classes, depending on the energy of the neutrons that sustain the fission chain reaction:

- Thermal reactors (the most common type of nuclear reactor) use slowed or thermal neutrons to keep up the fission of their fuel. Almost all current reactors are of this type. These contain neutron moderator materials that slow neutrons until their neutron temperature is *thermalized*, that is, until their kinetic energy approaches the average kinetic energy of the surrounding particles. Thermal neutrons have a far higher cross-section (probability) of fissioning the fissile nuclei uranium-235, plutonium-239, and plutonium-241, and a relatively lower probability of neutron capture by uranium-238 (U-238) compared to the faster neutrons that originally result from fission, allowing use of low-enriched uranium or even natural uranium fuel. The moderator is often also the coolant, usually water under high pressure to increase the boiling point. These are surrounded by a reactor vessel, instrumentation to monitor and control the reactor, radiation shielding, and a containment building.

- Fast neutron reactors use fast neutrons to cause fission in their fuel. They do not have a neutron moderator, and use less-moderating coolants. Maintaining a chain reaction requires the fuel to be more highly enriched in fissile material (about 20% or more) due to the relatively lower probability of fission versus capture by U-238. Fast reactors have the potential to produce less transuranic waste because all actinides are fissionable with fast neutrons, but they are more difficult to build and more expensive to operate. Overall, fast reactors are less common than thermal reactors in most applications. Some early power stations were fast reactors, as are some Russian naval propulsion units. Construction of prototypes is continuing.

Nuclear Fusion

Fusion power is an experimental technology, generally with hydrogen as fuel. While not suitable for power production, Farnsworth-Hirsch fusors are used to produce neutron radiation.

Classification by Moderator Material

Used by thermal reactors:

- Graphite-moderated reactors

- Water moderated reactors

    o   Heavy-water reactors (Used in Canada, India, Argentina, China, Pakistan, Romania and South Korea).

    o   Light-water-moderated reactors (LWRs). Light-water reactors (the most common type of thermal reactor) use ordinary water to moderate and cool the reactors. When at operating temperature, if the temperature of the water increases, its density drops, and fewer neutrons passing through it are slowed enough to trigger further reactions. That negative feedback stabilizes the reaction rate. Graphite and heavy-water reactors tend to be more thoroughly thermalized than light water reactors. Due to the extra thermalization, these types can use natural uranium/unenriched fuel.

- Light-element-moderated reactors.

    o   Molten salt reactors (MSRs) are moderated by light elements such as lithium or beryllium, which are constituents of the coolant/fuel matrix salts LiF and BeF_2.

    o   Liquid metal cooled reactors, such as those whose coolant is a mixture of lead and bismuth, may use BeO as a moderator.

- Organically moderated reactors (OMR) use biphenyl and terphenyl as moderator and coolant.

Classification By Coolant

- Water cooled reactor. There are 104 operating reactors in the United States. Of these, 69 are pressurized water reactors (PWR), and 35 are boiling water reactors (BWR).

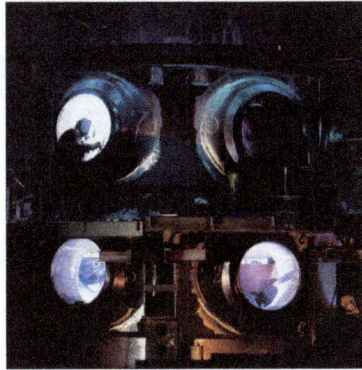

Treatment of the interior part of a VVER-1000 reactor frame on Atommash.

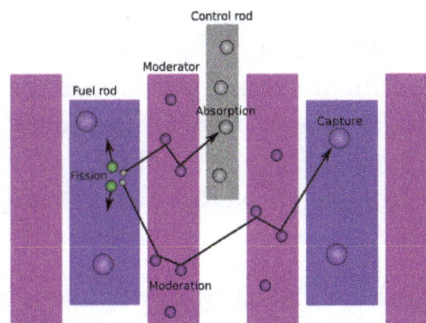

In thermal nuclear reactors (LWRs in specific), the coolant acts as a moderator that must slow down the neutrons before they can be efficiently absorbed by the fuel.

o   Pressurized water reactor (PWR) Pressurized water reactors constitute the large majority of all Western nuclear power plants.

- A primary characteristic of PWRs is a pressurizer, a specialized pressure vessel. Most commercial PWRs and naval reactors use pressurizers. During normal operation, a pressurizer is partially filled with water, and a steam bubble is maintained above it by heating the water with submerged heaters. During normal operation, the pressurizer is connected to the primary reactor pressure vessel (RPV) and the pressurizer "bubble" provides an expansion space for changes in water volume in the reactor. This arrangement also provides a means of pressure control for the reactor by increasing or decreasing the steam pressure in the pressurizer using the pressurizer heaters.

- Pressurised heavy water reactors are a subset of pressurized water reactors, sharing the use of a pressurized, isolated heat transport loop, but using heavy water as coolant and moderator for the greater neutron economies it offers.

o   Boiling water reactor (BWR)

- BWRs are characterized by boiling water around the fuel rods in the lower portion of a primary reactor pressure vessel. A boiling water reactor uses ^{235}U, enriched as uranium dioxide, as its fuel. The fuel is assembled into rods housed in a steel vessel that is submerged in water. The nuclear fission causes the water to boil, generating steam. This steam flows through pipes into turbines. The turbines are driven by the steam, and this process generates electricity. During normal operation, pressure is controlled by the amount of steam flowing from the reactor pressure vessel to the turbine.

- o   Pool-type reactor

- Liquid metal cooled reactor. Since water is a moderator, it cannot be used as a coolant in a fast reactor. Liquid metal coolants have included sodium, NaK, lead, lead-bismuth eutectic, and in early reactors, mercury.

  - o   Sodium-cooled fast reactor

  - o   Lead-cooled fast reactor

- Gas cooled reactors are cooled by a circulating inert gas, often helium in high-temperature designs, while carbon dioxide has been used in past British and French nuclear power plants. Nitrogen has also been used. Utilization of the heat varies, depending on the reactor. Some reactors run hot enough that the gas can directly power a gas turbine. Older designs usually run the gas through a heat exchanger to make steam for a steam turbine.

- Molten salt reactors (MSRs) are cooled by circulating a molten salt, typically a eutectic mixture of fluoride salts, such as FLiBe. In a typical MSR, the coolant is also used as a matrix in which the fissile material is dissolved.

Classification By Generation

- Generation I reactor (early prototypes, research reactors, non-commercial power producing reactors)

- Generation II reactor (most current nuclear power plants 1965–1996)

- Generation III reactor (evolutionary improvements of existing designs 1996-now)

- Generation IV reactor (technologies still under development unknown start date, possibly 2030)

In 2003, the French Commissariat à l'Énergie Atomique (CEA) was the first to refer to "Gen II" types in Nucleonics Week.

The first mentioning of "Gen III" was in 2000, in conjunction with the launch of the Generation IV International Forum (GIF) plans.

"Gen IV" was named in 2000, by the United States Department of Energy (DOE) for developing new plant types.

Classification By Phase of Fuel

- Solid fueled

- Fluid fueled

  o Aqueous homogeneous reactor

  o Molten salt reactor

- Gas fueled (theoretical)

Classification By Use

- Electricity

  o Nuclear power plants including small modular reactors

- Propulsion

  o Nuclear marine propulsion

  o Various proposed forms of rocket propulsion

- Other uses of heat

  o Desalination

  o Heat for domestic and industrial heating

  o Hydrogen production for use in a hydrogen economy

- Production reactors for transmutation of elements

  o Breeder reactors are capable of producing more fissile material than they con-sume during the fission chain reaction (by converting fertile U-238 to Pu-239, or Th-232 to U-233). Thus, a uranium breeder reactor, once running, can be re-fueled with natural or even depleted uranium, and a thorium breeder reactor can be re-fueled with thorium; however, an initial stock of fissile material is required.

  o Creating various radioactive isotopes, such as americium for use in smoke detec-tors, and cobalt-60, molybdenum-99 and others, used for imaging and medical treatment.

  o Production of materials for nuclear weapons such as weapons-grade plutonium

- Providing a source of neutron radiation (for example with the pulsed Godiva device) and positron radiation (e.g. neutron activation analysis and potassium-argon dating)

- Research reactor: Typically reactors used for research and training, materials testing, or the production of radioisotopes for medicine and industry. These are much small-er than power reactors or those propelling ships, and many are on university cam-puses. There are about 280 such reactors operating, in 56 countries. Some operate with high-enriched uranium fuel, and international efforts are underway to substitute low-enriched fuel.

Current Technologies

Diablo Canyon — a PWR

Pressurized water reactors (PWR)

These reactors use a pressure vessel to contain the nuclear fuel, control rods, moderator, and coolant. They are cooled and moderated by high-pressure liquid water. The hot radioactive water that leaves the pressure vessel is looped through a steam generator, which in turn heats a secondary (non-radioactive) loop of water to steam that can run turbines. They are the majority of current reactors. This is a thermal neutron reactor design, the newest of which are the VVER-1200, Advanced Pressurized Water Reactor and the European Pressurized Reactor. United States Naval reactors are of this type.

Boiling water reactors (BWR)

A BWR is like a PWR without the steam generator. A boiling water reactor is cooled and moderated by water like a PWR, but at a lower pressure, which allows the water to boil inside the pressure vessel producing the steam that runs the turbines. Unlike a PWR, there is no primary and secondary loop. The thermal efficiency of these reactors can be higher, and they can be simpler, and even potentially more stable and safe. This is a thermal neutron reactor design, the newest of which are the Advanced Boiling Water Reactor and the Economic Simplified Boiling Water Reactor.

The CANDU Qinshan Nuclear Power Plant

Pressurized Heavy Water Reactor (PHWR)

A Canadian design (known as CANDU), these reactors are heavy-water-cooled and -moderated pressurized-water reactors. Instead of using a single large pressure vessel as in a

PWR, the fuel is contained in hundreds of pressure tubes. These reactors are fueled with natural uranium and are thermal neutron reactor designs. PHWRs can be refueled while at full power, which makes them very efficient in their use of uranium (it allows for precise flux control in the core). CANDU PHWRs have been built in Canada, Argentina, China, India, Pakistan, Romania, and South Korea. India also operates a number of PHWRs, often termed 'CANDU-derivatives', built after the Government of Canada halted nuclear dealings with India following the 1974 Smiling Buddha nuclear weapon test.

The Ignalina Nuclear Power Plant — a RBMK type (closed 2009)

Reaktor Bolshoy Moschnosti Kanalniy (High Power Channel Reactor) (RBMK)

A Soviet design, built to produce plutonium as well as power. RBMKs are water cooled with a graphite moderator. RBMKs are in some respects similar to CANDU in that they are refuelable during power operation and employ a pressure tube design instead of a PWR-style pressure vessel. However, unlike CANDU they are very unstable and large, making containment buildings for them expensive. A series of critical safety flaws have also been identified with the RBMK design, though some of these were corrected following the Chernobyl disaster. Their main attraction is their use of light water and un-enriched uranium. As of 2010, 11 remain open, mostly due to safety improvements and help from international safety agencies such as the DOE. Despite these safety improvements, RBMK reactors are still considered one of the most dangerous reactor designs in use. RBMK reactors were deployed only in the former Soviet Union.

The Magnox Sizewell A nuclear power station

The Torness nuclear power station — an AGR

Gas-cooled reactor (GCR) and advanced gas-cooled reactor (AGR)

These are generally graphite moderated and CO_2 cooled. They can have a high thermal efficiency compared with PWRs due to higher operating temperatures. There are a number of operating reactors of this design, mostly in the United Kingdom, where the concept was developed. Older designs (i.e. Magnox stations) are either shut down or will be in the near future. However, the AGCRs have an anticipated life of a further 10 to 20 years. This is a thermal neutron reactor design. Decommissioning costs can be high due to large volume of reactor core.

Liquid-metal fast-breeder reactor (LMFBR)

This is a reactor design that is cooled by liquid metal, totally unmoderated, and produces more fuel than it consumes. They are said to "breed" fuel, because they produce fissionable fuel during operation because of neutron capture. These reactors can function much like a PWR in terms of efficiency, and do not require much high-pressure containment, as the liquid metal does not need to be kept at high pressure, even at very high temperatures. BN-350 and BN-600 in USSR and Superphénix in France were a reactor of this type, as was Fermi-I in the United States. The Monju reactor in Japan suffered a sodium leak in 1995 and was restarted in May 2010. All of them use/used liquid sodium. These reactors are fast neutron, not thermal neutron designs. These reactors come in two types:

The Superphénix, one of the few FBRs

Lead-cooled

Using lead as the liquid metal provides excellent radiation shielding, and allows for operation at very high temperatures. Also, lead is (mostly) transparent to neutrons, so fewer neutrons are lost in the coolant, and the coolant does not become radioactive. Unlike sodium, lead is mostly inert, so there is less risk of explosion or accident, but such large quantities of lead may be problematic from toxicology and disposal points of view. Often a reactor of this type would use a lead-bismuth eutectic mixture. In this case, the bismuth would present some minor radiation problems, as it is not quite as transparent to neutrons, and can be transmuted to a radioactive isotope more readily than lead. The Russian Alfa class submarine uses a lead-bismuth-cooled fast reactor as its main power plant.

Sodium-cooled

Most LMFBRs are of this type. The sodium is relatively easy to obtain and work with, and it also manages to actually prevent corrosion on the various reactor parts immersed in it. However, sodium explodes violently when exposed to water, so care must be taken, but such explosions would not be vastly more violent than (for example) a leak of superheated fluid from a SCWR or PWR. EBR-I, the first reactor to have a core meltdown, was of this type.

Pebble-bed reactors (PBR)

These use fuel molded into ceramic balls, and then circulate gas through the balls. The result is an efficient, low-maintenance, very safe reactor with inexpensive, standardized fuel. The prototype was the AVR.

Molten salt reactors

These dissolve the fuels in fluoride salts, or use fluoride salts for coolant. These have many safety features, high efficiency and a high power density suitable for vehicles. Notably, they have no high pressures or flammable components in the core. The prototype was the MSRE, which also used the Thorium fuel cycle. As a breeder reactor type, it reprocesses the spent fuel, extracting both Uranium and transuranics, leaving only 0.1% of transuranic waste compared to conventional once-through uranium-fueled light water reactors currently in use. A separate issue are the radioactive fission products, which are not reprocessable and need to be disposed of as with conventional reactors.

Aqueous Homogeneous Reactor (AHR)

These reactors use soluble nuclear salts dissolved in water and mixed with a coolant and a neutron moderator.

Future and Developing Technologies

Advanced Reactors

More than a dozen advanced reactor designs are in various stages of development. Some are evolutionary from the PWR, BWR and PHWR designs above, some are more radical departures. The former include the advanced boiling water reactor (ABWR), two of which are now operating with others under construction, and the planned passively safe Economic Simplified Boiling Water Reactor (ESBWR) and AP1000 units.

- The Integral Fast Reactor (IFR) was built, tested and evaluated during the 1980s and then retired under the Clinton administration in the 1990s due to nuclear non-proliferation policies of the administration. Recycling spent fuel is the core of its design and it therefore produces only a fraction of the waste of current reactors.

- The pebble-bed reactor, a high-temperature gas-cooled reactor (HTGCR), is designed so high temperatures reduce power output by Doppler broadening of the fuel's neutron cross-section. It uses ceramic fuels so its safe operating temperatures exceed the power-reduction temperature range. Most designs are cooled by inert helium. Helium is not subject to steam explosions, resists neutron absorption leading to radioactivity, and does not dissolve contaminants that can become radioactive. Typical designs have more layers (up to 7) of passive containment than light water reactors (usually 3). A unique feature that may aid safety is that the fuel-balls actually form the core's mechanism, and are replaced one-by-one as they age. The design of the fuel makes fuel reprocessing expensive.

- The Small, sealed, transportable, autonomous reactor (SSTAR) is being primarily researched and developed in the US, intended as a fast breeder reactor that is passively safe and could be remotely shut down in case the suspicion arises that it is being tampered with.

- The Clean And Environmentally Safe Advanced Reactor (CAESAR) is a nuclear reactor concept that uses steam as a moderator – this design is still in development.

- The Reduced moderation water reactor builds upon the Advanced boiling water reactor(ABWR) that is presently in use, it is not a complete fast reactor instead using mostly epithermal neutrons, which are between thermal and fast neutrons in speed.

- The hydrogen-moderated self-regulating nuclear power module (HPM) is a reactor design emanating from the Los Alamos National Laboratory that uses uranium hydride as fuel.

- Subcritical reactors are designed to be safer and more stable, but pose a number of engineering and economic difficulties. One example is the Energy amplifier.

- Thorium-based reactors. It is possible to convert Thorium-232 into U-233 in reactors specially designed for the purpose. In this way, thorium, which is four times more abundant than uranium, can be used to breed U-233 nuclear fuel. U-233 is also believed to have favourable nuclear properties as compared to traditionally used U-235, including better neutron economy and lower production of long lived transuranic waste.

  o Advanced heavy-water reactor (AHWR)— A proposed heavy water moderated nuclear power reactor that will be the next generation design of the PHWR type. Under development in the Bhabha Atomic Research Centre (BARC), India.

  o KAMINI — A unique reactor using Uranium-233 isotope for fuel. Built in India by BARC and Indira Gandhi Center for Atomic Research (IGCAR).

  o India is also planning to build fast breeder reactors using the thorium – Uranium-233 fuel cycle. The FBTR (Fast Breeder Test Reactor) in operation at Kalpakkam (India) uses Plutonium as a fuel and liquid sodium as a coolant.

Generation IV Reactors

Generation IV reactors are a set of theoretical nuclear reactor designs currently being researched. These designs are generally not expected to be available for commercial construction before 2030. Current reactors in operation around the world are generally considered second- or third-generation systems, with the first-generation systems having been retired some time ago. Research into these reactor types was officially started by the Generation IV International Forum (GIF) based on eight technology goals. The primary goals being to improve nuclear safety, improve proliferation resistance, minimize waste and natural resource utilization, and to decrease the cost to build and run such plants.

- Gas-cooled fast reactor

- Lead-cooled fast reactor

- Molten salt reactor

- Sodium-cooled fast reactor

- Supercritical water reactor

- Very-high-temperature reactor

Generation V+ Reactors

Generation V reactors are designs which are theoretically possible, but which are not being actively considered or researched at present. Though such reactors could be built with current or near term technology, they trigger little interest for reasons of economics, practicality, or safety.

- Liquid-core reactor. A closed loop liquid-core nuclear reactor, where the fissile material is molten uranium or uranium solution cooled by a working gas pumped in through holes in the base of the containment vessel.

- Gas-core reactor. A closed loop version of the nuclear lightbulb rocket, where the fissile material is gaseous uranium-hexafluoride contained in a fused silica vessel. A working gas (such as hydrogen) would flow around this vessel and absorb the UV light produced by the reaction. This reactor design could also function as a rocket engine, as featured in Harry Harrison's 1976 science-fiction novel 'Skyfall'. In theory, using UF_6 as a working fuel directly (rather than as a stage to one, as is done now) would mean lower processing costs, and very small reactors. In practice, running a reactor at such high power densities would probably produce unmanageable neutron flux, weakening most reactor materials, and therefore as the flux would be similar to that expected in fusion reactors, it would require similar materials to those selected by the International Fusion Materials Irradiation Facility.

  - Gas core EM reactor. As in the gas core reactor, but with photovoltaic arrays converting the UV light directly to electricity.

- Fission fragment reactor

- Hybrid nuclear fusion. Would use the neutrons emitted by fusion to fission a blanket of fertile material, like U-238 or Th-232 and transmutate other reactor's spent nuclear fuel/ nuclear waste into relatively more benign isotopes.

Fusion Reactors

Controlled nuclear fusion could in principle be used in fusion power plants to produce power without the complexities of handling actinides, but significant scientific and technical obstacles remain. Several fusion reactors have been built, but only recently reactors have been able to release more energy than the amount of energy used in the process. Despite research having started in the 1950s, no commercial fusion reactor is expected before 2050. The ITER project is currently leading the effort to harness fusion power.

Nuclear Fuel Cycle

Thermal reactors generally depend on refined and enriched uranium. Some nuclear reactors can operate with a mixture of plutonium and uranium. The process by which uranium ore is mined, processed, enriched, used, possibly reprocessed and disposed of is known as the nuclear fuel cycle.

Under 1% of the uranium found in nature is the easily fissionable U-235 isotope and as a result most reactor designs require enriched fuel. Enrichment involves increasing the percentage of U-235 and is usually done by means of gaseous diffusion or gas centrifuge. The enriched result is then converted into uranium dioxide powder, which is pressed and fired into pellet form. These pellets are stacked into tubes which are then sealed and called fuel rods. Many of these fuel rods are used in each nuclear reactor.

Most BWR and PWR commercial reactors use uranium enriched to about 4% U-235, and some commercial reactors with a high neutron economy do not require the fuel to be enriched at all (that is, they can use natural uranium). According to the International Atomic Energy Agency there are at least 100 research reactors in the world fueled by highly enriched (weapons-grade/90% enrichment uranium). Theft risk of this fuel (potentially used in the production of a nuclear weapon) has led to campaigns advocating conversion of this type of reactor to low-enrichment uranium (which poses less threat of proliferation).

Fissile U-235 and non-fissile but fissionable and fertile U-238 are both used in the fission process. U-235 is fissionable by thermal (i.e. slow-moving) neutrons. A thermal neutron is one which is moving about the same speed as the atoms around it. Since all atoms vibrate proportionally to their absolute temperature, a thermal neutron has the best opportunity to fission U-235 when it is moving at this same vibrational speed. On the other hand, U-238 is more likely to capture a neutron when the neutron is moving very fast. This U-239 atom will soon decay into plutonium-239, which is another fuel. Pu-239 is a viable fuel and must be accounted for even when a highly enriched uranium fuel is used. Plutonium fissions will dominate the U-235 fissions in some reactors, especially after the initial loading of U-235 is spent. Plutonium is fissionable with both fast and thermal neutrons, which make it ideal for either nuclear reactors or nuclear bombs.

Most reactor designs in existence are thermal reactors and typically use water as a neutron moderator (moderator means that it slows down the neutron to a thermal speed) and as a coolant.

But in a fast breeder reactor, some other kind of coolant is used which will not moderate or slow the neutrons down much. This enables fast neutrons to dominate, which can effectively be used to constantly replenish the fuel supply. By merely placing cheap unenriched uranium into such a core, the non-fissionable U-238 will be turned into Pu-239, "breeding" fuel.

In thorium fuel cycle thorium-232 absorbs a neutron in either a fast or thermal reactor. The thorium-233 beta decays to protactinium-233 and then to uranium-233, which in turn is used as fuel. Hence, like uranium-238, thorium-232 is a fertile material.

Fueling of Nuclear Reactors

The amount of energy in the reservoir of nuclear fuel is frequently expressed in terms of "full-power days," which is the number of 24-hour periods (days) a reactor is scheduled for operation at full power output for the generation of heat energy. The number of full-power days in a reactor's operating cycle (between refueling outage times) is related to the amount of fissile uranium-235 (U-235) contained in the fuel assemblies at the beginning of the cycle. A higher percentage of U-235 in the core at the beginning of a cycle will permit the reactor to be run for a greater number of full-power days.

At the end of the operating cycle, the fuel in some of the assemblies is "spent" and is discharged and replaced with new (fresh) fuel assemblies, although in practice it is the buildup of reaction poisons in nuclear fuel that determines the lifetime of nuclear fuel in a reactor. Long before all possible fission has taken place, the buildup of long-lived neutron absorbing fission byproducts impedes the chain reaction. The fraction of the reactor's fuel core replaced during refueling is typically one-fourth for a boiling-water reactor and one-third for a pressurized-water reactor. The disposition and storage of this spent fuel is one of the most challenging aspects of the operation of a commercial nuclear power plant. This nuclear waste is highly radioactive and its toxicity presents a danger for thousands of years.

Not all reactors need to be shut down for refueling; for example, pebble bed reactors, RBMK reactors, molten salt reactors, Magnox, AGR and CANDU reactors allow fuel to be shifted through the reactor while it is running. In a CANDU reactor, this also allows individual fuel elements to be situated within the reactor core that are best suited to the amount of U-235 in the fuel element.

The amount of energy extracted from nuclear fuel is called its burnup, which is expressed in terms of the heat energy produced per initial unit of fuel weight. Burn up is commonly expressed as megawatt days thermal per metric ton of initial heavy metal.

Safety

Nuclear safety covers the actions taken to prevent nuclear and radiation accidents or to limit their consequences. The nuclear power industry has improved the safety and performance of reactors, and has proposed new safer (but generally untested) reactor designs but there is no guarantee that the reactors will be designed, built and operated correctly. Mistakes do occur and the designers of reactors at Fukushima in Japan did not anticipate that a tsunami generated by an earthquake would disable the backup systems that were supposed to stabilize the reactor after the earthquake, despite multiple warnings by the NRG and the Japanese nuclear safety administration. According to UBS AG, the Fukushima

I nuclear accidents have cast doubt on whether even an advanced economy like Japan can master nuclear safety. Catastrophic scenarios involving terrorist attacks are also conceivable. An interdisciplinary team from MIT has estimated that given the expected growth of nuclear power from 2005–2055, at least four serious nuclear accidents would be expected in that period.

Accidents

Some serious nuclear and radiation accidents have occurred. Nuclear power plant accidents include the SL-1 accident (1961), the Three Mile Island accident (1979), Chernobyl disaster (1986), and the Fukushima Daiichi nuclear disaster (2011). Nuclear-powered submarine mishaps include the K-19 reactor accident (1961), the K-27 reactor accident (1968), and the K-431 reactor accident (1985).

Three of the reactors at Fukushima I overheated, causing meltdowns that eventually led to explosions, which released large amounts of radioactive material into the air.

Nuclear reactors have been launched into Earth orbit at least 34 times. A number of incidents connected with the unmanned nuclear-reactor-powered Soviet RORSAT radar satellite program resulted in spent nuclear fuel re-entering the Earth's atmosphere from orbit.

Natural Nuclear Reactors

Although nuclear fission reactors are often thought of as being solely a product of modern technology, the first nuclear fission reactors were in fact naturally occurring. A natural nuclear fission reactor can occur under certain circumstances that mimic the conditions in a constructed reactor. Fifteen natural fission reactors have so far been found in three separate ore deposits at the Oklo uranium mine in Gabon, West Africa. First discovered in 1972 by French physicist Francis Perrin, they are collectively known as the Oklo Fossil Reactors. Self-sustaining nuclear fission reactions took place in these reactors approximately 1.5 billion years ago, and ran for a few hundred thousand years, averaging 100 kW of power output during that time. The concept of a natural nuclear reactor was theorized as early as 1956 by Paul Kuroda at the University of Arkansas.

Such reactors can no longer form on Earth: radioactive decay over this immense time span has reduced the proportion of U-235 in naturally occurring uranium to below the amount required to sustain a chain reaction.

The natural nuclear reactors formed when a uranium-rich mineral deposit became inundated with groundwater that acted as a neutron moderator, and a strong chain reaction took place. The water moderator would boil away as the reaction increased, slowing it back down again and preventing a meltdown. The fission reaction was sustained for hundreds of thousands of years.

These natural reactors are extensively studied by scientists interested in geologic radioactive waste disposal. They offer a case study of how radioactive isotopes migrate through the Earth's crust. This is a significant area of controversy as opponents of geologic waste disposal fear that isotopes from stored waste could end up in water supplies or be carried into the environment.

Types of Nuclear Reactors

Light-water Reactor

The light-water reactor (LWR) is a type of thermal-neutron reactor that uses normal water, as opposed to heavy water, as both its coolant and neutron moderator – furthermore a solid form of fissile elements is used as fuel. Thermal-neutron reactors are the most common type of nuclear reactor, and light-water reactors are the most common type of thermal-neutron reactor.

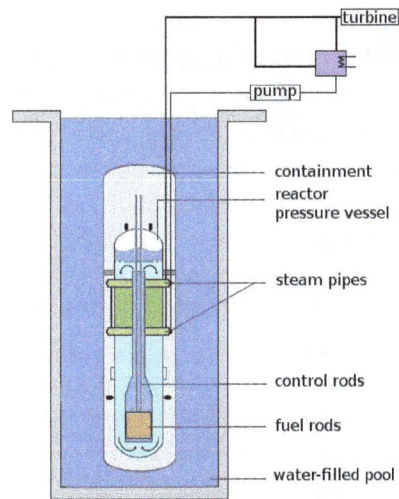

A simple light-water reactor

There are three varieties of light-water reactors: the pressurized water reactor (PWR), the boiling water reactor (BWR), and (most designs of) the supercritical water reactor (SCWR).

History

Early Concepts and Experiments

After the discoveries of fission, moderation and of the theoretical possibility of a nuclear chain reaction, early experimental results rapidly showed that natural uranium could only undergo a sustained chain reaction using graphite or heavy water as a moderator. While the world's first reactors (CP-1, X10 etc.) were successfully reaching criticality, uranium enrichment began to develop from theoretical concept to practical applications in order to meet the goal of the Manhattan Project, to build a nuclear explosive.

In May 1944, the first grams of enriched uranium ever produced reached criticality in the LOPO reactor at Los Alamos, which was used to estimate the critical mass of U235 to produce the atomic bomb. LOPO cannot be considered as the first light-water reactor because its fuel was not a solid uranium compound cladded with corrosion-resistant material, but was composed of uranyl sulfate salt dissolved in water. It is however the first aqueous homogeneous reactor and the first reactor using enriched uranium as fuel and ordinary water as a moderator.

By the end of the war, following an idea of Alvin Weinberg, natural uranium fuel elements were arranged in a lattice in ordinary water at the top of the X10 reactor to evaluate the neutron multiplication factor. The purpose of this experience was to determine the feasibility of a nuclear reactor using light water as a moderator and coolant, and cladded solid uranium as fuel. The results showed that, with a lightly enriched uranium, criticality could be reached. This experience was the first practical step toward light-water reactor.

After World War II and with the availability of enriched uranium, new concepts of reactor became feasible. In 1946, Eugene Wigner and Alvin Weinberg proposed and developed the concept of a reactor using enriched uranium as a fuel, and light water as a moderator and coolant. This concept was proposed for a reactor whose purpose was to test the behavior of materials under neutron flux. This reactor, the Material Testing Reactor (MTR), was built in Idaho at INL and reached criticality on March 31, 1952. For the design of this reactor, experiments were necessary, so a mock-up of the MTR was built at ORNL, to assess the hydraulic performances of the primary circuit and then to test its neutronic characteristics. This MTR mock-up, later called the Low Intensity Test Reactor (LITR), reached criticality on February 4, 1950 and was the world's first light-water reactor.

First Pressurized Water Reactors

Immediately after the end of World War II the United States Navy started a program under the direction of Captain (later Admiral) Hyman Rickover, with the goal of nuclear propulsion for ships. It developed the first pressurized water reactors in the early 1950s, and led to the successful deployment of the first nuclear submarine, the USS *Nautilus* (SSN-571).

The Soviet Union independently developed a version of the PWR in the late 1950s, under the name of VVER. While functionally very similar to the American effort, it also has certain design distinctions from Western PWRs.

First Boiling Water Reactor

Researcher Samuel Untermyer II led the effort to develop the BWR at the US National Reactor Testing Station (now the Idaho National Laboratory) in a series of tests called the BORAX experiments.

Overview

The family of nuclear reactors known as light-water reactors (LWR), cooled and moderated using ordinary water, tend to be simpler and cheaper to build than other types of nuclear reactor; due to these factors, they make up the vast majority of civil nuclear reactors and naval propulsion reactors in service throughout the world as of 2009. LWRs can be subdivided into three categories – pressurized water reactors (PWRs), boiling water reactors (BWRs), and supercritical water reactors (SCWRs). The SCWR remains hypothetical as of 2009; it is a Generation IV design that is still a light-water reactor, but it is only partially moderated by light water and exhibits certain characteristics of a fast neutron reactor.

The Koeberg nuclear power station, consisting of two pressurized water reactors fueled with uranium

The leaders in national experience with PWRs, offering reactors for export, are the United States (which offers the passively-safe AP1000, a Westinghouse design, as well as several smaller, modular, passively-safe PWRs, such as the Babcock & Wilcox MPower, and the NuScale MASLWR), the Russian Federation (offering both the VVER-1000 and the VVER-1200 for export), the Republic of France (offering the AREVA EPR for export), and Japan (offering the Mitsubishi Advanced Pressurized Water Reactor for export); in addition, both the People's Republic of China and the Republic of Korea are both noted to be rapidly ascending into the front rank of PWR-constructing nations as well, with the Chinese being engaged in a massive program of nuclear power expansion, and the Koreans currently designing and constructing their second generation of indigenous designs. The leaders in national experience with BWRs, offering reactors for export, are the United States and Japan, with the alliance of General Electric (of the US) and Hitachi (of Japan), offering both the Advanced Boiling Water Reactor (ABWR) and the Economic Simplified Boiling Water Reactor (ESBWR) for construction and export; in addition, Toshiba offers an ABWR variant for construction in Japan, as well. West Germany was also once a major player with BWRs. The other types of nuclear reactor in use for power generation are the heavy water moderated reactor, built by Canada (CANDU) and the Republic of India (AHWR), the advanced gas cooled reactor (AGCR), built by the United Kingdom, the liquid metal cooled reactor (LMFBR), built by the Russian Federation, the Republic of France, and Japan, and the graphite-moderated, water-cooled reactor (RBMK or LWGR), found exclusively within the Russian Federation and former Soviet states.

Though electricity generation capabilities are comparable between all these types of reactor, due to the aforementioned features, and the extensive experience with operations of the LWR, it is favored in the vast majority of new nuclear power plants. In addition, light-water reactors make up the vast majority of reactors that power naval nuclear-powered vessels. Four out of the five great powers with nuclear naval propulsion capacity use light-water reactors exclusively: the British Royal Navy, the Chinese People's Liberation Army Navy, the French Marine nationale, and the United States Navy. Only the Russian Federation's Navy has used a relative handful of liquid-metal cooled reactors in production vessels, specifically the Alfa class submarine, which used lead-bismuth eutectic as a reactor moderator and coolant, but the vast majority of Russian nuclear-powered boats and ships use light-water reactors exclusively. The reason for near exclusive LWR use aboard nuclear naval vessels is the level of inherent safety built into these types of reactors. Since light water is used as both a coolant and a neutron moderator in these reactors, if one of these reactors suffers damage due to military action, leading to a compromise of the reactor core's integrity, the resulting release of the light-water moderator will act to stop the nuclear reaction and shut the reactor down. This capability is known as a negative void coefficient of reactivity.

Currently-offered LWRs include the following

- ABWR

- AP1000

- APR-1400

- CPR-1000

- EPR

- VVER

LWR Statistics

Data from the International Atomic Energy Agency in 2009:

Reactors in operation.	359
Reactors under construction.	27
Number of countries with LWRs.	27
Generating capacity (Gigawatt).	328.4

Reactor Design

The light-water reactor produces heat by controlled nuclear fission. The nuclear reactor core is the portion of a nuclear reactor where the nuclear reactions take place. It mainly consists of nuclear fuel and control elements. The pencil-thin nuclear fuel rods, each about 12 feet (3.7 m) long, are grouped by the hundreds in bundles called fuel assemblies. Inside each fuel rod, pellets of uranium, or more commonly uranium oxide, are stacked end to end. The control elements, called control rods, are filled with pellets of substances like hafnium or cadmium that readily capture neutrons. When the control rods are lowered into the core, they absorb neutrons, which thus cannot take part in the chain reaction. On the converse, when the control rods are lifted out of the way, more neutrons strike the fissile uranium-235 or plutonium-239 nuclei in nearby fuel rods, and the chain reaction intensifies. All of this is enclosed in a water-filled steel pressure vessel, called the reactor vessel.

In the boiling water reactor, the heat generated by fission turns the water into steam, which directly drives the power-generating turbines. But in the pressurized water reactor, the heat generated by fission is transferred to a secondary loop via a heat exchanger. Steam is produced in the secondary loop, and the secondary loop drives the power-generating turbines. In either case, after flowing through the turbines, the steam turns back into water in the condenser.

The water required to cool the condenser is taken from a nearby river or ocean. It is then pumped back into the river or ocean, in warmed condition. The heat could also be dissipated via a cooling tower into the atmosphere. The United States uses LWR reactors for electric power production, in comparison to the heavy water reactors used in Canada.

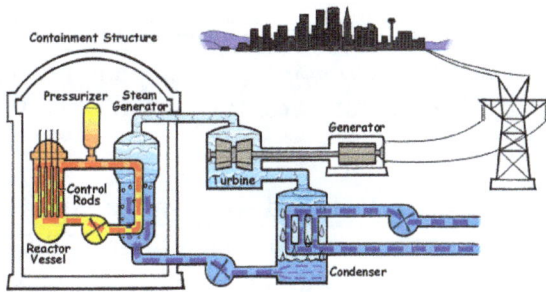

Animated diagram of a pressurized water reactor   Animated diagram of a boiling water reactor

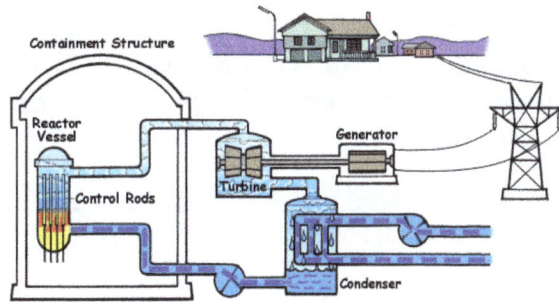

Control

Control rods are usually combined into control rod assemblies — typically 20 rods for a commercial pressurized water reactor assembly — and inserted into guide tubes within a fuel element. A control rod is removed from or inserted into the central core of a nuclear reactor in order to control the number of neutrons which will split further uranium atoms. This in turn affects the thermal power of the reactor, the amount of steam generated, and hence the electricity produced. The control rods are partially removed from the core to allow a chain reaction to occur. The number of control rods inserted and the distance by which they are inserted can be varied to control the reactivity of the reactor.

A pressurized water reactor head, with the control rods visible on the top

Usually there are also other means of controlling reactivity. In the PWR design a soluble neutron absorber, usually boric acid, is added to the reactor coolant allowing the complete extraction of the control rods during stationary power operation ensuring an even power and flux distribution over the entire core. Operators of the BWR design use the coolant flow through the core to control reactivity by varying the speed of the reactor recirculation pumps. An increase in the coolant flow through the core improves the removal of steam bubbles, thus increasing the density of the coolant/moderator with the result of increasing power.

Coolant

The light-water reactor also uses ordinary water to keep the reactor cooled. The cooling source, light water, is circulated past the reactor core to absorb the heat that it generates. The heat is carried away from the reactor and is then used to generate steam. Most reactor systems employ a cooling system that is physically separate from the water that will be boiled to produce pressurized steam for the turbines, like the pressurized-water reactor. But in some reactors the water for the steam turbines is boiled directly by the reactor core, for example the boiling-water reactor.

Many other reactors are also light-water cooled, notably the RBMK and some military plutonium-production reactors. These are not regarded as LWRs, as they are moderated by graphite, and as a result their nuclear characteristics are very different. Although the coolant flow rate in commercial PWRs is constant, it is not in nuclear reactors used on U.S. Navy ships.

Fuel

The use of ordinary water makes it necessary to do a certain amount of enrichment of the uranium fuel before the necessary criticality of the reactor can be maintained. The light-water reactor uses uranium 235 as a fuel, enriched to approximately 3 percent. Although this is its major fuel, the uranium 238 atoms also contribute to the fission process by converting to plutonium 239; about one-half of which is consumed in the reactor. Light-water reactors are generally refueled every 12 to 18 months, at which time, about 25 percent of the fuel is replaced.

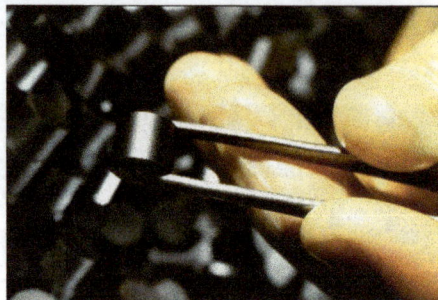

A nuclear fuel pellet

Nuclear fuel pellets that are ready for fuel assembly completion

The enriched UF_6 is converted into uranium dioxide powder that is then processed into pellet form. The pellets are then fired in a high-temperature, sintering furnace to create hard, ceramic pellets of enriched uranium. The cylindrical pellets then undergo a grinding process to achieve a uniform pellet size. The uranium oxide is dried before inserting into the tubes to try to eliminate moisture in the ceramic fuel that can lead to corrosion and hydrogen embrittlement. The pellets are stacked, according to each nuclear core's design specifications, into tubes of corrosion-resistant metal alloy. The tubes are sealed to contain the fuel pellets: these tubes are called fuel rods.

The finished fuel rods are grouped in special fuel assemblies that are then used to build up the nuclear fuel core of a power reactor. The metal used for the tubes depends on the design of the reactor – stainless steel was used in the past, but most reactors now use a zirconium alloy. For the most common types of reactors the tubes are assembled into bundles with the tubes spaced precise distances apart. These bundles are then given a unique identification number, which enables them to be tracked from manufacture through use and into disposal.

Pressurized water reactor fuel consists of cylindrical rods put into bundles. A uranium oxide ceramic is formed into pellets and inserted into zirconium alloy tubes that are bundled together. The zirconium alloy tubes are about 1 cm in diameter, and the fuel cladding gap is filled with helium gas to improve the conduction of heat from the fuel to the cladding. There are about 179-264 fuel rods per fuel bundle and about 121 to 193 fuel bundles are loaded into a reactor core. Generally, the fuel bundles consist of fuel rods bundled 14x14 to 17x17. PWR fuel bundles are about 4 meters in length. The zirconium alloy tubes are pressurized with helium to try to minimize pellet cladding interaction which can lead to fuel rod failure over long periods.

In boiling water reactors, the fuel is similar to PWR fuel except that the bundles are "canned"; that is, there is a thin tube surrounding each bundle. This is primarily done to prevent local density variations from effecting neutronics and thermal hydraulics of the nuclear core on a global scale. In modern BWR fuel bundles, there are either 91, 92, or 96 fuel rods per assembly depending on the manufacturer. A range between 368 assemblies for the smallest and 800 assemblies for the largest U.S. BWR forms the reactor core. Each BWR fuel rod is back filled with helium to a pressure of about three atmospheres (300 kPa).

Moderator

A neutron moderator is a medium which reduces the velocity of fast neutrons, thereby turning them into thermal neutrons capable of sustaining a nuclear chain reaction involving uranium-235. A good neutron moderator is a material full of atoms with light nuclei which do not easily absorb neutrons. The neutrons strike the nuclei and bounce off. After sufficient impacts, the velocity of the neutron will be comparable to the thermal velocities of the nuclei; this neutron is then called a thermal neutron.

The light-water reactor uses ordinary water, also called light water, as its neutron moderator. The light water absorbs too many neutrons to be used with unenriched natural uranium, and therefore uranium enrichment or nuclear reprocessing becomes necessary to operate such reactors, increasing overall costs. This differentiates it from a heavy water reactor, which uses heavy water as a neutron moderator. While ordinary water has some heavy water molecules in it, it is not enough to be important in most applications. In pressurized water reactors the coolant water is used as a moderator by letting the neutrons undergo multiple collisions with light hydrogen atoms in the water, losing speed in the process. This moderating of neutrons will happen more often when the water is denser, because more collisions will occur.

The use of water as a moderator is an important safety feature of PWRs, as any increase in temperature causes the water to expand and become less dense; thereby reducing the extent to which neutrons are slowed down and hence reducing the reactivity in the reactor. Therefore, if reactivity increases beyond normal, the reduced moderation of neutrons will cause the chain reaction to slow down, producing less heat. This property, known as the negative temperature coefficient of reactivity, makes PWR reactors very stable. In event of a loss-of-coolant accident, the moderator is also lost and the active fission reaction will stop. Heat is still produced after the chain reaction stops from the radioactive byproducts of fission, at about 5% of rated power. This "decay heat" will continue for 1 to 3 years after shut down, whereupon the reactor finally reaches "full cold shut-down". Decay heat, while dangerous and strong enough to melt the core, is not nearly as intense as an active fission reaction. During the post shutdown period the reactor requires cooling water to be pumped or the reactor will overheat. If the temperature exceeds 2200 degrees Celsius, cooling water will break down to hydrogen and oxygen, which can form a (chemically) explosive mixture. Decay heat is a major risk factor in LWR safety record.

PIUS reactor

PIUS, standing for *Process Inherent Ultimate Safety*, was a Swedish design concept for a light-water reactor system. It relied on passive measures, not requiring operator actions or external energy supplies, to provide safe operation. No units were ever built.

Boiling Water Reactor

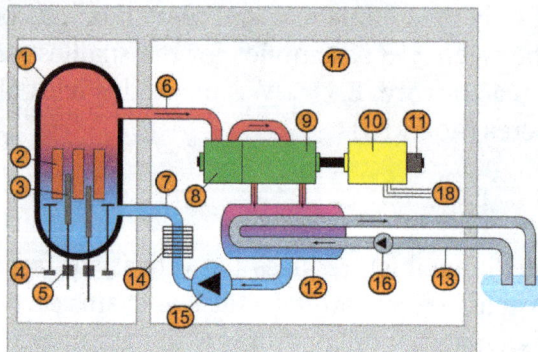

Schematic diagram of a *boiling water reactor* (BWR):

- 1. Reactor pressure vessel
- 2. Nuclear fuel element
- 3. Control rods
- 4. Recirculation pumps
- 5. Control rod drives
- 6. Steam
- 7. Feedwater
- 8. High pressure turbine
- 9. Low pressure turbine
- 10. Generator

- 11. Exciter

- 12. Condenser

- 13. Coolant

- 14. Pre-heater

- 15. Feedwater pump

- 16. Cold water pump

- 17. Concrete enclosure

- 18. Connection to electricity grid

The boiling water reactor (BWR) is a type of light water nuclear reactor used for the generation of electrical power. It is the second most common type of electricity-generating nuclear reactor after the pressurized water reactor (PWR), also a type of light water nuclear reactor. The main difference between a BWR and PWR is that in a BWR, the reactor core heats water, which turns to steam and then drives a steam turbine. In a PWR, the reactor core heats water, which does not boil. This hot water then exchanges heat with a lower pressure water system, which turns to steam and drives the turbine. The BWR was developed by the Idaho National Laboratory and General Electric (GE) in the mid-1950s. The main present manufacturer is GE Hitachi Nuclear Energy, which specializes in the design and construction of this type of reactor.

Overview

The *boiling water reactor* (BWR) uses demineralized water as a coolant and neutron moderator. Heat is produced by nuclear fission in the reactor core, and this causes the cooling water to boil, producing steam. The steam is directly used to drive a turbine, after which it is cooled in a condenser and converted back to liquid water. This water is then returned to the reactor core, completing the loop. The cooling water is maintained at about 75 atm (7.6 MPa, 1000–1100 psi) so that it boils in the core at about 285 °C (550 °F). In comparison, there is no significant boiling allowed in a pressurized water reactor (PWR) because of the high pressure maintained in its primary loop—approximately 158 atm (16 MPa, 2300 psi). The core damage frequency of the reactor was estimated to be between 10^{-4} and 10^{-7} (i.e., one core damage accident per every 10,000 to 10,000,000 reactor years).

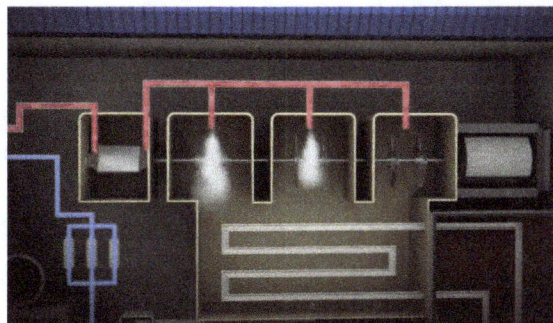

Animation of a BWR with cooling towers.

Components

Condensate and Feedwater

Steam exiting the turbine flows into condensers located underneath the low pressure turbines where the steam is cooled and returned to the liquid state (condensate). The condensate is then pumped through feedwater heaters that raise its temperature using extraction steam from various turbine stages. Feedwater from the feedwater heaters enters the reactor pressure vessel (RPV) through nozzles high on the vessel, well above the top of the nuclear fuel assemblies (these nuclear fuel assemblies constitute the "core") but below the water level.

The feedwater enters into the downcomer or annulus region and combines with water exiting the moisture separators. The feedwater subcools the saturated water from the moisture separators. This water now flows down the downcomer or annulus region, which is separated from the core by a tall shroud. The water then goes through either jet pumps or internal recirculation pumps that provide additional pumping power (hydraulic head). The water now makes a 180 degree turn and moves up through the lower core plate into the nuclear core where the fuel elements heat the water. Water exiting the fuel channels at the top guide is saturated with a steam quality of about 15%. Typical core flow may be 45,000,000 kg/h (100,000,000 lb/h) with 6,500,000 kg/h (14,500,000 lb/h) steam flow. However, core-average void fraction is a significantly higher fraction (~40%). These sort of values may be found in each plant's publicly available Technical Specifications, Final Safety Analysis Report, or Core Operating Limits Report.

The heating from the core creates a thermal head that assists the recirculation pumps in recirculating the water inside of the RPV. A BWR can be designed with no recirculation pumps and rely entirely on the thermal head to recirculate the water inside of the RPV. The forced recirculation head from the recirculation pumps is very useful in controlling power, however, and allows achieving higher power levels that would not otherwise be possible. The thermal power level is easily varied by simply increasing or decreasing the forced recirculation flow through the recirculation pumps.

The two phase fluid (water and steam) above the core enters the riser area, which is the upper region contained inside of the shroud. The height of this region may be increased to increase the thermal natural recirculation pumping head. At the top of the riser area is the moisture separator. By swirling the two phase flow in cyclone separators, the steam is separated and rises upwards towards the steam dryer while the water remains behind and flows horizontally out into the downcomer or annulus region. In the downcomer or annulus region, it combines with the feedwater flow and the cycle repeats.

The saturated steam that rises above the separator is dried by a chevron dryer structure. The "wet" steam goes through a tortuous path where the water droplets are slowed down and directed out into the downcomer or annulus region. The "dry" steam then exits the RPV through four main steam lines and goes to the turbine.

Control Systems

Reactor power is controlled via two methods: by inserting or withdrawing control rods and by changing the water flow through the reactor core.

Positioning (withdrawing or inserting) control rods is the normal method for controlling power when starting up a BWR. As control rods are withdrawn, neutron absorption decreases in the control material and increases in the fuel, so reactor power increases. As control rods are inserted, neutron absorption increases in the control material and decreases in the fuel, so reactor power decreases. Differently from the PWR, in a BWR the control rods (boron carbide plates) are inserted from below to give a more homogeneous distribution of the power: in the upper side the density of the water is lower due to vapour formation, making the neutron moderation less efficient and the fission probability lower. In normal operation, the control rods are only used to keep a homogeneous power distribution in the reactor and compensate the consumption of the fuel, while the power is controlled through the water flow. Some early BWRs and the proposed ESBWR (Economic Simplified BWR made by General Electric Hitachi) designs use only natural circulation with control rod positioning to control power from zero to 100% because they do not have reactor recirculation systems.

Changing (increasing or decreasing) the flow of water through the core is the normal and convenient method for controlling power from approximately 30% to 100% reactor power. When operating on the so-called "100% rod line," power may be varied from approximately 30% to 100% of rated power by changing the reactor recirculation system flow by varying the speed of the recirculation pumps or modulating flow control valves. As flow of water through the core is increased, steam bubbles ("voids") are more quickly removed from the core, the amount of liquid water in the core increases, neutron moderation increases, more neutrons are slowed down to be absorbed by the fuel, and reactor power increases. As flow of water through the core is decreased, steam voids remain longer in the core, the amount of liquid water in the core decreases, neutron moderation decreases, fewer neutrons are slowed down to be absorbed by the fuel, and reactor power decreases.

Reactor pressure in a BWR is controlled by the main turbine or main steam bypass valves. Unlike a PWR, where the turbine steam demand is set manually by the operators, in a BWR, the turbine valves will modulate to maintain reactor pressure at a setpoint. Under this control mode, the turbine will automatically follow reactor power changes. When the turbine is offline or trips, the main steam bypass/dump valves will open to direct steam directly to the condenser. These bypass valves will automatically or manually modulate as necessary to maintain reactor pressure and control the reactor's heatup and cooldown rates while steaming is still in progress.

Reactor water level is controlled by the main feedwater system. From about 0.5% power to 100% power, feedwater will automatically control the water level in the reactor. At low power conditions, the feedwater controller acts as a simple PID control by watching reactor water level. At high power conditions, the controller is switched to a "Three-Element" control mode, where the controller looks at the current water level in the reactor, as well as the amount of water going in and the amount of steam leaving the reactor. By using the water injection and steam flow rates, the feed water control system can rapidly anticipate water level deviations and respond to maintain water level within a few inches of set point. If one of the two feedwater pumps fails during operation, the feedwater system will command the recirculation system to rapidly reduce core flow, effectively reducing reactor power from 100% to 50% in a few seconds. At this power level a single feedwater pump can maintain the core water level. If all feedwater is lost, the reactor will scram and the Emergency Core Cooling System is used to restore reactor water level.

Steam Turbines

Steam produced in the reactor core passes through steam separators and dryer plates above the core and then directly to the turbine, which is part of the reactor circuit. Because the water around the core of a reactor is always contaminated with traces of radionuclides, the turbine must be shielded during normal operation, and radiological protection must be provided during maintenance. The increased cost related to operation and maintenance of a BWR tends to balance the savings due to the simpler design and greater thermal efficiency of a BWR when compared with a PWR. Most of the radioactivity in the water is very short-lived (mostly N-16, with a 7-second half-life), so the turbine hall can be entered soon after the reactor is shut down.

BWR steam turbines employ a high-pressure turbine designed to handle saturated steam, and multiple low-pressure turbines. The high pressure turbine receives steam directly from the reactor. The high pressure turbine exhaust passes through a steam reheater which superheats the steam to over 400 degrees F for the low pressure turbines to use. The exhaust of the low pressure turbines is sent to the main condenser. The steam reheaters take some of the reactor's steam and use it as a heating source to reheat what comes out of the high pressure turbine exhaust. While the reheaters take steam away from the turbine, the net-result is that the reheaters improve the thermodynamic efficiency of the plant.

Reactor Core

A modern BWR fuel assembly comprises 74 to 100 fuel rods, and there are up to approximately 800 assemblies in a reactor core, holding up to approximately 140 short tons of low-enriched uranium. The number of fuel assemblies in a specific reactor is based on considerations of desired reactor power output, reactor core size and reactor power density.

Safety Systems

A modern reactor has many safety systems that are designed with a defence in depth philosophy, which is a design philosophy that is integrated throughout construction and commissioning.

A BWR is similar to a pressurized water reactor (PWR) in that the reactor will continue to produce heat even after the fission reactions have stopped, which could make a core damage incident possible. This heat is produced by the radioactive decay of fission products and materials that have been activated by neutron absorption. BWRs contain multiple safety systems for cooling the core after emergency shut down.

Refueling Systems

The reactor fuel rods are occasionally replaced by removing them from the top of the containment vessel. A typical fuel cycle lasts 18–24 months, with about one third of fuel assemblies being replaced during a refueling outage. The remaining fuel assemblies are shuffled to new core locations to maximize the efficiency and power produced in the next fuel cycle.

Because they are hot both radioactively and thermally, this is done via cranes and under water. For this reason the spent fuel storage pools are above the reactor in typical installations. They are shielded by water several times their height, and stored in rigid arrays in which their geometry is controlled to avoid criticality. In the Fukushima reactor incident this became problematic because water was lost from one or more spent fuel pools and the earthquake could have altered the geometry. The fact that the fuel rods' cladding is a zirconium alloy was also problematic since this element can react with steam at extreme temperatures to produce hydrogen, which can ignite with oxygen in the air. Normally the fuel rods are kept sufficiently cool in the reactor and spent fuel pools that this is not a concern, and the cladding remains intact for the life of the rod.

Evolution

Early Concepts

The BWR concept was developed slightly later than the PWR concept. Development of the BWR started in the early 1950s, and was a collaboration between General Electric (GE) and several US national laboratories.

Research into nuclear power in the US was led by the 3 military services. The Navy, seeing the possibility of turning submarines into full-time underwater vehicles, and ships that could steam around the world without refueling, sent their man in engineering, Captain Hyman Rickover to run their nuclear power program. Rickover decided on the PWR route for the Navy, as the early researchers in the field of nuclear power feared that the direct production of steam within a reactor would cause instability, while they knew that the use of pressurized water would definitively work as a means of heat transfer. This concern led to the US's first research effort in nuclear power being devoted to the PWR, which was highly suited for naval vessels (submarines, especially), as space was at a premium, and PWRs could be made compact and high-power enough to fit in such, in any event.

But other researchers wanted to investigate whether the supposed instability caused by boiling water in a reactor core would really cause instability. During early reactor development, a small group of engineers accidentally increased the reactor power level on an experimental reactor to such an extent that the water quickly boiled, this shut down the reactor, indicating the useful self-moderating property in emergency circumstances. In particular, Samuel Untermyer II, a researcher at Argonne National Laboratory, proposed and oversaw a series of experiments: the BORAX experiments—to see if a *boiling water reactor* would be feasible for use in energy production. He found that it was, after subjecting his reactors to quite strenuous tests, proving the safety principles of the BWR.

Following this series of tests, GE got involved and collaborated with INL to bring this technology to market. Larger-scale tests were conducted through the late 1950s/early/mid-1960s that only partially used directly-generated (primary) nuclear boiler system steam to feed the turbine and incorporated heat exchangers for the generation of secondary steam to drive separate parts of the turbines. The literature does not indicate why this was the case, but it was eliminated on production models of the BWR.

First Series of Production

The first generation of production boiling water reactors saw the incremental development of the unique and distinctive features of the BWR: the torus (used to quench steam in the event of a transient requiring the quenching of steam), as well as the drywell, the elimination of the heat exchanger, the steam dryer, the distinctive general layout of the reactor building, and the standardization of reactor control and safety systems. The first, General Electric (GE), series of production BWRs evolved through 6 iterative design phases, each termed BWR/1 through BWR/6. (BWR/4s, BWR/5s, and BWR/6s are the most common types in service today.) The vast majority of BWRs in service throughout the world belong to one of these design phases.

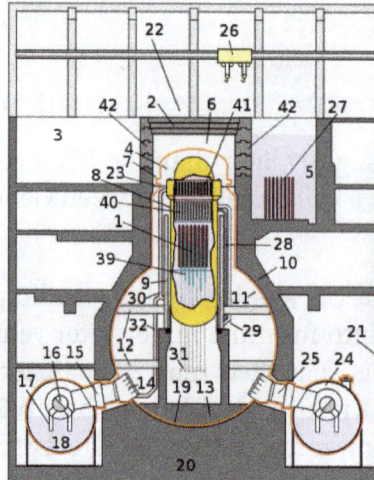

Cross-section sketch of a typical BWR Mark I containment

- 1st generation BWR: BWR/1 with Mark I containment.

- 2nd generation BWRs: BWR/2, BWR/3 and some BWR/4 with Mark I containment. Other BWR/4, and BWR/5 with Mark-II containment.

- 3rd generation BWRs: BWR/6 with Mark-III containment.

Browns Ferry Unit 1 drywell and wetwell under construction, a BWR/4 using the Mark I containment

Containment variants were constructed using either concrete or steel for the Primary Containment, Drywell and Wetwell in various combinations.

Apart from the GE designs there were others by ABB, MITSU, Toshiba and KWU.

Advanced Boiling Water Reactor

A newer design of BWR is known as the Advanced Boiling Water Reactor (ABWR). The ABWR was developed in the late 1980s and early 1990s, and has been further improved to the present day. The ABWR incorporates advanced technologies in the design, including computer control, plant automation, control rod removal, motion, and insertion, in-core pumping, and nuclear safety to deliver improvements over the original series of production BWRs, with a high power output (1350 MWe per reactor), and a significantly lowered probability of core damage. Most significantly, the ABWR was a completely standardized design, that could be made for series production.

The ABWR was approved by the United States Nuclear Regulatory Commission for production as a standardized design in the early 1990s. Subsequently, numerous ABWRs were built in Japan. One development spurred by the success of the ABWR in Japan is that General Electric's nuclear energy division merged with Hitachi Corporation's nuclear energy division, forming GE Hitachi Nuclear Energy, which is now the major worldwide developer of the BWR design.

Simplified Boiling Water Reactor

Parallel to the development of the ABWR, General Electric also developed a different concept, known as the *simplified boiling water reactor* (SBWR). This smaller 600 megawatt electrical reactor was notable for its incorporation—for the first time ever in a light water reactor—of "passive safety" design principles. The concept of passive safety means that the reactor, rather than requiring the intervention of active systems, such as emergency injection pumps, to keep the reactor within safety margins, was instead designed to return to a safe state solely through operation of natural forces if a safety-related contingency developed.

For example, if the reactor got too hot, it would trigger a system that would release soluble neutron absorbers (generally a solution of borated materials, or a solution of borax), or materials that greatly hamper a chain reaction by absorbing neutrons, into the reactor core. The tank containing the soluble neutron absorbers would be located above the reactor, and the absorption solution, once the system was triggered, would flow into the core through force of gravity, and bring the reaction to a near-complete stop. Another example was the Isolation Condenser system, which relied on the principle of hot water/steam rising to bring hot coolant into large heat exchangers located above the reactor in very deep tanks of water, thus accomplishing residual heat removal. Yet another example was the omission of recirculation pumps within the core; these pumps were used in other BWR designs to keep cooling water moving; they were expensive, hard to reach to repair, and could occasionally fail; so as to improve reliability, the ABWR incorporated no less than 10 of these recirculation pumps, so that even if several failed, a sufficient number would remain serviceable so that an unscheduled shutdown would not be necessary, and the pumps could be repaired during the next refueling outage. Instead, the designers of the *simplified boiling water reactor* used thermal analysis to design the reactor core such that natural circulation (cold water falls, hot water rises) would bring water to the center of the core to be boiled.

The ultimate result of the passive safety features of the SBWR would be a reactor that would not require human intervention in the event of a major safety contingency for at least 48 hours following the safety contingency; thence, it would only require periodic refilling of cooling water tanks located completely outside of the reactor, isolated from the cooling system, and designed to remove reactor waste heat through evaporation. The *simplified boiling water reactor* was submitted to the United States Nuclear Regulatory Commission, however, it was withdrawn prior to approval; still, the concept remained intriguing to General Electric's designers, and served as the basis of future developments.

Economic Simplified Boiling Water Reactor

During a period beginning in the late 1990s, GE engineers proposed to combine the features of the *advanced boiling water reactor* design with the distinctive safety features of the *simplified boiling water reactor* design, along with scaling up the resulting design to a larger size of 1,600 MWe (4,500 MWth). This Economic Simplified Boiling Water Reactor (ESBWR) design was submitted to the US Nuclear Regulatory Commission for approval in April 2005, and design certification was granted by the NRC in September 2014.

Reportedly, this design has been advertised as having a core damage probability of only 3×10^{-8} core damage events per reactor-year. That is, there would need to be 3 million ESBWRs operating before one would expect a single core-damaging event during their 100-year lifetimes. Earlier designs of the BWR, the BWR/4, had core damage probabilities as high as 1×10^{-5} core-damage events per reactor-year. This extraordinarily low CDP for the ESBWR far exceeds the other large LWRs on the market.

Advantages and Disadvantages

Advantages

- The reactor vessel and associated components operate at a substantially lower pressure of about 70–75 bars (1,020–1,090 psi) compared to about 155 bars (2,250 psi) in a PWR.

- Pressure vessel is subject to significantly less irradiation compared to a PWR, and so does not become as brittle with age.

- Operates at a lower nuclear fuel temperature.

- Fewer components due to no steam generators and no pressurizer vessel. (Older BWRs have external recirculation loops, but even this piping is eliminated in modern BWRs, such as the ABWR.) This also makes BWRs simpler to operate.

- Lower risk (probability) of a rupture causing loss of coolant compared to a PWR, and lower risk of core damage should such a rupture occur. This is due to fewer pipes, fewer large diameter pipes, fewer welds and no steam generator tubes.

- NRC assessments of limiting fault potentials indicate if such a fault occurred, the average BWR would be less likely to sustain core damage than the average PWR due to the robustness and redundancy of the Emergency Core Cooling System (ECCS).

- Measuring the water level in the pressure vessel is the same for both normal and emergency operations, which results in easy and intuitive assessment of emergency conditions.

- Can operate at lower core power density levels using natural circulation without forced flow.

- A BWR may be designed to operate using only natural circulation so that recirculation pumps are eliminated entirely. (The new ESBWR design uses natural circulation.)

- BWRs do not use boric acid to control fission burn-up to avoid the production of tritium (contamination of the turbines), leading to less possibility of corrosion within the reactor vessel and piping. (Corrosion from boric acid must be carefully monitored in PWRs; it has been demonstrated that reactor vessel head corrosion can occur if the reactor vessel head is not properly maintained. Since BWRs do not utilize boric acid, these contingencies are eliminated.)

- The power control by reduction of the moderator density (vapour bubbles in the water) instead of by addition of neutron absorbers (boric acid in PWR) leads to breeding of U-238 by fast neutrons, producing fissile Pu-239.

- BWRs generally have N-2 redundancy on their major safety-related systems, which normally consist of four "trains" of components. This generally means that up to two of the four components of a safety system can fail and the system will still perform if called upon.

- Due to their single major vendor (GE/Hitachi), the current fleet of BWRs have predictable, uniform designs that, while not completely standardized, generally are very similar to one another. The ABWR/ESBWR designs are completely standardized. Lack of standardization remains a problem with PWRs, as, at least in the United States, there are three design families represented among the current PWR fleet (Combustion Engineering, Westinghouse, and Babcock & Wilcox), within these families, there are quite divergent designs. Still, some countries could reach a high level of standardisation with PWRs, like France.

  o Additional families of PWRs are being introduced. For example, Mitsubishi's APWR, Areva's US-EPR, and Westinghouse's AP1000/AP600 will add diversity and complexity to an already diverse crowd, and possibly cause customers seeking stability and predictability to seek other designs, such as the BWR.

- BWRs are overrepresented in imports, when the importing nation does not have a nuclear navy (PWRs are favored by nuclear naval states due to their compact, high-power design used on nuclear-powered vessels; since naval reactors are generally not exported, they cause national skill to be developed in PWR design, construction, and operation). This may be due to the fact that BWRs are ideally suited for peaceful uses like power generation, process/industrial/district heating, and desalinization, due to low cost, simplicity, and safety focus, which come at the expense of larger size and slightly lower thermal efficiency.

  o Sweden is standardized mainly on BWRs.

  o Mexico's two reactors are BWRs.

  o Japan experimented with both PWRs and BWRs, but most builds as of late have been of BWRs, specifically ABWRs.

  o In the CEGB open competition in the early 1960s for a standard design for UK 2nd-generation power reactors, the PWR didn't even make it to the final round, which was a showdown between the BWR (preferred for its easily understood design as well as for being predictable and "boring") and the AGR, a uniquely British design; the indigenous design won, possibly on technical merits, possibly due to the proximity of a general election. In the 1980s the CEGB built a PWR, Sizewell B.

Disadvantages

- BWRs require more complex calculations for managing consumption of nuclear fuel during operation due to "two phase (water and steam) fluid flow" in the upper part of the core. This also requires more instrumentation in the reactor core.

- Larger pressure vessel than for a PWR of similar power, with correspondingly higher cost, in particular for older models that still use a main steam generator and associated piping.

- Contamination of the turbine by short-lived activation products. This means that shielding and access control around the steam turbine are required during normal operations due to the radiation levels arising from the steam entering directly from the reactor core. This is a moderately minor concern, as most of the radiation flux is due to Nitrogen-16 (activation of oxygen in the water), which has a half-life of 7 seconds, allowing the turbine chamber to be entered into within minutes of shutdown.

- Though the present fleet of BWRs is said to be less likely to suffer core damage from the "1 in 100,000 reactor-year" limiting fault than the present fleet of PWRs, (due to increased ECCS robustness and redundancy) there have been concerns raised about the pressure containment ability of the as-built, unmodified Mark I containment – that such may be insufficient to contain pressures generated by a limiting fault combined with complete ECCS failure that results in extremely severe core damage. In this double failure scenario, assumed to be extremely unlikely prior to the Fukushima I nuclear accidents, an unmodified Mark I containment can allow some degree of radioactive release to occur. This is supposed to be mitigated by the modification of the Mark I containment; namely, the addition of an outgas stack system that, if containment pressure exceeds critical setpoints, is supposed to allow the orderly discharge of pressurizing gases after the gases pass through activated carbon filters designed to trap radionuclides.

- Control rods are inserted from below for current BWR designs. There are two available hydraulic power sources that can drive the control rods into the core for a BWR under emergency conditions. There is a dedicated high pressure hydraulic accumulator and also the pressure inside of the reactor pressure vessel available to each control rod. Either the dedicated accumulator (one per rod) or reactor pressure is capable of fully inserting each rod. Most other reactor types use top entry control rods that are held up in the withdrawn position by electromagnets, causing them to fall into the reactor by gravity if power is lost.

Technical and Background Information

Start-up ("Going Critical")

Reactor start up (criticality) is achieved by withdrawing control rods from the core to raise core reactivity to a level where it is evident that the nuclear chain reaction is self-sustaining. This is known as "going critical". Control rod withdrawal is performed slowly, as to carefully monitor core conditions as the reactor approaches criticality. When the reactor is observed to become slightly super-critical, that is, reactor power is increasing on its own, the reactor is declared critical.

Rod motion is performed using rod drive control systems. Newer BWRs such as the ABWR and ESBWR as well as all German and Swedish BWRs use the Fine Motion Control Rod Drive system, which allows multiple rods to be controlled with very smooth motions. This allows a reactor operator to evenly increase the core's reactivity until the reactor is critical. Older BWR designs use a manual control system, which is usually limited to controlling one or four control rods at a time, and only through a series of notched positions with fixed intervals between these positions. Due to the limitations of the manual control system it is possible while starting-up that the core can be placed into a condition where a single control rod can cause a large uneven reactivity change which can potentially challenge the fuel's thermal design margins. As a result, GE developed a set of rules in 1977 called BPWS (Banked Position Withdrawal Sequence) which help minimize the worth of any single control rod and prevent fuel damage in the case of a control rod drop accident. BPWS separates control rods into four groups, A1, A2, B1, and B2. Then, either all of the A control rods or B control rods are pulled full out in a defined sequence to create a "checkboard" pattern. Next the opposing group (B or A) is pulled in a defined sequence to positions 02, then 04, 08, 16, and finally full out (48), until the reactor enters the power operation range where thermal limits are no longer bounding. By following a BPWS compliant start-up sequence, the manual control system can be used to evenly and safely raise the entire core to critical, and prevent any fuel rods from exceeding 280 cal/gm energy release during any postulated event which could potentially damage the fuel.

Thermal Margins

Several calculated/measured quantities are tracked while operating a BWR:

- Maximum Fraction Limiting Critical Power Ratio, or MFLCPR;

- Fraction Limiting Linear Heat Generation Rate, or FLLHGR;

- Average Planar Linear Heat Generation Rate, or APLHGR;

- Pre-Conditioning Interim Operating Management Recommendation, or PCIOMR;

MFLCPR, FLLHGR, and APLHGR must be kept less than 1.0 during normal operation; administrative controls are in place to assure some margin of error and margin of safety to these licensed limits. Typical computer simulations divide the reactor core into 24–25 axial planes; relevant quantities (margins, burnup, power, void history) are tracked for each "node" in the reactor core (764 fuel assemblies x 25 nodes/assembly = 19100 nodal calculations/quantity).

Maximum Fraction Limiting Critical Power Ratio (Mflcpr)

Specifically, MFLCPR represents how close the leading fuel bundle is to "dry-out" (or "departure from nucleate boiling" for a PWR). Transition boiling is the unstable transient region where nucleate boiling tends toward film boiling. A water drop dancing on a hot frying pan is an example of film boiling. During film boiling a volume of insulating vapor separates the heated surface from the cooling fluid; this causes the temperature of the heated surface to increase drastically to once again reach equilibrium heat transfer with the cooling fluid. In other words, steam semi-insulates the heated surface and surface temperature rises to allow heat to get to the cooling fluid (through convection and radiative heat transfer).

MFLCPR is monitored with an empirical correlation that is formulated by vendors of BWR fuel (GE, Westinghouse, AREVA-NP). The vendors have test rigs where they simulate nuclear heat with resistive heating and determine experimentally what conditions of coolant flow, fuel assembly power, and reactor pressure will be in/out of the transition boiling region for a particular fuel design. In essence, the vendors make a model of the fuel assembly but power it with resistive heaters. These mock fuel assemblies are put into a test stand where data points are taken at specific powers, flows, pressures. It is obvious that nuclear fuel could be damaged by film boiling; this would cause the fuel cladding to overheat and fail. Experimental data is conservatively applied to BWR fuel to ensure that the transition to film boiling does not occur during normal or transient operation. Typical SLMCPR/MCPRSL (Safety Limit MCPR) licensing limit for a BWR core is substantiated by a calculation that proves that 99.9% of fuel rods in a BWR core will not enter the transition to film boiling during normal operation or anticipated operational occurrences. Since the BWR is boiling water, and steam does not transfer heat as well as liquid water, MFLCPR typically occurs at the top of a fuel assembly, where steam volume is the highest.

Fraction Limiting Linear Heat Generation Rate (Fllhgr)

FLLHGR (FDLRX, MFLPD) is a limit on fuel rod power in the reactor core. For new fuel, this limit is typically around 13 kW/ft (43 kW/m) of fuel rod. This limit ensures that the centerline temperature of the fuel pellets in the rods will not exceed the melting point of the fuel material (uranium/gadolinium oxides) in the event of the worst possible plant transient/scram anticipated to occur. To illustrate the response of LHGR in transient imagine the rapid closure of the valves that admit steam to the turbines at full power. This causes the immediate cessation of steam flow and an immediate rise in BWR pressure. This rise in pressure effectively subcools the reactor coolant instantaneously; the voids (vapor) collapse into solid water. When the voids collapse in the reactor, the fission reaction is encouraged (more thermal neutrons); power increases drastically (120%) until it is terminated by the automatic insertion of the control rods. So, when the reactor is isolated from the turbine rapidly, pressure in the vessel rises rapidly, which collapses the water vapor, which causes a power excursion which is terminated by the Reactor Protection System. If a fuel pin was operating at 13.0 kW/ft prior to the transient, the void collapse would cause its power to rise. The FLLHGR limit is in place to ensure that the highest powered fuel rod will not melt if its power was rapidly increased following a pressurization transient. Abiding by the LHGR limit precludes melting of fuel in a pressurization transient.

Average Planar Linear Heat Generation Rate (Aplhgr)

APLHGR, being an average of the Linear Heat Generation Rate (LHGR), a measure of the decay heat present in the fuel bundles, is a margin of safety associated with the potential for fuel failure to occur during a LBLOCA (large-break loss-of-coolant accident – a massive pipe rupture leading to catastrophic loss of coolant pressure within the reactor, considered the most threatening "design basis accident" in probabilistic risk assessment and nuclear safety), which is anticipated to lead to the temporary exposure of the core; this core drying-out event is termed core "uncovery", for the core loses its heat-removing cover of coolant, in the case of a BWR, light water. If the core is uncovered for too long, fuel failure can occur; for the purpose of design, fuel failure is assumed to occur when the temperature of the uncovered fuel reaches a critical temperature (1100 °C, 2200 °F). BWR designs incorporate failsafe protection systems to rapidly cool and make safe the uncovered fuel prior to it reaching this temperature; these failsafe systems are known as the Emergency Core Cooling System. The ECCS is designed to rapidly flood the reactor pressure vessel, spray water on the core itself, and sufficiently cool the reactor fuel in this event. However, like any system, the ECCS has limits, in this case, to its cooling capacity, and there is a possibility that fuel could be designed that produces so much decay heat that the ECCS would be overwhelmed and could not cool it down successfully.

So as to prevent this from happening, it is required that the decay heat stored in the fuel assemblies at any one time does not overwhelm the ECCS. As such, the measure of decay heat generation known as LHGR was developed by GE's engineers, and from this measure, APLHGR is derived. APLHGR is monitored to ensure that the reactor is not operated at an average power level that would defeat the primary containment systems. When a refueled core is licensed to operate, the fuel vendor/licensee simulate events with computer models. Their approach is to simulate worst case events when the reactor is in its most vulnerable state.

Aplhgr Is Commonly Pronounced As "Apple Hugger" In The Industry.

Pre-Conditioning Interim Operating Management Recommendation (PCIOMR)

PCIOMR is a set of rules and limits to prevent cladding damage due to pellet-clad interaction. During the first nuclear heatup, nuclear fuel pellets can crack. The jagged edges of the pellet can rub and interact with the inner cladding wall. During power increases in the fuel pellet, the ceramic fuel material expands faster than the fuel cladding, and the jagged edges of the fuel pellet begin to press into the cladding, potentially causing a perforation. To prevent this from occurring, two corrective actions were taken. The first is the inclusion of a thin barrier layer against the inner walls of the fuel cladding which are resistant to perforation due to pellet-clad interactions, and the second is a set of rules created under PCIOMR.

The PCIOMR rules require initial "conditioning" of new fuel. This means, for the first nuclear heatup of each fuel element, that local bundle power must be ramped very slowly to prevent cracking of the fuel pellets and limit the differences in the rates of thermal expansion of the fuel. PCIOMR rules also limit the maximum local power change (in kW/ft*hr), prevent pulling control rods below the tips of adjacent control rods, and require control rod sequences to be analyzed against core modelling software to prevent pellet-clad interactions. PCIOMR analysis look at local power peaks and xenon transients which could be caused by control rod position changes or rapid power changes to ensure that local power rates never exceed maximum ratings.

Pressurized Heavy-water Reactor

A pressurized heavy-water reactor (PHWR) is a nuclear reactor, commonly using unenriched natural uranium as its fuel, that uses heavy water (deuterium oxide D_2O) as its coolant and neutron moderator. The heavy water coolant is kept under pressure, allowing it to be heated to higher temperatures without boiling, much as in a pressurized water reactor. While heavy water is significantly more expensive than ordinary light water, it creates greatly enhanced neutron economy, allowing the reactor to operate without fuel-enrichment facilities (offsetting the additional expense of the heavy water) and enhancing the ability of the reactor to make use of alternate fuel cycles.

Purpose of Using Heavy Water

The key to maintaining a nuclear reaction within a nuclear reactor is to use the neutrons released during fission to stimulate fission in other nuclei. With careful control over the geometry and reaction rates, this can lead to a self-sustaining chain reaction, a state known as "criticality".

Natural uranium consists of a mixture of various isotopes, primarily ^{238}U and a much smaller amount (about 0.72% by weight) of ^{235}U. ^{238}U can only be fissioned by neutrons that are relatively energetic, about 1 MeV or above. No amount of ^{238}U can be made "critical", however, since it will tend to parasitically absorb more neutrons than it releases by the fission process. ^{235}U, on the other hand, can support a self-sustained chain reaction, but due to the low natural abundance of ^{235}U, natural uranium cannot achieve criticality by itself.

The "trick" to making a working reactor is to slow some of the neutrons to the point where their probability of causing nuclear fission in ^{235}U increases to a level that permits a sustained chain reaction in the uranium as a whole. This requires the use of a neutron moderator, which absorbs some of the neutrons' kinetic energy, slowing them down to an energy comparable to the thermal energy of the moderator nuclei themselves (leading to the terminology of "thermal neutrons" and "thermal reactors"). During this slowing-down process it is beneficial to physically separate the neutrons from the uranium, since ^{238}U nuclei have an enormous parasitic affinity for neutrons in this intermediate energy range (a reaction known as "resonance" absorption). This is a fundamental reason for designing reactors with discrete solid fuel separated by moderator, rather than employing a more homogeneous mixture of the two materials.

Water makes an excellent moderator; the hydrogen atoms in the water molecules are very close in mass to a single neutron, and the collisions thus have a very efficient momentum transfer, similar conceptually to the collision of two billiard balls. However, despite being a good moderator, water is relatively effective at absorbing neutrons. Using water as a moderator will absorb so many neutrons that there will be too few left to react with the small amount of ^{235}U in the fuel, again precluding criticality in natural uranium. Instead, in order to fuel a light-water reactor, first the amount of ^{235}U in the uranium must be increased, producing enriched uranium, which generally contains between 3% and 5% ^{235}U by weight (the waste from this process is known as depleted uranium, consisting primarily of ^{238}U). In this enriched form there *is* enough ^{235}U to react with the water-moderated neutrons to maintain criticality.

One complication of this approach is the requirement to build a uranium enrichment facility, which are generally expensive to build and operate. They also present a nuclear proliferation concern; the same systems used to enrich the ^{235}U can also be used to produce much more "pure" weapons-grade material (90% or more ^{235}U), suitable for producing a nuclear bomb. This is not a trivial exercise by any means, but feasible enough that enrichment facilities present a significant nuclear proliferation risk.

An alternative solution to the problem is to use a moderator that does *not* absorb neutrons as readily as water. In this case potentially all of the neutrons being released can be moderated and used in reactions with the ^{235}U, in which case there *is* enough ^{235}U in natural uranium to sustain criticality. One such moderator is heavy water, or deuterium-oxide. Although it reacts dynamically with the neutrons in a similar fashion to light water (albeit with less energy transfer on average, given that heavy hydrogen, or deuterium, is about twice the mass of hydrogen), it already has the extra neutron that light water would normally tend to absorb.

Advantages and Disadvantages

The use of heavy water as the moderator is the key to the PHWR (pressurized heavy water reactor) system, enabling the use of natural uranium as the fuel (in the form of ceramic UO_2), which means that it can be operated without expensive uranium enrichment facilities. The mechanical arrangement of the PHWR, which places most of the moderator at lower temperatures, is particularly efficient because the resulting thermal neutrons are "more thermal" than in traditional designs, where the moderator normally is much hotter. These features mean that a PHWR can use natural uranium and other fuels, and does so more efficiently than light water reactors (LWRs).

Pressurised heavy-water reactors do have some drawbacks. Heavy water generally costs hundreds of dollars per kilogram, though this is a trade-off against reduced fuel costs. The reduced energy content of natural uranium as compared to enriched uranium necessitates more frequent replacement of fuel; this is normally accomplished by use of an on-power refuelling system. The increased rate of fuel movement through the reactor also results in higher volumes of spent fuel than in LWRs employing enriched uranium. However, since unenriched uranium fuel accumulates a lower density of fission products than enriched uranium fuel, it generates less heat, allowing more compact storage.

Nuclear Proliferation

Opponents of heavy-water reactors suggest that such reactors pose a much greater risk of nuclear proliferation than comparable light water reactors. These concerns stem from the fact that during normal reactor operation, uranium-238 in the natural uranium fuel of a heavy-water reactor is converted into plutonium-239, a fissile material suitable for use in nuclear weapons, via neutron capture followed by two β^- decays. As a result, if the fuel of a heavy-water reactor is changed frequently, significant amounts of weapons-grade plutonium can be chemically extracted from the irradiated natural uranium fuel by nuclear reprocessing. In this way, the materials necessary to construct a nuclear weapon can be obtained without any uranium enrichment.

In addition, the use of heavy water as a moderator results in the production of small amounts of tritium when the deuterium nuclei in the heavy water absorb neutrons, a very inefficient reaction. Tritium is essential for the production of boosted fission weapons, which in turn enable the easier production of thermonuclear weapons, including neutron bombs. It is unclear whether it is possible to use this method to produce tritium on a practical scale.

The proliferation risk of heavy-water reactors was demonstrated when India produced the plutonium for Operation Smiling Buddha, its first nuclear weapon test, by extraction from the spent fuel of a heavy-water research reactor known as the CIRUS reactor.

Graphite-moderated Reactor

A graphite reactor is a nuclear reactor that uses carbon as a neutron moderator, which allows un-enriched uranium to be used as nuclear fuel.

The very first artificial nuclear reactor, the Chicago Pile-1, used graphite as moderator. Two graphite moderated reactors were involved in major accidents: An untested graphite annealing process contributed to the Windscale fire (but the graphite itself did not catch fire), and a graphite fire during the Chernobyl disaster contributed to the spread of radioactive material (but was not a cause of the accident itself).

Types

There are several types of graphite-moderated nuclear reactors that have been used in commercial electricity generation:

- Gas-cooled reactors
    - Magnox
    - UNGG reactor
    - Advanced gas-cooled reactor (AGR)
- Water-cooled reactors
    - RBMK
    - EGP-6
- High-temperature gas-cooled reactors (past)
    - Dragon reactor
    - AVR
    - Peach Bottom Nuclear Generating Station, Unit 1
    - THTR-300
    - Fort St. Vrain Generating Station
- High temperature gas-cooled reactors (in development or construction)
    - Pebble-bed reactor
    - Prismatic fuel reactor
    - UHTREX Ultra-high-temperature reactor experiment

Research Reactors

There have been a number of Research or Test Reactors built that use graphite as the moderator.

- Chicago Pile-1, described more below

- Chicago Pile-2

- Transient Reactor Test Facility (TREAT)

- Molten Salt Reactor Experiment (MSRE)

History

The first artificial nuclear reactor, Chicago Pile-1, a graphite-moderated device that produced a microscopic amount of heat, was constructed by a team led by Enrico Fermi in 1942. The construction and testing of this reactor (an "atomic pile") was part of the Manhattan Project. This work led to the construction of the X-10 Graphite Reactor at Oak Ridge National Laboratory, which was the first nuclear reactor designed and built for continuous operation, and began operation in 1943.

Accidents

There have been several major accidents in graphite moderated reactors, with the Windscale fire and the Chernobyl disaster probably the best known.

In the Windscale fire, an untested annealing process for the graphite was used, and that contributed to the accident – however it was the uranium fuel rather than the graphite in the reactor that caught fire. The only graphite moderator damage was found to be localized around burning fuel elements.

In the Chernobyl disaster the graphite was a contributing factor to the cause of the accident. Due to overheating from lack of adequate cooling the fuel rods began to deteriorate. After the AZ5 button was pressed to shut down the reactor, the control rods jammed in the middle of the core causing a positive loop since the nuclear fuel reacted to graphite. This is what has been dubbed the "final trigger" of events before the rupture. A graphite fire after the main event contributed to the spread of radioactive material. The massive power excursion in Chernobyl during a mishandled test led to the rupture of the reactor vessel and a series of steam explosions, which destroyed the reactor building. Now exposed to both air and the heat from the reactor core, the graphite moderator in the reactor core caught fire, and this fire sent a plume of highly radioactive fallout into the atmosphere and over an extensive geographical area.

In addition, the French Saint-Laurent Nuclear Power Plant and the Spanish Vandellòs Nuclear Power Plant – both UNGG graphite-moderated natural uranium reactors – suffered major accidents. Particularly noteworthy are an partial core meltdown on 17. October 1969 and an heat excursion during graphite annealing on 13. March 1980 in Saint-Laurent, which were both classified as INES 4. The Vandellòs NPP was damaged on 19. October 1989, and a repair was considered not economic.

Gas-cooled Reactor

A gas-cooled reactor (GCR) is a nuclear reactor that uses graphite as a neutron moderator and carbon dioxide (helium can also be used) as coolant. Although there are many other types of reactor cooled by gas, the terms *GCR* and to a lesser extent *gas cooled reactor* are particularly used to refer to this type of reactor.

The GCR was able to use natural uranium as fuel, enabling the countries that developed them to fabricate their own fuel without relying on other countries for supplies of enriched uranium, which was at the time of their development only available from the United States or Soviet Union.

Generation I GCR

There were two main types of generation I GCR:

- The Magnox reactors developed by the United Kingdom.

- The UNGG reactors developed by France.

The main difference between these two types is in the fuel cladding material. Both types were mainly constructed in their countries of origin, with a few export sales: Magnox plants to Italy and Japan, and a UNGG to Spain. Both types used fuel cladding materials that were unsuitable for medium term storage under water, making reprocessing an essential part of the nuclear fuel cycle.

Both types were, in their countries of origin, also designed and used to produce weapons-grade plutonium, but at the cost of major interruption to their use for power generation despite the provision of online refuelling.

In the UK, the Magnox was replaced by the advanced gas-cooled reactor (AGR), an improved Generation II gas cooled reactors. In France, the UNGG was replaced by the pressurized water reactor (PWR). More recently, GCRs based on the declassified drawings of the early Magnox reactors have been constructed by North Korea at the Yongbyon Nuclear Scientific Research Center.

Types

Gas-cooled reactor types include:

- Heavy Water Gas Cooled Reactor (HWGCR), e.g. KS 150 and Brennilis Nuclear Power Plant

- Gas-cooled reactor (graphite moderated)

  o Magnox

  o UNGG reactor (French design, 10 built, last retired in 1994)

- Advanced gas-cooled reactor

- Gas-cooled fast reactor

- Gas turbine modular helium reactor, General Atomics design : He-cooled, Graphite moderated

- High temperature gas cooled reactor

- Pebble bed reactor

  o AVR reactor

  o THTR-300

      ○ Pebble bed modular reactor

- Very high temperature reactor

Breeder Reactor

A breeder reactor is a nuclear reactor that generates more fissile material than it consumes. These devices achieve this because their neutron economy is high enough to breed more fissile fuel than they use from fertile material, such as uranium-238 or thorium-232. Breeders were at first found attractive because their fuel economy was better than light water reactors, but interest declined after the 1960s as more uranium reserves were found, and new methods of uranium enrichment reduced fuel costs.

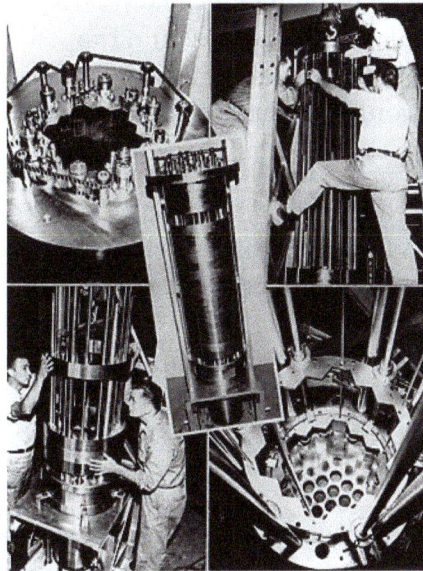

Assembly of the core of Experimental Breeder Reactor I in Idaho, United States, 1951

Fuel Efficiency and Types of Nuclear Waste

Fission Probabilities of Selected Actinides, Thermal vs. Fast Neutrons				
Isotope	Thermal Fission Cross Section	Thermal Fission %	Fast Fission Cross Section	Fast Fission %
Th-232	nil	1 (non-fissile)	0.350 barn	3 (non-fissile)
U-232	76.66 barn	59	2.370 barn	95
U-233	531.2 barn	89	2.450 barn	93
U-235	584.4 barn	81	2.056 barn	80
U-238	11.77 microbarn	1 (non-fissile)	1.136 barn	11
Np-237	0.02249 barn	3 (non-fissile)	2.247 barn	27
Pu-238	17.89 barn	7	2.721 barn	70
Pu-239	747.4 barn	63	2.338 barn	85
Pu-240	58.77 barn	1 (non-fissile)	2.253 barn	55
Pu-241	1012 barn	75	2.298 barn	87

Pu-242	0.002557 barn	1 (non-fissile)	2.027 barn	53
Am-241	600.4 barn	1 (non-fissile)	0.2299 microbarn	21
Am-242m	6409 barn	75	2.550 barn	94
Am-243	0.1161 barn	1 (non-fissile)	2.140 barn	23
Cm-242	5.064 barn	1 (non-fissile)	2.907 barn	10
Cm-243	617.4 barn	78	2.500 barn	94
Cm-244	1.037 barn	4 (non-fissile)	0.08255 microbarn	33

Breeder reactors could, in principle, extract almost all of the energy contained in uranium or thorium, decreasing fuel requirements by a factor of 100 compared to widely used once-through light water reactors, which extract less than 1% of the energy in the uranium mined from the earth. The high fuel efficiency of breeder reactors could greatly reduce concerns about fuel supply or energy used in mining. Adherents claim that with seawater uranium extraction, there would be enough fuel for breeder reactors to satisfy our energy needs for 5 billion years at 1983's total energy consumption rate, thus making nuclear energy effectively a renewable energy.

Nuclear waste became a greater concern by the 1990s. In broad terms, spent nuclear fuel has two main components. The first consists of fission products, the leftover fragments of fuel atoms after they have been split to release energy. Fission products come in dozens of elements and hundreds of isotopes, all of them lighter than uranium. The second main component of spent fuel is transuranics (atoms heavier than uranium), which are generated from uranium or heavier atoms in the fuel when they absorb neutrons but do not undergo fission. All transuranic isotopes fall within the actinide series on the periodic table, and so they are frequently referred to as the actinides.

The physical behavior of the fission products is markedly different from that of the transuranics. In particular, fission products do not themselves undergo fission, and therefore cannot be used for nuclear weapons. Furthermore, only seven long-lived fission product isotopes have half-lives longer than a hundred years, which makes their geological storage or disposal less problematic than for transuranic materials.

With increased concerns about nuclear waste, breeding fuel cycles became interesting again because they can reduce actinide wastes, particularly plutonium and minor actinides. Breeder reactors are designed to fission the actinide wastes as fuel, and thus convert them to more fission products.

After "spent nuclear fuel" is removed from a light water reactor, it undergoes a complex decay profile as each nuclide decays at a different rate. Due to a physical oddity referenced below, there is a large gap in the decay half-lives of fission products compared to transuranic isotopes. If the transuranics are left in the spent fuel, after 1,000 to 100,000 years, the slow decay of these transuranics would generate most of the radioactivity in that spent fuel. Thus, removing the transuranics from the waste eliminates much of the long-term radioactivity of spent nuclear fuel.

Today's commercial light water reactors do breed some new fissile material, mostly in the form of plutonium. Because commercial reactors were never designed as breeders, they do not convert enough uranium-238 into plutonium to replace the uranium-235 consumed. Nonetheless, at least one-third of the power produced by commercial nuclear reactors comes from fission of plutonium generated within the fuel. Even with this level of plutonium consumption, light water reactors consume only part of the plutonium and minor actinides they produce, and nonfissile isotopes of plutonium build up, along with significant quantities of other minor actinides.

Conversion Ratio, Breakeven, Breeding Ratio, Doubling Time, and Burnup

One measure of a reactor's performance is the "conversion ratio" (the average number of new fissile atoms created per fission event). All proposed nuclear reactors except specially designed and operated actinide burners experience some degree of conversion. As long as there is any amount of a fertile material within the neutron flux of the reactor, some new fissile material is always created.

The ratio of new fissile material in spent fuel to fissile material consumed from the fresh fuel is known as the "conversion ratio" or "breeding ratio" of a reactor.

For example, commonly used light water reactors have a conversion ratio of approximately 0.6. Pressurized heavy water reactors (PHWR) running on natural uranium have a conversion ratio of 0.8. In a breeder reactor, the conversion ratio is higher than 1. "Breakeven" is achieved when the conversion ratio becomes 1: the reactor produces as much fissile material as it uses.

"Doubling time" is the amount of time it would take for a breeder reactor to produce enough new fissile material to create a starting fuel load for another nuclear reactor. This was considered an important measure of breeder performance in early years, when uranium was thought to be scarce. However, since uranium is more abundant than thought, and given the amount of plutonium available in spent reactor fuel, doubling time has become a less important metric in modern breeder reactor design.

"Burnup" is a measure of how much energy has been extracted from a given mass of heavy metal in fuel, often expressed (for power reactors) in terms of gigawatt-days per ton of heavy metal. Burnup is an important factor in determining the types and abundances of isotopes produced by a fission reactor. Breeder reactors, by design, have extremely high burnup compared to a conventional reactor, as breeder reactors produce much more of their waste in the form of fission products, while most or all of the actinides are meant to be fissioned and destroyed.

In the past breeder reactor development focused on reactors with low breeding ratios, from 1.01 for the Shippingport Reactor running on thorium fuel and cooled by conventional light water to over 1.2 for the Soviet BN-350 liquid-metal-cooled reactor. Theoretical models of breeders with liquid sodium coolant flowing through tubes inside fuel elements ("tube-in-shell" construction) suggest breeding ratios of at least 1.8 are possible on an industrial scale. The Soviet BR-1 test reactor achieved a breeding ratio of 2.5 under non-commercial conditions.

Types of Breeder Reactor

Many types of breeder reactor are possible:

A 'breeder' is simply a reactor designed for very high neutron economy with an associated conversion rate higher than 1.0. In principle, almost any reactor design could possibly be tweaked to become a breeder. An example of this process is the evolution of the Light Water Reactor, a very heavily moderated thermal design, into the Super Fast Reactor concept, using light water in an extremely low-density supercritical form to increase the neutron economy high enough to allow breeding.

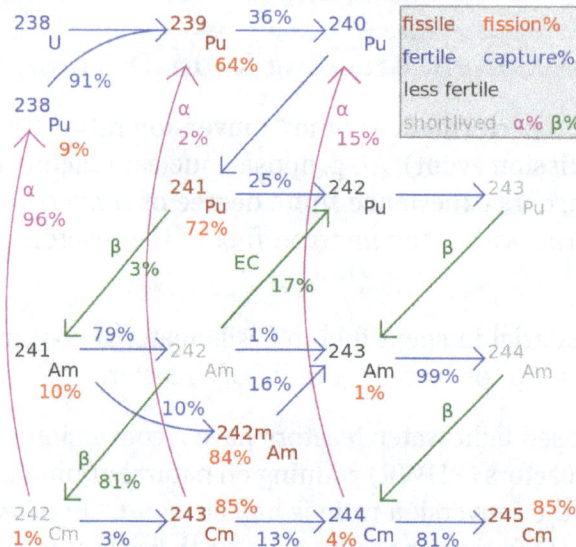

Production of heavy transuranic actinides in current thermal-neutron fission reactors through neutron capture and decays. Starting at uranium-238, isotopes of plutonium, americium, and curium are all produced. In a fast neutron breeder reactor, all these isotopes may be burned as fuel.

Aside from water cooled, there are many other types of breeder reactor currently envisioned as possible. These include molten-salt cooled, gas cooled, and liquid metal cooled designs in many variations. Almost any of these basic design types may be fueled by uranium, plutonium, many minor actinides, or thorium, and they may be designed for many different goals, such as creating more fissile fuel, long-term steady-state operation, or active burning of nuclear wastes.

For convenience, it is perhaps simplest to divide the extant reactor designs into two broad categories based upon their neutron spectrum, which has the natural effect of dividing the reactor designs into those designed to use primarily uranium and transuranics, and those designed to use thorium and avoid transuranics.

- Fast breeder reactor or FBR uses fast (unmoderated) neutrons to breed fissile plutonium and possibly higher transuranics from fertile uranium-238. The fast spectrum is flexible enough that it can also breed fissile uranium-233 from thorium, if desired.

- Thermal breeder reactor use thermal spectrum (moderated) neutrons to breed fissile uranium-233 from thorium (thorium fuel cycle). Due to the behavior of the various nuclear fuels, a thermal breeder is thought commercially feasible only with thorium fuel, which avoids the buildup of the heavier transuranics.

Reprocessing

Fission of the nuclear fuel in any reactor produces neutron-absorbing fission products. Because of this unavoidable physical process, it is necessary to reprocess the fertile material from a breeder reactor to remove those neutron poisons. This step is required if one is to fully utilize the ability to breed as much or more fuel than is consumed. All reprocessing can present a proliferation concern, since it extracts weapons usable material from spent fuel. The most common reprocessing technique, PUREX, presents a particular concern, since it was expressly designed to separate pure plutonium. Early proposals for the breeder reactor fuel cycle posed an even greater proliferation concern because they would use PUREX to separate plutonium in a highly attractive isotopic form for use in nuclear weapons.

Several countries are developing reprocessing methods that do not separate the plutonium from the other actinides. For instance, the non-water based pyrometallurgical electrowinning process, when used to reprocess fuel from an integral fast reactor, leaves large amounts of radioactive actinides in the reactor fuel. More conventional advanced, water-based reprocessing systems like PUREX, include SANEX, UNEX, DIAMEX, COEX, and TRUEX—as well as proposals to combine PUREX with co-processes.

All these systems have better proliferation resistance than PUREX, though their adoption rate is low.

In the thorium cycle, thorium-232 breeds by converting first to protactinium-233, which then decays to uranium-233. If the protactinium remains in the reactor, small amounts of U-232 are also produced, which has the strong gamma emitter Tl-208 in its decay chain. Similar to uranium-fueled designs, the longer the fuel and fertile material remain in the reactor, the more of these undesirable elements build up. Inside the envisioned commercial thorium reactors high levels of U232 would be allowed to accumulate, leading to extremely high gamma radiation doses from any uranium derived from thorium. These gamma rays complicate the safe handling of a weapon and the design of its electronics; this explains why U-233 has never been pursued for weapons beyond proof-of-concept demonstrations.

Waste Reduction

Actinides and fission products by half-life								
Actinides by decay chain				Half-life range (y)		Fission products of ^{235}U by yield		
$4n$	$4n+1$	$4n+2$	$4n+3$			4.5–7%	0.04–1.25%	<0.001%
^{228}Ra^{No}				4–6	†		^{155}Eu^{b}	
^{244}Cm^{f}	^{241}Pu^{f}	^{250}Cf	^{227}Ac^{No}	10–29		^{90}Sr	^{85}Kr	^{113m}Cd^{b}
^{232}U^{f}		^{238}Pu^{fNo}	^{243}Cm^{f}	29–97		^{137}Cs	^{151}Sm^{b}	^{121m}Sn
^{248}Bk	^{249}Cf^{f}	^{242m}Am^{f}		141–351				
	^{241}Am^{f}		^{251}Cf^{f}	430–900		No fission products have a half-life in the range of 100–210 k years …		
		^{226}Ra^{No}	^{247}Bk	1.3 k – 1.6 k				
^{240}Pu^{fNo}	^{229}Th^{No}	^{246}Cm^{f}	^{243}Am^{f}	4.7 k – 7.4 k				
	^{245}Cm^{f}	^{250}Cm		8.3 k – 8.5 k				
			^{239}Pu^{fNo}	24.1 k				
		^{230}Th^{No}	^{231}Pa^{No}	32 k – 76 k				

					Half-life					
$^{236}\text{Np}^f$	$^{233}\text{U}^{f№}$	$^{234}\text{U}^{№}$			150 k – 250 k	‡	$^{99}\text{Tc}^{□}$		^{126}Sn	
^{248}Cm		$^{242}\text{Pu}^f$			327 k – 375 k				$^{79}\text{Se}^{□}$	
					1.53 M		^{93}Zr			
	$^{237}\text{Np}^{f№}$				2.1 M – 6.5 M		$^{135}\text{Cs}^{□}$		^{107}Pd	
$^{236}\text{U}^{№}$			$^{247}\text{Cm}^f$		15 M – 24 M				$^{129}\text{I}^{□}$	
$^{244}\text{Pu}^{№}$					80 M		... nor beyond 15.7 M years			
$^{232}\text{Th}^{№}$		$^{238}\text{U}^{№}$	$^{235}\text{U}^{f№}$		0.7 G – 14.1 G					

Legend for superscript symbols
□ has thermal neutron capture cross section in the range of 8–50 barns
f fissile
m metastable isomer
№ naturally occurring radioactive material (NORM)
þ neutron poison (thermal neutron capture cross section greater than 3k barns)
† range 4–97 y: Medium-lived fission product
‡ over 200,000 y: Long-lived fission product

Nuclear waste became a greater concern by the 1990s. Breeding fuel cycles attracted renewed interest because of their potential to reduce actinide wastes, particularly plutonium and minor actinides. Since breeder reactors on a closed fuel cycle would use nearly all of the actinides fed into them as fuel, their fuel requirements would be reduced by a factor of about 100. The volume of waste they generate would be reduced by a factor of about 100 as well. While there is a huge reduction in the *volume* of waste from a breeder reactor, the *activity* of the waste is about the same as that produced by a light water reactor.

In addition, the waste from a breeder reactor has a different decay behavior, because it is made up of different materials. Breeder reactor waste is mostly fission products, while light water reactor waste has a large quantity of transuranics. After spent nuclear fuel has been removed from a light water reactor for longer than 100,000 years, these transuranics would be the main source of radioactivity. Eliminating them would eliminate much of the long-term radioactivity from the spent fuel.

In principle, breeder fuel cycles can recycle and consume all actinides, leaving only fission products. As the graphic in this section indicates, fission products have a peculiar 'gap' in their aggregate half-lives, such that no fission products have a half-life longer than 91 years and shorter than two hundred thousand years. As a result of this physical oddity, after several hundred years in storage, the activity of the radioactive waste from a Fast Breeder Reactor would quickly drop to the low level of the long-lived fission products. However, to obtain this benefit requires the highly efficient separation of transuranics from spent fuel. If the fuel reprocessing methods used leave a large fraction of the transuranics in the final waste stream, this advantage would be greatly reduced.

Both types of breeding cycles can reduce actinide wastes:

- The fast breeder reactor's fast neutrons can fission actinide nuclei with even numbers of both protons and neutrons. Such nuclei usually lack the low-speed "thermal neutron" resonances of fissile fuels used in LWRs.

- The thorium fuel cycle inherently produces lower levels of heavy actinides. The fertile material in the thorium fuel cycle has an atomic weight of 232, while the fertile material in the uranium fuel cycle has an atomic weight of 238. That mass difference means that thorium-232 requires six more neutron capture events per nucleus before the transuranic elements can be produced. In addition to this simple mass difference, the reactor gets two chances to fission the nuclei as the mass increases: First as the effective fuel nuclei U233, and as it absorbs two more neutrons, again as the fuel nuclei U235.

A reactor whose main purpose is to destroy actinides, rather than increasing fissile fuel stocks, is sometimes known as a burner reactor. Both breeding and burning depend on good neutron economy, and many designs can do either. Breeding designs surround the core by a breeding blanket of fertile material. Waste burners surround the core with non-fertile wastes to be destroyed. Some designs add neutron reflectors or absorbers.

Breeder Reactor Concepts

There are several concepts for breeder reactors; the two main ones are:

- Reactors with a fast neutron spectrum are called fast breeder reactors (FBR) – these typically utilize uranium-238 as fuel.

- Reactors with a thermal neutron spectrum are called thermal breeder reactors – these typically utilize thorium-232 as fuel.

Fast Breeder Reactor

In 2006 all large-scale fast breeder reactor (FBR) power stations were liquid metal fast breeder reactors (LMFBR) cooled by liquid sodium. These have been of one of two designs:

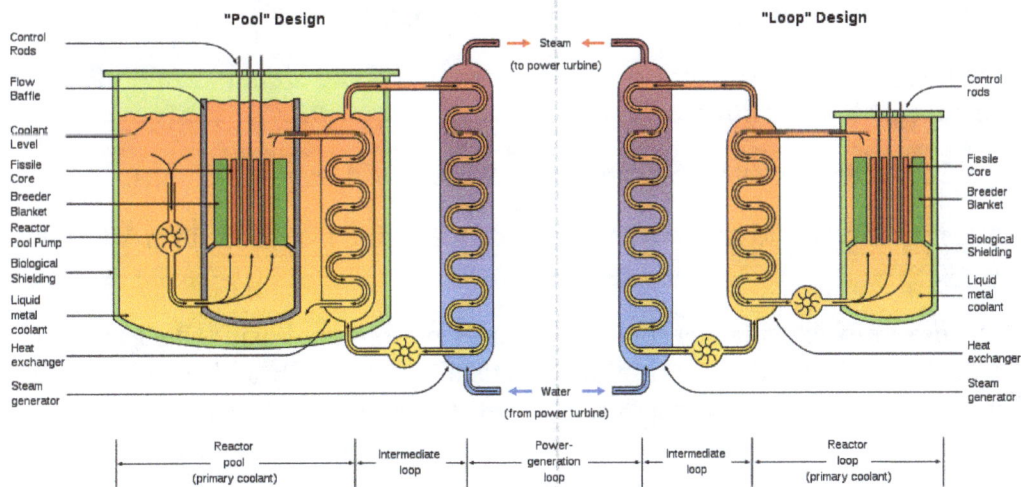

Schematic diagram showing the difference between the Loop and Pool types of LMFBR.

- *Loop* type, in which the primary coolant is circulated through primary heat exchangers outside the reactor tank (but inside the biological shield due to radioactive sodium-24 in the primary coolant)

- *Pool* type, in which the primary heat exchangers and pumps are immersed in the reactor tank

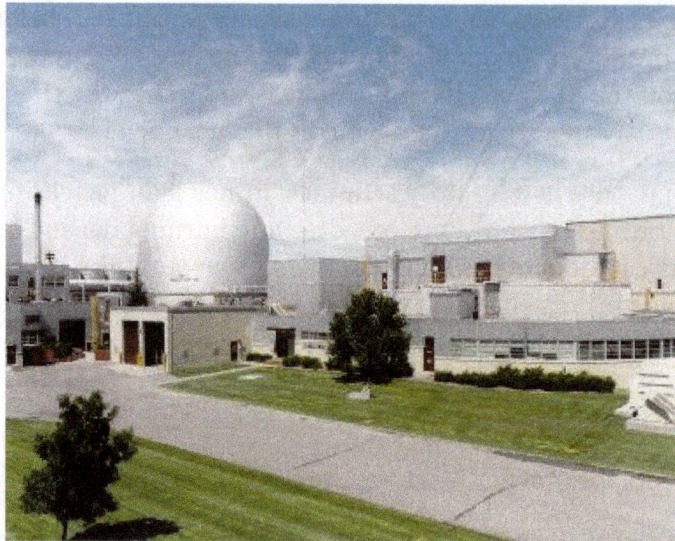

Experimental Breeder Reactor II, which served as the prototype for the Integral Fast Reactor

All current fast neutron reactor designs use liquid metal as the primary coolant, to transfer heat from the core to steam used to power the electricity generating turbines. FBRs have been built cooled by liquid metals other than sodium—some early FBRs used mercury, other experimental reactors have used a sodium-potassium alloy called NaK. Both have the advantage that they are liquids at room temperature, which is convenient for experimental rigs but less important for pilot or full scale power stations. Lead and lead-bismuth alloy have also been used. The relative merits of lead vs sodium are discussed here. Looking further ahead, four of the proposed generation IV reactor types are FBRs:

- Gas-Cooled Fast Reactor (GFR) cooled by helium.

- Sodium-Cooled Fast Reactor (SFR) based on the existing Liquid Metal FBR (LMFBR) and Integral Fast Reactor designs.

- Lead-Cooled Fast Reactor (LFR) based on Soviet naval propulsion units.

- Supercritical Water Reactor (SCWR) based on existing LWR and supercritical boiler technology.

FBRs usually use a mixed oxide fuel core of up to 20% plutonium dioxide (PuO_2) and at least 80% uranium dioxide (UO_2). Another fuel option is metal alloys, typically a blend of uranium, plutonium, and zirconium (used because it is "transparent" to neutrons). Enriched uranium can also be used on its own.

Many designs surround the core in a blanket of tubes that contain non-fissile uranium-238, which, by capturing fast neutrons from the reaction in the core, converts to fissile plutonium-239 (as is some of the uranium in the core), which is then reprocessed and used as nuclear fuel. Other FBR designs rely on the geometry of the fuel itself (which also contains uranium-238), arranged to attain sufficient fast neutron capture. The plutonium-239 (or the fissile uranium-235) fission cross-section is much smaller in a fast spectrum than in a thermal spectrum, as is the ratio between the $^{239}Pu/^{235}U$ fission cross-section and the ^{238}U absorption cross-section. This increases the concentration of ^{239}Pu/^{235}U needed to sustain a chain reaction, as well as the ratio of breeding to fission.

On the other hand, a fast reactor needs no moderator to slow down the neutrons at all, taking advantage of the fast neutrons producing a greater number of neutrons per fission than slow neutrons. For this reason ordinary liquid water, being a moderator as well as a neutron absorber, is an undesirable primary coolant for fast reactors. Because large amounts of water in the core are required to cool the reactor, the yield of neutrons and therefore breeding of ^{239}Pu are strongly affected. Theoretical work has been done on reduced moderation water reactors, which may have a sufficiently fast spectrum to provide a breeding ratio slightly over 1. This would likely result in an unacceptable power derating and high costs in an liquid-water-cooled reactor, but the supercritical water coolant of the SCWR has sufficient heat capacity to allow adequate cooling with less water, making a fast-spectrum water-cooled reactor a practical possibility.

The only commercially operating reactor as of 2015 is the BN-600 reactor in Russia, a 560MW sodium cooled reactor.

Integral Fast Reactor

One design of fast neutron reactor, specifically designed to address the waste disposal and plutonium issues, was the *integral fast reactor* (also known as an *integral fast breeder reactor*, although the original reactor was designed to not breed a net surplus of fissile material).

To solve the waste disposal problem, the IFR had an on-site electrowinning fuel reprocessing unit that recycled the uranium and all the transuranics (not just plutonium) via electroplating, leaving just short half-life fission products in the waste. Some of these fission products could later be separated for industrial or medical uses and the rest sent to a waste repository (where they would not have to be stored for anywhere near as long as wastes containing long half-life transuranics). The IFR pyroprocessing system uses molten cadmium cathodes and electrorefiners to reprocess metallic fuel directly on-site at the reactor. Such systems not only commingle all the minor actinides with both uranium and plutonium, they are compact and self-contained, so that no plutonium-containing material ever needs to be transported away from the site of the breeder reactor. Breeder reactors incorporating such technology would most likely be designed with breeding ratios very close to 1.00, so that after an initial loading of enriched uranium and/or plutonium fuel, the reactor would then be refueled only with small deliveries of natural uranium metal. A quantity of natural uranium metal equivalent to a block about the size of a milk crate delivered once per month would be all the fuel such a 1 gigawatt reactor would need. Such self-contained breeders are currently envisioned as the final self-contained and self-supporting ultimate goal of nuclear reactor designers. The project was canceled in 1994 by United States Secretary of Energy Hazel O'Leary.

Other Fast Reactors

Another proposed fast reactor is a fast molten salt reactor, in which the molten salt's moderating properties are insignificant. This is typically achieved by replacing the light metal fluorides (e.g. LiF, BeF$_2$) in the salt carrier with heavier metal chlorides (e.g., KCl, RbCl, ZrCl$_4$).

The graphite core of the Molten Salt Reactor Experiment

Several prototype FBRs have been built, ranging in electrical output from a few light bulbs' equivalent (EBR-I, 1951) to over 1,000 MWe. As of 2006, the technology is not economically competitive to thermal reactor technology—but India, Japan, China, South Korea and Russia are all committing substantial research funds to further development of Fast Breeder reactors, anticipating that rising uranium prices will change this in the long term. Germany, in contrast, abandoned the technology due to safety concerns. The SNR-300 fast breeder reactor was finished after 19 years despite cost overruns summing up to a total of 3.6 billion Euros, only to then be abandoned.

As well as their thermal breeder program, India is also developing FBR technology, using both uranium and thorium feedstocks.

Thermal Breeder Reactor

The advanced heavy water reactor (AHWR) is one of the few proposed large-scale uses of thorium. India is developing this technology, their interest motivated by substantial thorium reserves; almost a third of the world's thorium reserves are in India, which also lacks significant uranium reserves.

The Shippingport Reactor, used as a prototype Light Water Breeder for five years beginning in August, 1977

The third and final core of the Shippingport Atomic Power Station 60 MWe reactor was a light water thorium breeder, which began operating in 1977. It used pellets made of thorium dioxide and uranium-233 oxide; initially, the U-233 content of the pellets was 5–6% in the seed region, 1.5–3% in the blanket region and none in the reflector region. It operated at 236 MWt, generating 60 MWe and ultimately produced over 2.1 billion kilowatt hours of electricity. After five years, the core was removed and found to contain nearly 1.4% more fissile material than when it was installed, demonstrating that breeding from thorium had occurred.

The liquid fluoride thorium reactor (LFTR) is also planned as a thorium thermal breeder. Liquid-fluoride reactors may have attractive features, such as inherent safety, no need to manufacture fuel rods and possibly simpler reprocessing of the liquid fuel. This concept was first investigated at the Oak Ridge National Laboratory Molten-Salt Reactor Experiment in the 1960s. From 2012 it became the subject of renewed interest worldwide. Japan, China, the UK, as well as private US, Czech and Australian companies have expressed intent to develop and commercialize the technology.

Breeder Reactor Controversy

Like many aspects of nuclear power, fast breeder reactors have been subject to much controversy over the years. In 2010 the International Panel on Fissile Materials said "After six decades and the expenditure of the equivalent of tens of billions of dollars, the promise of breeder reactors remains largely unfulfilled and efforts to commercialize them have been steadily cut back in most countries". In Germany, the United Kingdom, and the United States, breeder reactor development programs have been abandoned. The rationale for pursuing breeder reactors—sometimes explicit and sometimes implicit—was based on the following key assumptions:

- It was expected that uranium would be scarce and high-grade deposits would quickly become depleted if fission power were deployed on a large scale; the reality, however, is that since the end of the cold war, uranium has been much cheaper and more abundant than early designers expected.

- It was expected that breeder reactors would quickly become economically competitive with the light-water reactors that dominate nuclear power today, but the reality is that capital costs are at least 25% more than water cooled reactors.

- It was thought that breeder reactors could be as safe and reliable as light-water reactors, but safety issues are cited as a concern with fast reactors that use a sodium coolant, where a leak could lead to a sodium fire.

- It was expected that the proliferation risks posed by breeders and their "closed" fuel cycle, in which plutonium would be recycled, could be managed. But since plutonium breeding reactors produce plutonium from U238, and thorium reactors produce fissile U233 from thorium, all breeding cycles could theoretically pose proliferation risks.

There are some past anti-nuclear advocates that have become pro-nuclear power as a clean source of electricity since breeder reactors effectively recycle most of their waste. This solves one of the most important negative issues of nuclear power. In the documentary *Pandora's Promise*, a case is made for breeder reactors because they provide a real, high kW alternative to fossil fuel energy. According to the movie, one pound of uranium provides as much energy as 5000 barrels of oil.

FBRs have been built and operated in the United States, the United Kingdom, France, the former USSR, India and Japan. The experimental FBR SNR-300 was built in Germany but never operated and eventually shut down amid political controversy following the Chernobyl disaster. As of 2014 one such reactor was being used for power generation, with another scheduled for early 2015. Several reactors are planned, many for research related to the Generation IV reactor initiative.

Breeder Reactor Development and Notable Breeder Reactors

Notable Breeder reactors											
Reactor	Country when built	Started	Shutdown	Design MWe	Final MWe	Thermal Power MWt	Capacity factor	No of leaks	Neutron temperature	Coolant	Reactor class
DFR	UK	1962	1977	14	11	65	34%	7	Fast	NaK	Test
BN-350	Soviet Union	1973	1999	135	52	750	43%	15	Fast	Sodium	Prototype
Rapsodie	France	1967	1983	0	-	40	-	2	Fast	Sodium	Test
Phénix	France	1975	2010	233	130	563	40.5%	31	Fast	Sodium	Prototype
PFR	UK	1976	1994	234	234	650	26.9%	20	Fast	Sodium	Prototype
KNK II	Germany	1977	1991	18	17	58	17.1%	21	Fast	Sodium	Research/Test
SNR-300	Germany	1985 (partial operation)	1991	327	-	-	-	-	Fast	Sodium	Prototype/Commercial
BN-600	Soviet Union	1981	operating	560	560	1470	74.2%	27	Fast	Sodium	Prototype/Commercial(Gen2)
FFTF	USA	1982	1993	0	-	400	-	1	Fast	Sodium	Test
Superphénix	France	1985	1998	1200	1200	3000	7.9%	7	Fast	Sodium	Prototype/Commercial(Gen2)
FBTR	India	1985	operating	13	-	40	-	6	Fast	Sodium	Test
PFBR	India	commissioning	commissioning	500	-	1250	-	-	Fast	Sodium	Prototype/Commercial(Gen3)
Jōyō	Japan	1977	operating	0	-	150	-	-	Fast	Sodium	Test
Monju	Japan	1995	dormant	246	246	714	trial only	1	Fast	Sodium	Prototype
BN-800	Russia	2015	operating	789	880	2100	-	-	Fast	Sodium	Prototype/Commercial(Gen3)
MSRE	USA	1965	1969	0	-	7.4	-	-	Epithermal	Molten Salt(FLiBe)	Test
Clementine	USA	1946	1952	0	-	0.025	-	-	Fast	Mercury	World's First Fast Reactor
EBR-1	USA	1951	1964	0.2	0.2	1.4	-	-	Fast	NaK	World's First Power Reactor
Fermi-1	USA	1963	1972	66	66	200	-	-	Fast	Sodium	Prototype
EBR-2	USA	1964	1994	19	19	62.5	-	-	Fast	Sodium	Experimental/Test
Shippingport	USA	1977 as breeder	1982	60	60	236	-	-	Thermal	Light Water	Experimental-Core3

The Soviet Union (comprising Russia and other countries, dissolved in 1991) constructed a series of fast reactors, the first being mercury-cooled and fueled with plutonium metal, and the later plants sodium-cooled and fueled with plutonium oxide.

BR-1 (1955) was 100W (thermal) was followed by BR-2 at 100 kW and then the 5MW BR-5.

BOR-60 (first criticality 1969) was 60 MW, with construction started in 1965.

Future Plants

In 2012 an FBR called the Prototype Fast Breeder Reactor was under construction in India, due to be completed that year, with commissioning date known by mid-year. The FBR program of India includes the concept of using fertile thorium-232 to breed fissile uranium-233. India is also pursuing the thorium thermal breeder reactor. A thermal breeder is not possible with purely uranium/plutonium based technology. Thorium fuel is the strategic direction of the power program of India, owing to the nation's large reserves of thorium, but worldwide known reserves of thorium are also some four times those of uranium. India's Department of Atomic Energy (DAE) said in 2007 that it would simultaneously construct four more breeder reactors of 500 MWe each including two at Kalpakkam.

The new Indian FBR-600 is being used to advance these plans. The FBR-600 is a pool-type sodium cooled reactor with a rating of 600MWe and advanced active and passive safety features.

The Chinese Experimental Fast Reactor is a 65 MW (thermal), 20 MW (electric), sodium-cooled, pool-type reactor with a 30-year design lifetime and a target burnup of 100 MWd/kg.

The China Experimental Fast Reactor (CEFR) is a 25 MW(e) prototype for the planned China Prototype Fast Reactor (CFRP). It started generating power on 21 July 2011.

China also initiated a research and development project in thorium molten-salt thermal breeder reactor technology (Liquid fluoride thorium reactor), formally announced at the Chinese Academy of Sciences (CAS) annual conference in January 2011. Its ultimate target is to investigate and develop a thorium-based molten salt nuclear system over about 20 years.

Kirk Sorensen, former NASA scientist and Chief Nuclear Technologist at Teledyne Brown Engineering, has long been a promoter of thorium fuel cycle and particularly liquid fluoride thorium reactors. In 2011, Sorensen founded Flibe Energy, a company aimed to develop 20-50 MW LFTR reactor designs to power military bases.

South Korea is developing a design for a standardized modular FBR for export, to complement the standardized PWR (Pressurized Water Reactor) and CANDU designs they have already developed and built, but has not yet committed to building a prototype.

Russia has a plan for increasing its fleet of fast breeder reactors significantly. A BN-800 reactor (800 MWe) at Beloyarsk was completed in 2012, succeeding a smaller BN-600. In June 2014 the

BN-800 was started in the minimum power mode. The reactor working at 35% of the nominal efficiency, was plugged-in the energy network on 10 December 2015.

A cutaway model of the BN-600 reactor, superseded by the BN-800 reactor family.

Plans for the construction of an even larger BN-1200 reactor (1,200 MWe) initially anticipated completion in 2018, with two additional BN-1200 reactors built by the end of 2030. However, in 2015 Rosenergoatom postponed construction indefinitely to allow fuel design to be improved after more experience of operating the BN-800 reactor, and amongst cost concerns.

An experimental lead-cooled fast reactor, BREST-300 will be built at the Siberian Chemical Combine (SCC) in Seversk. The BREST design is seen as a successor to the BN series and the 300 MWe unit at the SCC could be the forerunner to a 1,200 MWe version for wide deployment as a commercial power generation unit. The development program is as part of an Advanced Nuclear Technologies Federal Program 2010-2020 that seeks to exploit fast reactors as a way to be vastly more efficient in the use of uranium while 'burning' radioactive substances that otherwise would have to be disposed of as waste. BREST refers to *bystry reaktor so svintsovym teplonositelem*, Russian for 'fast reactor with lead coolant'. Its core would measure about 2.3 metres in diameter by 1.1 metres in height and contain 16 tonnes of fuel. The unit would be refuelled every year, with each fuel element spending five years in total within the core. Lead coolant temperature would be around 540 °C, giving a high efficiency of 43%, primary heat production of 700 MWt yielding electrical power of 300 MWe. The operational lifespan of the unit could be 60 years. The design is expected to be completed by NIKIET in 2014 for construction between 2016 and 2020.

Construction of the BN-800 reactor

On February 16, 2006, the U.S., France and Japan signed an "arrangement" to research and develop sodium-cooled fast reactors in support of the Global Nuclear Energy Partnership. In April 2007 the Japanese government selected Mitsubishi Heavy Industries as the "core company in FBR development in Japan". Shortly thereafter, MHI started a new company, Mitsubishi FBR Systems (MFBR) to develop and eventually sell FBR technology.

The Marcoule Nuclear Site in France, location of the Phénix (on the left) and possible future site of the ASTRID Gen-IV reactor.

In September 2010 the French government allocated 651.6 million euros to the *Commissariat à l'énergie atomique* to finalize the design of "Astrid" (Advanced Sodium Technological Reactor for Industrial Demonstration), a 600 MW reactor design of the 4th generation to be operational in 2020. As of 2013 the UK had shown interest in the PRISM reactor and was working in concert with France to develop ASTRID.

In October 2010 GE Hitachi Nuclear Energy signed a memorandum of understanding with the operators of the US Department of Energy's Savannah River site, which should allow the construction of a demonstration plant based on the company's S-PRISM fast breeder reactor prior to the design receiving full NRC licensing approval. In October 2011 The Independent reported that the UK Nuclear Decommissioning Authority (NDA) and senior advisers within the Department for Energy and Climate Change (DECC) had asked for technical and financial details of the PRISM, partly as a means of reducing the country's plutonium stockpile.

The traveling wave reactor proposed in a patent by Intellectual Ventures is a fast breeder reactor designed to not need fuel reprocessing during the decades-long lifetime of the reactor. The breed-burn wave in the TWR design does not move from one end of the reactor to the other but gradually from the inside out. Moreover, as the fuel's composition changes through nuclear transmutation, fuel rods are continually reshuffled within the core to optimize the neutron flux and fuel usage at any given point in time. Thus, instead of letting the wave propagate through the fuel, the fuel itself is moved through a largely stationary burn wave. This is contrary to many media reports, which have popularized the concept as a candle-like reactor with a burn region that moves down a stick of fuel. By replacing a static core configuration with an actively managed "standing wave" or "soliton" core, TerraPower's design avoids the problem of cooling a highly variable burn region. Under this scenario, the reconfiguration of fuel rods is accomplished remotely by robotic devices; the containment vessel remains closed during the procedure, and there is no associated downtime.

Steam Turbine

A steam turbine is a device that extracts thermal energy from pressurized steam and uses it to do mechanical work on a rotating output shaft. Its modern manifestation was invented by Sir Charles Parsons in 1884.

The rotor of a modern steam turbine used in a power plant

Because the turbine generates rotary motion, it is particularly suited to be used to drive an electrical generator – about 90% of all electricity generation in the United States (1996) is by use of steam turbines. The steam turbine is a form of heat engine that derives much of its improvement in thermodynamic efficiency from the use of multiple stages in the expansion of the steam, which results in a closer approach to the ideal reversible expansion process.

History

The first device that may be classified as a reaction steam turbine was little more than a toy, the classic Aeolipile, described in the 1st century by Greek mathematician Hero of Alexandria in Roman Egypt. In 1551, Taqi al-Din in Ottoman Egypt described a steam turbine with the practical application of rotating a spit. Steam turbines were also described by the Italian Giovanni Branca (1629) and John Wilkins in England (1648). The devices described by Taqi al-Din and Wilkins are today known as steam jacks. In 1672 an impulse steam turbine driven car was designed by Ferdinand Verbiest. A more modern version of this car was produced some time in the late 18th century by an unknown German mechanic.

A 250 kW industrial steam turbine from 1910 (right) directly linked to a generator (left).

The modern steam turbine was invented in 1884 by Sir Charles Parsons, whose first model was connected to a dynamo that generated 7.5 kW (10 hp) of electricity. The invention of Parsons' steam turbine made cheap and plentiful electricity possible and revolutionized marine transport and naval warfare. Parsons' design was a reaction type. His patent was licensed and the turbine scaled-up shortly after by an American, George Westinghouse. The Parsons turbine also turned out to be easy to scale up. Parsons had the satisfaction of seeing his invention adopted for all major world power stations, and the size of generators had increased from his first 7.5 kW set up to units of 50,000 kW capacity. Within Parson's lifetime, the generating capacity of a unit was scaled up by about 10,000 times, and the total output from turbo-generators constructed by his firm C. A. Parsons and Company and by their licensees, for land purposes alone, had exceeded thirty million horse-power.

A number of other variations of turbines have been developed that work effectively with steam. The *de Laval turbine* (invented by Gustaf de Laval) accelerated the steam to full speed before running it against a turbine blade. De Laval's impulse turbine is simpler, less expensive and does not need to be pressure-proof. It can operate with any pressure of steam, but is considerably less efficient. fr:Auguste Rateau developed a pressure compounded impulse turbine using the de Laval principle as early as 1896, obtained a US patent in 1903, and applied the turbine to a French torpedo boat in 1904. He taught at the École des mines de Saint-Étienne for a decade until 1897, and later founded a successful company that was incorporated into the Alstom firm after his death. One of the founders of the modern theory of steam and gas turbines was Aurel Stodola, a Slovak physicist and engineer and professor at the Swiss Polytechnical Institute (now ETH) in Zurich. His work *Die Dampfturbinen und ihre Aussichten als Wärmekraftmaschinen* (English: The Steam Turbine and its prospective use as a Heat Engine) was published in Berlin in 1903. A further book *Dampf und Gas-Turbinen* (English: Steam and Gas Turbines) was published in 1922.

The *Brown-Curtis turbine*, an impulse type, which had been originally developed and patented by the U.S. company International Curtis Marine Turbine Company, was developed in the 1900s in conjunction with John Brown & Company. It was used in John Brown-engined merchant ships and warships, including liners and Royal Navy warships.

Manufacturing

The present-day manufacturing industry for steam turbines is dominated by Chinese power equipment makers. Harbin Electric, Shanghai Electric, and Dongfang Electric, the top three power equipment makers in China, collectively hold a majority stake in the worldwide market share for steam turbines in 2009-10 according to Platts. Other manufacturers with minor market share include Bhel, Siemens, Alstom, GE, Doosan Škoda Power, Mitsubishi Heavy Industries, and Toshiba. The consulting firm Frost & Sullivan projects that manufacturing of steam turbines will become more consolidated by 2020 as Chinese power manufacturers win increasing business outside of China.

Types

Steam turbines are made in a variety of sizes ranging from small <0.75 kW (<1 hp) units (rare) used as mechanical drives for pumps, compressors and other shaft driven equipment, to 1 500 000 kW (1.5 GW; 2 000 000 hp) turbines used to generate electricity. There are several classifications for modern steam turbines.

Blade and Stage Design

Turbine blades are of two basic types, blades and nozzles. Blades move entirely due to the impact of steam on them and their profiles do not converge. This results in a steam velocity drop and essentially no pressure drop as steam moves through the blades. A turbine composed of blades alternating with fixed nozzles is called an impulse turbine, Curtis turbine, Rateau turbine, or Brown-Curtis turbine. Nozzles appear similar to blades, but their profiles converge near the exit. This results in a steam pressure drop and velocity increase as steam moves through the nozzles. Nozzles move due to both the impact of steam on them and the reaction due to the high-velocity steam at the exit. A turbine composed of moving nozzles alternating with fixed nozzles is called a reaction turbine or Parsons turbine.

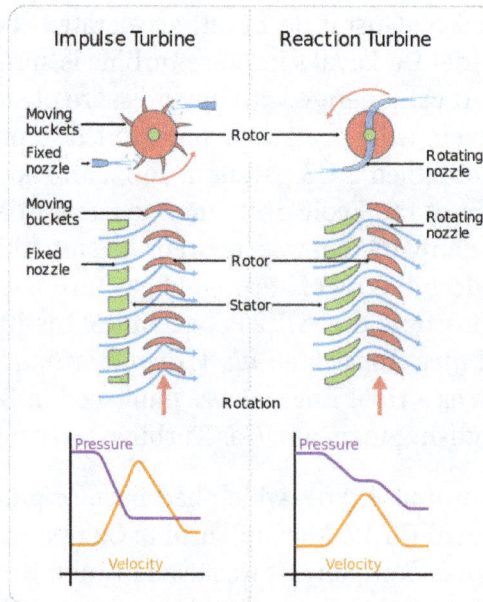

Schematic diagram outlining the difference between an impulse and a 50% reaction turbine

Except for low-power applications, turbine blades are arranged in multiple stages in series, called compounding, which greatly improves efficiency at low speeds. A reaction stage is a row of fixed nozzles followed by a row of moving nozzles. Multiple reaction stages divide the pressure drop between the steam inlet and exhaust into numerous small drops, resulting in a pressure-compounded turbine. Impulse stages may be either pressure-compounded, velocity-compounded, or pressure-velocity compounded. A pressure-compounded impulse stage is a row of fixed nozzles followed by a row of moving blades, with multiple stages for compounding. This is also known as a Rateau turbine, after its inventor. A velocity-compounded impulse stage (invented by Curtis and also called a "Curtis wheel") is a row of fixed nozzles followed by two or more rows of moving blades alternating with rows of fixed blades. This divides the velocity drop across the stage into several smaller drops. A series of velocity-compounded impulse stages is called a pressure-velocity compounded turbine.

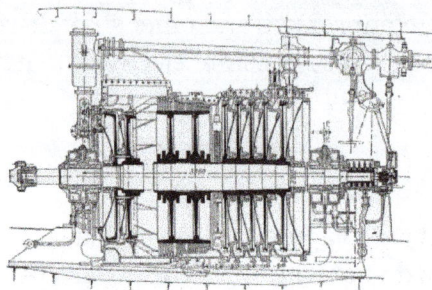

Diagram of an AEG marine steam turbine circa 1905

By 1905, when steam turbines were coming into use on fast ships (such as HMS *Dreadnought*) and in land-based power applications, it had been determined that it was desirable to use one or more Curtis wheels at the beginning of a multi-stage turbine (where the steam pressure is highest), followed by reaction stages. This was more efficient with high-pressure steam due to reduced leakage between the turbine rotor and the casing. This is illustrated in the drawing of the German 1905 AEG marine steam turbine. The steam from the boilers enters from the right at high pressure through a throttle, controlled manually by an operator (in this case a sailor known as the throttleman). It passes through five Curtis wheels and numerous reaction stages (the small blades at the edges of the two large rotors in the middle) before exiting at low pressure, almost certainly to a condenser. The condenser provides a vacuum that maximizes the energy extracted from the steam, and condenses the steam into feedwater to be returned to the boilers. On the left are several additional reaction stages (on two large rotors) that rotate the turbine in reverse for astern operation, with steam admitted by a separate throttle. Since ships are rarely operated in reverse, efficiency is not a priority in astern turbines, so only a few stages are used to save cost.

Blade Design Challenges

A major challenge facing turbine design is reducing the creep experienced by the blades. Because of the high temperatures and high stresses of operation, steam turbine materials become damaged through these mechanisms. As temperatures are increased in an effort to improve turbine efficiency, creep becomes more significant. To limit creep, thermal coatings and superalloys with solid-solution strengthening and grain boundary strengthening are used in blade designs.

Protective coatings are used to reduce the thermal damage and to limit oxidation. These coatings are often stabilized zirconium dioxide-based ceramics. Using a thermal protective coating limits the temperature exposure of the nickel superalloy. This reduces the creep mechanisms experienced in the blade. Oxidation coatings limit efficiency losses caused by a buildup on the outside of the blades, which is especially important in the high-temperature environment.

The nickel-based blades are alloyed with aluminum and titanium to improve strength and creep resistance. The microstructure of these alloys is composed of different regions of composition. A uniform dispersion of the gamma-prime phase – a combination of nickel, aluminum, and titanium – promotes the strength and creep resistance of the blade due to the microstructure.

Refractory elements such as rhenium and ruthenium can be added to the alloy to improve creep strength. The addition of these elements reduces the diffusion of the gamma prime phase, thus preserving the fatigue resistance, strength, and creep resistance.

Steam Supply and Exhaust Conditions

These types include condensing, non-condensing, reheat, extraction and induction.

Condensing turbines are most commonly found in electrical power plants. These turbines receive steam from a boiler and exhaust it to a condenser. The exhausted steam is at a pressure well below atmospheric, and is in a partially condensed state, typically of a quality near 90%.

Non-condensing or back pressure turbines are most widely used for process steam applications. The exhaust pressure is controlled by a regulating valve to suit the needs of the process steam pressure. These are commonly found at refineries, district heating units, pulp and paper plants, and desalination facilities where large amounts of low pressure process steam are needed.

A low-pressure steam turbine in a nuclear power plant. These turbines exhaust steam at a pressure below atmospheric.

Reheat turbines are also used almost exclusively in electrical power plants. In a reheat turbine, steam flow exits from a high pressure section of the turbine and is returned to the boiler where additional superheat is added. The steam then goes back into an intermediate pressure section of the turbine and continues its expansion. Using reheat in a cycle increases the work output from the turbine and also the expansion reaches conclusion before the steam condenses, thereby minimizing the erosion of the blades in last rows. In most of the cases, maximum number of reheats employed in a cycle is 2 as the cost of super-heating the steam negates the increase in the work output from turbine.

Extracting type turbines are common in all applications. In an extracting type turbine, steam is released from various stages of the turbine, and used for industrial process needs or sent to boiler feedwater heaters to improve overall cycle efficiency. Extraction flows may be controlled with a valve, or left uncontrolled.

Induction turbines introduce low pressure steam at an intermediate stage to produce additional power.

Casing or Shaft Arrangements

These arrangements include single casing, tandem compound and cross compound turbines. Single casing units are the most basic style where a single casing and shaft are coupled to a generator. Tandem compound are used where two or more casings are directly coupled together to drive a single generator. A cross compound turbine arrangement features two or more shafts not in line driving two or more generators that often operate at different speeds. A cross compound turbine is typically used for many large applications. A typical 1930s-1960s naval installation is illustrated below; this shows high- and low-pressure turbines driving a common reduction gear, with a geared cruising turbine on one high-pressure turbine.

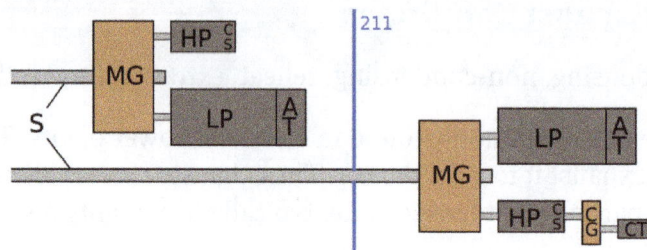

Starboard steam turbine machinery arrangement of Japanese *Furutaka*- and *Aoba*-class cruisers.

Two-flow Rotors

The moving steam imparts both a tangential and axial thrust on the turbine shaft, but the axial thrust in a simple turbine is unopposed. To maintain the correct rotor position and balancing, this force must be counteracted by an opposing force. Thrust bearings can be used for the shaft bearings, the rotor can use dummy pistons, it can be double flow- the steam enters in the middle of the shaft and exits at both ends, or a combination of any of these. In a double flow rotor, the blades in each half face opposite ways, so that the axial forces negate each other but the tangential forces act together. This design of rotor is also called two-flow, double-axial-flow, or double-exhaust. This arrangement is common in low-pressure casings of a compound turbine.

A two-flow turbine rotor. The steam enters in the middle of the shaft, and exits at each end, balancing the axial force.

Principle of Operation and Design

An ideal steam turbine is considered to be an isentropic process, or constant entropy process, in which the entropy of the steam entering the turbine is equal to the entropy of the steam leaving the turbine. No steam turbine is truly isentropic, however, with typical isentropic efficiencies ranging from 20–90% based on the application of the turbine. The interior of a turbine comprises several sets of blades or *buckets*. One set of stationary blades is connected to the casing and one set of rotating blades is connected to the shaft. The sets intermesh with certain minimum clearances, with the size and configuration of sets varying to efficiently exploit the expansion of steam at each stage.

Theoretical Turbine Efficiency

To maximize turbine efficiency the steam is expanded, doing work, in a number of stages. These stages are characterized by how the energy is extracted from them and are known as either impulse or reaction turbines. Most steam turbines use a mixture of the reaction and impulse designs: each stage behaves as either one or the other, but the overall turbine uses both. Typically, lower pressure sections are reaction type and higher pressure stages are impulse type.

Impulse Turbines

An impulse turbine has fixed nozzles that orient the steam flow into high speed jets. These jets contain significant kinetic energy, which is converted into shaft rotation by the bucket-like shaped rotor blades, as the steam jet changes direction. A pressure drop occurs across only the stationary blades, with a net increase in steam velocity across the stage. As the steam flows through the nozzle its pressure falls from inlet pressure to the exit pressure (atmospheric pressure, or more usually, the condenser vacuum). Due to this high ratio of expansion of steam, the steam leaves the nozzle with a very high velocity. The steam leaving the moving blades has a large portion of the maximum velocity of the steam when leaving the nozzle. The loss of energy due to this higher exit velocity is commonly called the carry over velocity or leaving loss.

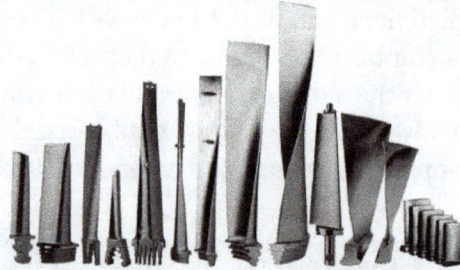

A selection of impulse turbine blades

The law of moment of momentum states that the sum of the moments of external forces acting on a fluid which is temporarily occupying the control volume is equal to the net time change of angular momentum flux through the control volume.

The swirling fluid enters the control volume at radius r_1 with tangential velocity V_{w1} and leaves at radius r_2 with tangential velocity V_{w2}.

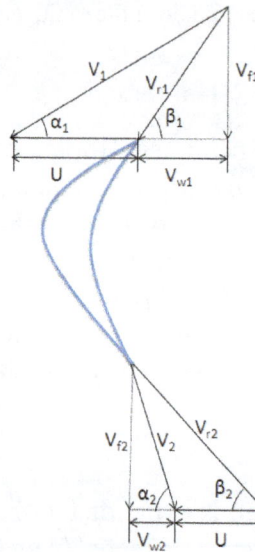

Velocity triangle

A velocity triangle paves the way for a better understanding of the relationship between the various velocities. In the adjacent figure we have:

V_1 and V_2 are the absolute velocities at the inlet and outlet respectively.

V_{f1} and V_{f2} are the flow velocities at the inlet and outlet respectively.

$V_{w1} + U$ and V_{w2} are the swirl velocities at the inlet and outlet respectively.

V_{r1} and V_{r2} are the relative velocities at the inlet and outlet respectively.

U_1 and U_2 are the velocities of the blade at the inlet and outlet respectively.

α is the guide vane angle and β is the blade angle.

Then by the law of moment of momentum, the torque on the fluid is given by:

$$T = \dot{m}(r_2 V_{w2} - r_1 V_{w1})$$

For an impulse steam turbine: $r_2 = r_1 = r.$. Therefore, the tangential force on the blades is $F_u = \dot{m}(V_{w1} - V_{w2})$. The work done per unit time or power developed: $W = T * \omega.$.

When ω is the angular velocity of the turbine, then the blade speed is $U = \omega * r.$. The power developed is then $W = \dot{m}U(\Delta V_w).$.

Blade Efficiency

Blade efficiency (η_b) can be defined as the ratio of the work done on the blades to kinetic energy supplied to the fluid, and is given by

$$\eta_b = \frac{Work\ Done}{Kinetic\ Energy\ Supplied} = \frac{2UV_w}{V_1^2}$$

Stage efficiency

A stage of an impulse turbine consists of a nozzle set and a moving wheel. The stage efficiency defines a relationship between enthalpy drop in the nozzle and work done in the stage.

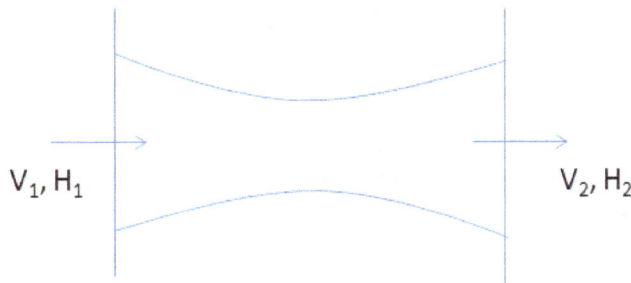

V_1, H_1 V_2, H_2

Convergent Divergent Nozzle

Convergent-divergent nozzle

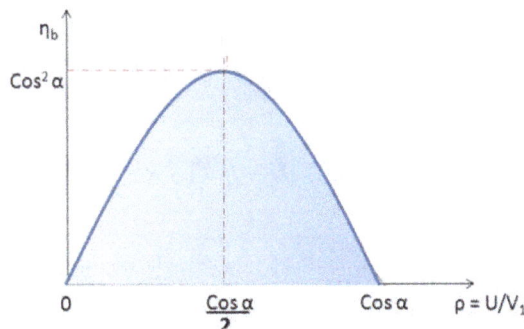

Graph depicting efficiency of Impulse turbine

$$\eta_{stage} = \frac{Work\ done\ on\ blade}{Energy\ supplied\ per\ stage} = \frac{U\Delta V_w}{\Delta h}$$

Where $\ddot{A}h = h_2 - h_1$ is the specific enthalpy drop of steam in the nozzle.

By the first law of thermodynamics: $h_1 + \frac{V_1^2}{2} = h_2 + \frac{V_2^2}{2}$

Assuming that V_1 is appreciably less than V_2, we get $\ddot{A}h \approx \frac{V_2^2}{2}$ Furthermore, stage efficiency is the product of blade efficiency and nozzle efficiency, or $\eta_{stage} = \eta_b * \eta_N$

Nozzle efficiency is given by $\eta_N = \frac{V_2^2}{2(h_1 - h_2)}$,, where the enthalpy (in J/Kg) of steam at the entrance of the nozzle is h_1 and the enthalpy of steam at the exit of the nozzle is h_2. $\Delta V_w = V_{w1} - (-V_{w2})$

$$\Delta V_w = V_{w1} + V_{w2}$$

$$\Delta V_w = V_{r1} \cos\beta_1 + V_{r2} \cos\beta_2$$

$$\Delta V_w = V_{r1} \cos\beta_1 (1 + \frac{V_{r2} \cos\beta_2}{V_{r1} \cos\beta_1})$$

The ratio of the cosines of the blade angles at the outlet and inlet can be taken and denoted

$c = \frac{\cos\beta_2}{\cos\beta_1}$. The ratio of steam velocities relative to the rotor speed at the outlet to the inlet of the

blade is defined by the friction coefficient $k = \frac{V_{r2}}{V_{r1}}$..

$k < 1$ and depicts the loss in the relative velocity due to friction as the steam flows around the blades ($k = 1$ for smooth blades).

$$\eta_b = \frac{2U\Delta V_w}{V_1^2} = \frac{2U(\cos\alpha_1 - U/V_1)(1 + kc)}{V_1}$$

The ratio of the blade speed to the absolute steam velocity at the inlet is termed the blade speed

ratio $\rho = \frac{U}{V_1}$

η_b is maximum when $\frac{d\eta_b}{d\rho} = 0$ or, $\frac{d}{d\rho}(2\cos\alpha_1 - \rho^2(1 + kc)) = 0$.. That implies $\rho = \frac{\cos\alpha_1}{2}$ and

therefore $\frac{U}{V_1} = \frac{\cos\alpha_1}{2}$.. Now $\rho_{opt} = \frac{U}{V_1} = \frac{\cos\alpha_1}{2}$ (for a single stage impulse turbine)

Therefore, the maximum value of stage efficiency is obtained by putting the value of $\dfrac{U}{V_1} = \dfrac{\cos\alpha_1}{2}$ in the expression of η_b /

We get: $(\eta_b)_{max} = 2(\rho\cos\alpha_1 - \rho^2)(1+kc) = \dfrac{\cos^2\alpha_1(1+kc)}{2}$.

For equiangular blades, $\beta_1 \quad \beta_2$, therefore $c = 1$, and we get $(\eta_b)_{max} = \dfrac{\cos^2\alpha_1(1+k)}{2}$. If the friction due to the blade surface is neglected then $(\eta_b)_{max} = \cos^2\alpha_1$.

Conclusions on Maximum Efficiency

$$(\eta_b)_{max} = \cos^2\alpha_1$$

1. For a given steam velocity work done per kg of steam would be maximum when $\cos^2\alpha_1 = 1$ or $\alpha_1 = 0$.

2. As α_1 increases, the work done on the blades reduces, but at the same time surface area of the blade reduces, therefore there are less frictional losses.

Reaction Turbines

In the *reaction turbine*, the rotor blades themselves are arranged to form convergent nozzles. This type of turbine makes use of the reaction force produced as the steam accelerates through the nozzles formed by the rotor. Steam is directed onto the rotor by the fixed vanes of the stator. It leaves the stator as a jet that fills the entire circumference of the rotor. The steam then changes direction and increases its speed relative to the speed of the blades. A pressure drop occurs across both the stator and the rotor, with steam accelerating through the stator and decelerating through the rotor, with no net change in steam velocity across the stage but with a decrease in both pressure and temperature, reflecting the work performed in the driving of the rotor.

Blade Efficiency

Energy input to the blades in a stage:

$E = \Delta h$ is equal to the kinetic energy supplied to the fixed blades (f) + the kinetic energy supplied to the moving blades (m).

Or, E = enthalpy drop over the fixed blades, $\ddot{A}h_f$ + enthalpy drop over the moving blades, $\ddot{A}h_m$.

The effect of expansion of steam over the moving blades is to increase the relative velocity at the exit. Therefore, the relative velocity at the exit V_{r2} is always greater than the relative velocity at the inlet V_{r2}.

In terms of velocities, the enthalpy drop over the moving blades is given by:

$\Delta h_m = \dfrac{V_{r2}^2 - V_{r1}^2}{2}$ (it contributes to a change in static pressure)

The enthalpy drop in the fixed blades, with the assumption that the velocity of steam entering the fixed blades is equal to the velocity of steam leaving the previously moving blades is given by:

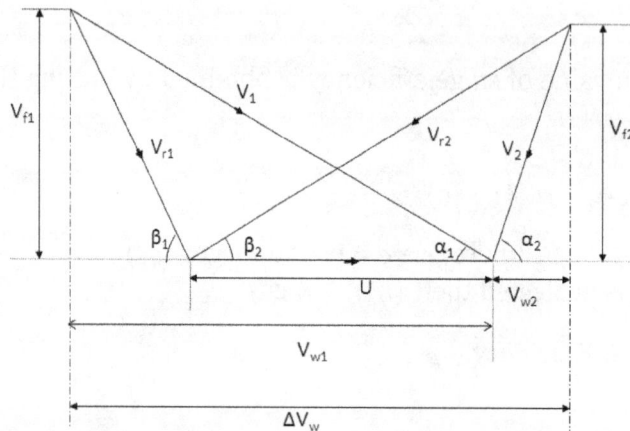

Velocity Diagram

$\Delta h_f = \dfrac{V_1^2 - V_0^2}{2}$ where V_o is the inlet velocity of steam in the nozzle

V_0 is very small and hence can be neglected

Therefore, $\Delta h_f = \dfrac{V_1^2}{2}$

$$E = \Delta h_f + \Delta h_m$$

$$E = \frac{V_1^2}{2} + \frac{V_{r2}^2 - V_{r1}^2}{2}$$

A very widely used design has half degree of reaction or 50% reaction and this is known as Parson's turbine. This consists of symmetrical rotor and stator blades. For this turbine the velocity triangle is similar and we have:

$$\alpha_1 = \beta_2, \quad \beta_1 = \alpha_2$$

$$V_1 = V_{r2}, \quad V_{r1} = V_2$$

Assuming *Parson's turbine* and obtaining all the expressions we get

$$E = V_1^2 - \frac{V_{r1}^2}{2}$$

From the inlet velocity triangle we have

$$V_{r1}^2 = V_1^2 + U^2 - 2UV_1 \cos \alpha_1$$

$$E = V_1^2 - \frac{V_1^2}{2} - \frac{U^2}{2} + \frac{2UV_1 \cos \alpha_1}{2}$$

$$E = \frac{V_1^2 - U^2 + 2UV_1 \cos \alpha_1}{2}$$

Work done (for unit mass flow per second): $W = U * \Delta V_w = U * (2 * V_1 \cos \alpha_1 - U)$

Therefore, the blade efficiency is given by

$$\eta_b = \frac{2U(2V_1 \cos\alpha_1 - U)}{V_1^2 - U^2 + 2V_1 U \cos\alpha_1}$$

Condition of Maximum Blade Efficiency

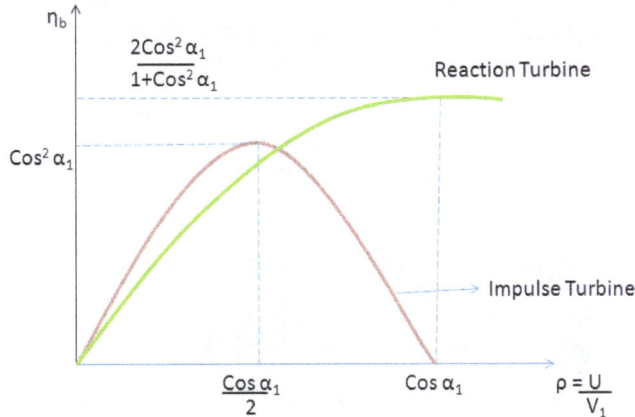

Comparing Efficiencies of Impulse and Reaction turbines

If $\rho = \dfrac{U}{V_1}$,, then

$$(\eta_b)_{max} = \frac{2\rho(\cos\alpha_1 - \rho)}{V_1^2 - U^2 + 2UV_1 \cos\alpha_1}$$

For maximum efficiency $\dfrac{d\eta_b}{d\rho} = 0$, we get

$$(1 - \rho^2 + 2\rho\cos\alpha_1)(4\cos\alpha_1 - 4\rho) - 2\rho(2\cos\alpha_1 - \rho)(-2\rho + 2\cos\alpha_1) = 0$$

and this finally gives $\rho_{opt} = \dfrac{U}{V_1} = \cos\alpha_1$

Therefore, $(\eta_b)_{max}$ is found by putting the value of $(\eta_b)_{max}$ in the expression of blade efficiency

$$(\eta_b)_{reaction} = \frac{2\cos^2\alpha_1}{1 + \cos^2\alpha_1}$$

$$(\eta_b)_{impulse} = \cos^2\alpha_1$$

Practical Turbine Efficiency

Practical thermal efficiency of a steam turbine varies with turbine size, load condition, gap losses and friction losses. They reach top values up to about 50% in a 1200MW turbine smaller ones have a lower efficiency.

Operation and Maintenance

Because of the high pressures used in the steam circuits and the materials used, steam turbines and their casings have high thermal inertia. When warming up a steam turbine for use, the main steam stop valves (after the boiler) have a bypass line to allow superheated steam to slowly bypass the valve and proceed to heat up the lines in the system along with the steam turbine. Also, a turning gear is engaged when there is no steam to slowly rotate the turbine to ensure even heating to prevent uneven expansion. After first rotating the turbine by the turning gear, allowing time for the rotor to assume a straight plane (no bowing), then the turning gear is disengaged and steam is admitted to the turbine, first to the astern blades then to the ahead blades slowly rotating the turbine at 10–15 RPM (0.17–0.25 Hz) to slowly warm the turbine. The warm up procedure for large steam turbines may exceed ten hours.

A modern steam turbine generator installation

During normal operation, rotor imbalance can lead to vibration, which, because of the high rotation velocities, could lead to a blade breaking away from the rotor and through the casing. To reduce this risk, considerable efforts are spent to balance the turbine. Also, turbines are run with high quality steam: either superheated (dry) steam, or saturated steam with a high dryness fraction. This prevents the rapid impingement and erosion of the blades which occurs when condensed water is blasted onto the blades (moisture carry over). Also, liquid water entering the blades may damage the thrust bearings for the turbine shaft. To prevent this, along with controls and baffles in the boilers to ensure high quality steam, condensate drains are installed in the steam piping leading to the turbine.

Maintenance requirements of modern steam turbines are simple and incur low costs (typically around $0.005 per kWh); their operational life often exceeds 50 years.

Speed Regulation

The control of a turbine with a governor is essential, as turbines need to be run up slowly to prevent damage and some applications (such as the generation of alternating current electricity) require precise speed control. Uncontrolled acceleration of the turbine rotor can lead to an overspeed trip, which causes the governor and throttle valves that control the flow of steam to the turbine to close. If these valves fail then the turbine may continue accelerating until it breaks apart, often catastrophically. Turbines are expensive to make, requiring precision manufacture and special quality materials.

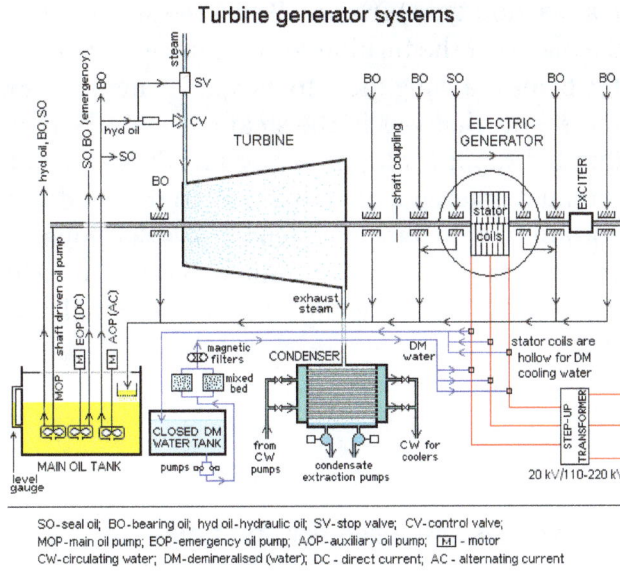

Diagram of a steam turbine generator system

During normal operation in synchronization with the electricity network, power plants are governed with a five percent droop speed control. This means the full load speed is 100% and the no-load speed is 105%. This is required for the stable operation of the network without hunting and drop-outs of power plants. Normally the changes in speed are minor. Adjustments in power output are made by slowly raising the droop curve by increasing the spring pressure on a centrifugal governor. Generally this is a basic system requirement for all power plants because the older and newer plants have to be compatible in response to the instantaneous changes in frequency without depending on outside communication.

Thermodynamics of Steam Turbines

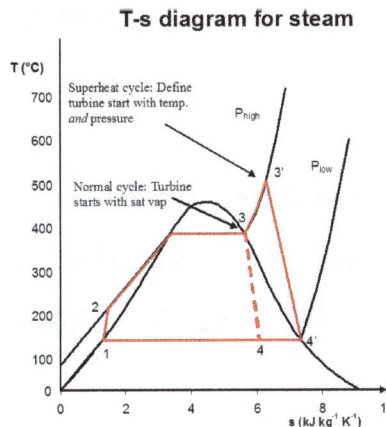

T-s diagram of a superheated Rankine cycle

The steam turbine operates on basic principles of thermodynamics using the part 3-4 of the Rankine cycle shown in the adjoining diagram. Superheated steam (or dry saturated steam, depending on application) leaves the boiler at high temperature and high pressure. At entry to the turbine, the steam gains kinetic energy by passing through a nozzle (a fixed nozzle in an impulse type turbine or the fixed blades in a reaction type turbine). When the steam leaves the nozzle it is moving at high velocity towards the blades of the turbine rotor. A force is created on the blades due to the pressure of the vapor on the blades causing them to move. A generator or other such device can be placed on the shaft, and the energy that was in the steam can now be stored and used. The steam leaves the turbine as a saturated vapor (or liquid-vapor mix depending on application) at a lower temperature and pressure than it entered with and is sent to the condenser to be cooled. The first law enables us to find an formula for the rate at which work is developed per unit mass. Assuming there is no heat transfer to the surrounding environment and that the changes in kinetic and potential energy are negligible compared to the change in specific enthalpy we arrive at the following equation

$$\frac{\dot{W}}{\dot{m}} = h_3 - h_4$$

where

- $\dot{W}$ is the rate at which work is developed per unit time

- $\dot{m}$ is the rate of mass flow through the turbine

Isentropic Efficiency

To measure how well a turbine is performing we can look at its isentropic efficiency. This compares the actual performance of the turbine with the performance that would be achieved by an ideal, isentropic, turbine. When calculating this efficiency, heat lost to the surroundings is assumed to be zero. The starting pressure and temperature is the same for both the actual and the ideal turbines, but at turbine exit the energy content ('specific enthalpy') for the actual turbine is greater than that for the ideal turbine because of irreversibility in the actual turbine. The specific enthalpy is evaluated at the same pressure for the actual and ideal turbines in order to give a good comparison between the two.

The isentropic efficiency is found by dividing the actual work by the ideal work.

$$\eta_t = \frac{h_3 - h_4}{h_3 - h_{4s}}$$

where

- h_3 is the specific enthalpy at state three

- h_4 is the specific enthalpy at state 4 for the actual turbine

- h_{4s} is the specific enthalpy at state 4s for the isentropic turbine

Direct Drive

Electrical power stations use large steam turbines driving electric generators to produce most (about 80%) of the world's electricity. The advent of large steam turbines made central-station electricity generation practical, since reciprocating steam engines of large rating became very bulky, and operated at slow speeds. Most central stations are fossil fuel power plants and nuclear power plants; some installations use geothermal steam, or use concentrated solar power (CSP) to create the steam. Steam turbines can also be used directly to drive large centrifugal pumps, such as feedwater pumps at a thermal power plant.

A direct-drive 5 MW steam turbine fuelled with biomass

The turbines used for electric power generation are most often directly coupled to their generators. As the generators must rotate at constant synchronous speeds according to the frequency of the electric power system, the most common speeds are 3,000 RPM for 50 Hz systems, and 3,600 RPM for 60 Hz systems. Since nuclear reactors have lower temperature limits than fossil-fired plants, with lower steam quality, the turbine generator sets may be arranged to operate at half these speeds, but with four-pole generators, to reduce erosion of turbine blades.

Marine Propulsion

Turbinia, 1894, the first steam turbine-powered ship

Fig. 5. Views of high and low pressure turbines completed and ready for installation

High and low pressure turbines for SS *Maui*.

The higher cost of turbines and the associated gears or generator/motor sets is offset by lower maintenance requirements and the smaller size of a turbine when compared to a reciprocating engine having an equivalent power, although the fuel costs are higher than a diesel engine because steam turbines have lower thermal efficiency. To reduce fuel costs the thermal efficiency of both types of engine have been improved over the years. Today, propulsion steam turbine cycle efficiencies have yet to break 50%, yet diesel engines routinely exceed 50%, especially in marine applications. Diesel power plants also have lower operating costs since fewer operators are required. Thus, conventional steam power is used in very few new ships. An exception is LNG carriers which often find it more economical to use boil-off gas with a steam turbine than to re-liquify it.

Parsons turbine from the 1928 Polish destroyer *Wicher*.

Nuclear-powered ships and submarines use a nuclear reactor to create steam for turbines. Nuclear power is often chosen where diesel power would be impractical (as in submarine applications) or the logistics of refuelling pose significant problems (for example, icebreakers). It has been estimated that the reactor fuel for the Royal Navy's *Vanguard*-class submarines is sufficient to last 40 circumnavigations of the globe – potentially sufficient for the vessel's entire service life. Nuclear propulsion has only been applied to a very few commercial vessels due to the expense of maintenance and the regulatory controls required on nuclear systems and fuel cycles.

Early Development

The development of steam turbine marine propulsion from 1894-1935 was dominated by the need to reconcile the high efficient speed of the turbine with the low efficient speed (less than 300 rpm) of the ship's propeller at an overall cost competitive with reciprocating engines. In 1894, efficient reduction gears were not available for the high powers required by ships, so direct drive was necessary. In *Turbinia*, which has direct drive to each propeller shaft, the efficient speed of the turbine was reduced after initial trials by directing the steam flow through all three direct drive turbines (one on each shaft) in series, probably totaling around 200 turbine stages operating in series. Also, there were three propellers on each shaft for operation at high speeds. The high shaft speeds of the era are represented by one of the first US turbine-powered destroyers, USS *Smith*, launched in 1909, which had direct drive turbines and whose three shafts turned at 724 rpm at 28.35 knots. The use of turbines in several casings exhausting steam to each other in series became standard in most subsequent marine propulsion applications, and is a form of cross-compounding. The first turbine was called the high pressure (HP) turbine, the last turbine was the low pressure (LP) turbine, and any turbine in between was an intermediate pressure (IP) turbine. A much later arrangement than *Turbinia* can be seen on RMS *Queen Mary* in Long Beach, California, launched in 1934, in which each shaft is powered by four turbines in series connected to the ends of the two input shafts of a single-reduction gearbox. They are the HP, 1st IP, 2nd IP, and LP turbines.

Cruising Machinery and Gearing

The quest for economy was even more important when cruising speeds were considered. Cruising speed is roughly 50% of a warship's maximum speed and 20-25% of its maximum power level. This would be a speed used on long voyages when fuel economy is desired. Although this brought the propeller speeds down to an efficient range, turbine efficiency was greatly reduced, and early turbine ships had poor cruising ranges. A solution that proved useful through most of the steam turbine propulsion era was the cruising turbine. This was an extra turbine to add even more stages, at first attached directly to one or more shafts, exhausting to a stage partway along the HP turbine, and not used at high speeds. As reduction gears became available around 1911, some ships, notably the battleship USS *Nevada*, had them on cruising turbines while retaining direct drive main turbines. Reduction gears allowed turbines to operate in their efficient range at a much higher speed than the shaft, but were expensive to manufacture.

Cruising turbines competed at first with reciprocating engines for fuel economy. An example of the retention of reciprocating engines on fast ships was the famous RMS *Titanic* of 1911, which along with her sisters RMS *Olympic* and HMHS *Britannic* had triple-expansion engines on the two outboard shafts, both exhausting to an LP turbine on the center shaft. After adopting turbines with the *Delaware*-class battleships launched in 1909, the United States Navy reverted to reciprocating machinery on the *New York*-class battleships of 1912, then went back to turbines on *Nevada* in 1914. The lingering fondness for reciprocating machinery was because the US Navy had no plans for capital ships exceeding 21 knots until after World War I, so top speed was less important than economical cruising. The United States had acquired the Philippines and Hawaii as territories in 1898, and lacked the British Royal Navy's worldwide network of coaling stations. Thus, the US Navy in 1900-1940 had the greatest need of any nation for fuel economy, especially as the prospect of war with Japan arose following World War I. This need was compounded by the US not launching any cruisers 1908-1920, so destroyers were required to perform long-range missions usually assigned to cruisers. So, various cruising solutions were fitted on US destroyers launched 1908-1916. These included small reciprocating engines and geared or ungeared cruising turbines on one or two shafts. However, once fully geared turbines proved economical in initial cost and fuel they were rapidly adopted, with cruising turbines also included on most ships. Beginning in 1915 all new Royal Navy destroyers had fully geared turbines, and the United States followed in 1917.

In the Royal Navy, speed was a priority until the Battle of Jutland in mid-1916 showed that in the battlecruisers too much armour had been sacrificed in its pursuit. The British used exclusively turbine-powered warships from 1906. Because they recognized that a significant cruising range would be desirable given their worldwide empire, some warships, notably the *Queen Elizabeth*-class battleships, were fitted with cruising turbines from 1912 onwards following earlier experimental installations.

In the US Navy, the *Mahan*-class destroyers, launched 1935-36, introduced double-reduction gearing. This further increased the turbine speed above the shaft speed, allowing smaller turbines than single-reduction gearing. Steam pressures and temperatures were also increasing progressively, from 300 psi/425 F (2.07 MPa/218 C)(saturation temperature) on the World War I-era *Wickes* class to 615 psi/850 F (4.25 MPa/454 C) superheated steam on some World War II *Fletcher*-class destroyers and later ships. A standard configuration emerged of an axial-flow high pressure turbine (sometimes with a cruising turbine attached) and a double-axial-flow low pressure turbine connected to a double-reduction gearbox. This arrangement continued throughout the steam era in the US Navy and was also used in some Royal Navy designs. Machinery of this configuration can be seen on many preserved World War II-era warships in several countries. When US Navy warship construction resumed in the early 1950s, most surface combatants and aircraft carriers used 1,200 psi/950 F (8.28 MPa/510 C) steam. This continued until the end of the US Navy steam-powered warship era with the *Knox*-class frigates of the early 1970s. Amphibious and auxiliary ships continued to use 600 psi (4.14 MPa) steam post-World War II, with USS *Iwo Jima*, launched in 2001, possibly the last non-nuclear steam-powered ship built for the US Navy.

Turbo-electric Drive

NS *50 Let Pobedy*, a nuclear icebreaker with nuclear-turbo-electric propulsion

Turbo-electric drive was introduced on the battleship USS *New Mexico*, launched in 1917. Over the next eight years the US Navy launched five additional turbo-electric-powered battleships and two aircraft carriers (initially ordered as *Lexington*-class battlecruisers). Ten more turbo-electric capital ships were planned, but cancelled due to the limits imposed by the Washington Naval Treaty. Although *New Mexico* was refitted with geared turbines in a 1931-33 refit, the remaining turbo-electric ships retained the system throughout their careers. This system used two large steam turbine generators to drive an electric motor on each of four shafts. The system was less costly initially than reduction gears and made the ships more maneuverable in port, with the shafts able to reverse rapidly and deliver more reverse power than with most geared systems. Some ocean liners were also built with turbo-electric drive, as were some troop transports and mass-production destroyer escorts in World War II. However, when the US designed the "treaty cruisers", beginning with USS *Pensacola* launched in 1927, geared turbines were used to conserve weight, and remained in use for all fast steam-powered ships thereafter.

Current Usage

Since the 1980s, steam turbines have been replaced by gas turbines on fast ships and by diesel engines on other ships; exceptions are nuclear-powered ships and submarines and LNG carriers. Some auxiliary ships continue to use steam propulsion. In the U.S. Navy, the conventionally powered steam turbine is still in use on all but one of the Wasp-class amphibious assault ships. The U.S. Navy also operates steam turbines on their nuclear powered Nimitz-class and Ford-class aircraft carriers along with all of their nuclear submarines (Ohio-, Los Angeles-, Seawolf-, and Virginia-classes). The Royal Navy decommissioned its last conventional steam-powered surface warship class, the *Fearless*-class landing platform dock, in 2002. In 2013, the French Navy ended its steam era with the decommissioning of its last *Tourville*-class frigate. Amongst the other blue-water navies, the Russian Navy currently operates steam-powered *Kuznetsov*-class aircraft carriers and *Sovremenny*-class destroyers. The Indian Navy currently operates two conventional steam-powered carriers, INS *Viraat*, a former British *Centaur*-class aircraft carrier (to be decommissioned in 2016), and INS *Vikramaditya*, a modified *Kiev*-class aircraft carrier; it also operates three *Brahmaputra*-class frigates commissioned in the early 2000s and two *Godavari*-class frigates currently scheduled for decommissioning.

Most other naval forces either retired or re-engined their steam-powered warships by 2010. The Chinese Navy currently operates steam-powered Russian *Kuznetsov*-class aircraft carriers and *Sovremenny*-class destroyers; it also operates steam-powered *Luda*-class destroyers. The JS *Kurama*, the last steam-powered JMSDF *Shirane*-class destroyer, will be decommissioned and replaced in 2017. As of 2016, the Brazilian Navy operates *São Paulo*, a former French *Clemenceau*-class aircraft carrier, while the Mexican Navy currently operates four former U.S. *Knox*-class frigates and two former U.S. *Bronstein*-class frigates. The Royal Thai Navy, Egyptian Navy and the Republic of China Navy respectively operate one, two and six former U.S. *Knox*-class frigates. The Peruvian Navy currently operates the former Dutch *De Zeven Provinciën*-class cruiser *BAP Almirante Grau*; the Ecuadorian Navy currently operates two *Condell*-class frigates (modified *Leander*-class frigates).

Locomotives

A steam turbine locomotive engine is a steam locomotive driven by a steam turbine.

The main advantages of a steam turbine locomotive are better rotational balance and reduced hammer blow on the track. However, a disadvantage is less flexible output power so that turbine locomotives were best suited for long-haul operations at a constant output power.

The first steam turbine rail locomotive was built in 1908 for the Officine Meccaniche Miani Silvestri Grodona Comi, Milan, Italy. In 1924 Krupp built the steam turbine locomotive T18 001, operational in 1929, for Deutsche Reichsbahn.

Testing

British, German, other national and international test codes are used to standardize the procedures and definitions used to test steam turbines. Selection of the test code to be used is an agreement between the purchaser and the manufacturer, and has some significance to the design of the turbine and associated systems. In the United States, ASME has produced several performance test codes on steam turbines. These include ASME PTC 6-2004, Steam Turbines, ASME PTC 6.2-2011, Steam Turbines in Combined Cycles, PTC 6S-1988, Procedures for Routine Performance Test of Steam Turbines. These ASME performance test codes have gained international recognition and acceptance for testing steam turbines. The single most important and differentiating characteristic of ASME performance test codes, including PTC 6, is that the test uncertainty of the measurement indicates the quality of the test and is not to be used as a commercial tolerance.

Nuclear Reactor Coolant

A nuclear reactor coolant is a coolant in a nuclear reactor used to remove heat from the nuclear reactor core and transfer it to electrical generators and the environment. Frequently, a chain of two coolant loops are used because the primary coolant loop takes on short-term radioactivity from the reactor.

Water

Almost all currently operating nuclear power plants are light water reactors using ordinary water under high pressure as coolant and neutron moderator. About 1/3 are boiling water reactors where the primary coolant undergoes phase transition to steam inside the reactor. About 2/3 are pressurized water reactors at even higher pressure. Current reactors stay under the critical point at around 374 °C and 218 bar where the distinction between liquid and gas disappears, which limits thermal efficiency, but the proposed supercritical water reactor would operate above this point.

Heavy water reactors use deuterium oxide which has similar properties to ordinary water but much lower neutron capture, allowing more thorough moderation.

Molten Metal

Fast reactors have a high power density and do not need neutron moderation. Most have been liquid metal cooled reactors using molten sodium. Lead, lead-bismuth eutectic, and other metals have also been proposed and occasionally used. Mercury was used in the first fast reactor.

Molten Salt

Molten salts share with metals the advantage of low vapor pressure even at high temperatures, and are less chemically reactive than sodium. Salts containing light elements like FLiBe can also provide moderation. In the Molten-Salt Reactor Experiment it even served as a solvent carrying the nuclear fuel.

Gas

Gases have also been used as coolant. Helium is extremely inert both chemically and with respect to nuclear reactions but has a low heat capacity, necessitating rapid circulation. Carbon dioxide has also been used in Magnox and AGR reactors. Gases, however, need to be under pressure for sufficient density at high temperatures.

Hydrocarbons

Organically moderated and cooled reactors was an early power reactor concept studied. They were not successful.

Nuclear Reprocessing

Nuclear reprocessing technology was developed to chemically separate and recover fissionable plutonium from irradiated nuclear fuel. Reprocessing serves multiple purposes, whose relative importance has changed over time. Originally, reprocessing was used solely to extract plutonium for producing nuclear weapons. With the commercialization of nuclear power, the reprocessed plutonium was recycled back into MOX nuclear fuel for thermal reactors. The reprocessed uranium, which constitutes the bulk of the spent fuel material, can in principle also be re-used as fuel, but that is only economical when uranium prices are high. Finally, a breeder reactor is not restricted to using recycled plutonium and uranium. It can employ all the actinides, closing the nuclear fuel cycle and potentially multiplying the energy extracted from natural uranium by about 60 times.

Nuclear reprocessing reduces the volume of high-level waste, but by itself does not reduce radioactivity or heat generation and therefore does not eliminate the need for a geological waste repository. Reprocessing has been politically controversial because of the potential to contribute to nuclear proliferation, the potential vulnerability to nuclear terrorism, the political challenges of repository siting (a problem that applies equally to direct disposal of spent fuel), and because of its high cost compared to the once-through fuel cycle. In the United States, the Obama administration stepped back from President Bush's plans for commercial-scale reprocessing and reverted to a program focused on reprocessing-related scientific research. Nuclear fuel reprocessing is performed routinely in Europe, Russia and Japan.

Separated Components and Disposition

The potentially useful components dealt with in nuclear reprocessing comprise specific actinides (plutonium, uranium, and some minor actinides). The lighter elements components include fission products, activation products, and cladding.

material	disposition
plutonium, minor actinides, reprocessed uranium	fission in fast, fusion, or subcritical reactor
reprocessed uranium, cladding, filters	less stringent storage as intermediate-level waste
long-lived fission and activation products	nuclear transmutation or geological repository
medium-lived fission products ^{137}Cs and ^{90}Sr	medium-term storage as high-level waste
useful radionuclides and noble metals	industrial and medical uses

History

The first large-scale nuclear reactors were built during World War II. These reactors were designed for the production of plutonium for use in nuclear weapons. The only reprocessing required, therefore, was the extraction of the plutonium (free of fission-product contamination) from the spent natural uranium fuel. In 1943, several methods were proposed for separating the relatively small quantity of plutonium from the uranium and fission products. The first method selected, a precipitation process called the bismuth phosphate process, was developed and tested at the Oak Ridge National Laboratory (ORNL) between 1943 and 1945 to produce quantities of plutonium for evaluation and use in the US weapons programs. ORNL produced the first macroscopic quantities (grams) of separated plutonium with these processes.

The bismuth phosphate process was first operated on a large scale at the Hanford Site, in the later part of 1944. It was successful for plutonium separation in the emergency situation existing then, but it had a significant weakness: the inability to recover uranium.

The first successful solvent extraction process for the recovery of pure uranium and plutonium was developed at ORNL in 1949. The PUREX process is the current method of extraction. Separation plants were also constructed at Savannah River Site and a smaller plant at West Valley Reprocessing Plant which closed by 1972 because of its inability to meet new regulatory requirements.

Reprocessing of civilian fuel has long been employed at the COGEMA La Hague site in France, the Sellafield site in the United Kingdom, the Mayak Chemical Combine in Russia, and at sites such as the Tokai plant in Japan, the Tarapur plant in India, and briefly at the West Valley Reprocessing Plant in the United States.

In October 1976, concern of nuclear weapons proliferation (especially after India demonstrated nuclear weapons capabilities using reprocessing technology) led President Gerald Ford to issue a Presidential directive to indefinitely suspend the commercial reprocessing and recycling of plutonium in the U.S. On 7 April 1977, President Jimmy Carter banned the reprocessing of commercial reactor spent nuclear fuel. The key issue driving this policy was the serious threat of nuclear weapons proliferation by diversion of plutonium from the civilian fuel cycle, and to encourage other nations to follow the USA lead. After that, only countries that already had large investments in reprocessing infrastructure continued to reprocess spent nuclear fuel. President Reagan lifted the ban in 1981, but did not provide the substantial subsidy that would have been necessary to start up commercial reprocessing.

In March 1999, the U.S. Department of Energy (DOE) reversed its policy and signed a contract with a consortium of Duke Energy, COGEMA, and Stone & Webster (DCS) to design and operate a mixed oxide (MOX) fuel fabrication facility. Site preparation at the Savannah River Site (South Carolina) began in October 2005. In 2011 the New York Times reported "...11 years after the government awarded a construction contract, the cost of the project has soared to nearly $5 billion. The vast concrete and steel structure is a half-finished hulk, and the government has yet to find a single customer, despite offers of lucrative subsidies." TVA (currently the most likely customer) said in April 2011 that it would delay a decision until it could see how MOX fuel performed in the nuclear accident at Fukushima Daiichi.

Separation Technologies

Water and Organic Solvents

PUREX

PUREX, the current standard method, is an acronym standing for *Plutonium and Uranium Recovery by EXtraction*. The PUREX process is a liquid-liquid extraction method used to reprocess spent nuclear fuel, to extract uranium and plutonium, independent of each other, from the fission products. This is the most developed and widely used process in the industry at present. When used on fuel from commercial power reactors the plutonium extracted typically contains too much Pu-240 to be considered "weapons-grade" plutonium, ideal for use in a nuclear weapon. Nevertheless, highly reliable nuclear weapons can be built at all levels of technical sophistication using reactor-grade plutonium. Moreover, reactors that are capable of refueling frequently can be used to produce weapon-grade plutonium, which can later be recovered using PUREX. Because of this, PUREX chemicals are monitored.

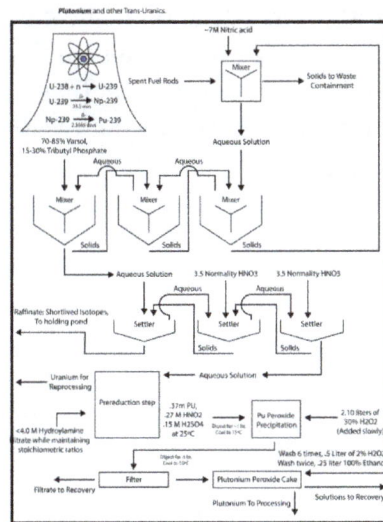

Plutonium Processing

Modifications of PUREX

UREX

The PUREX process can be modified to make a UREX (URanium EXtraction) process which could be used to save space inside high level nuclear waste disposal sites, such as the Yucca Mountain nuclear waste repository, by removing the uranium which makes up the vast majority of the mass and volume of used fuel and recycling it as reprocessed uranium.

The UREX process is a PUREX process which has been modified to prevent the plutonium from being extracted. This can be done by adding a plutonium reductant before the first metal extraction step. In the UREX process, ~99.9% of the uranium and >95% of technetium are separated from each other and the other fission products and actinides. The key is the addition of acetohydroxamic acid (AHA) to the extraction and scrub sections of the process. The addition of AHA greatly diminishes the extractability of plutonium and neptunium, providing somewhat greater proliferation resistance than with the plutonium extraction stage of the PUREX process.

TRUEX

Adding a second extraction agent, octyl(phenyl)-N, N-dibutyl carbamoylmethyl phosphine oxide(CMPO) in combination with tributylphosphate, (TBP), the PUREX process can be turned into the TRUEX (TRansUranic EXtraction) process. TRUEX was invented in the USA by Argonne National Laboratory and is designed to remove the transuranic metals (Am/Cm) from waste. The idea is that by lowering the alpha activity of the waste, the majority of the waste can then be disposed of with greater ease. In common with PUREX this process operates by a solvation mechanism.

DIAMEX

As an alternative to TRUEX, an extraction process using a malondiamide has been devised. The DIAMEX (DIAMideEXtraction) process has the advantage of avoiding the formation of organic waste which contains elements other than carbon, hydrogen, nitrogen, and oxygen. Such an organic waste can be burned without the formation of acidic gases which could contribute to acid rain (although the acidic gases could be recovered by a scrubber). The DIAMEX process is being worked on in Europe by the French CEA. The process is sufficiently mature that an industrial plant could be constructed with the existing knowledge of the process. In common with PUREX this process operates by a solvation mechanism.

SANEX

Selective ActiNide EXtraction. As part of the management of minor actinides it has been proposed that the lanthanides and trivalent minor actinides should be removed from the PUREX raffinate by a process such as DIAMEX or TRUEX. In order to allow the actinides such as americium to be either reused in industrial sources or used as fuel, the lanthanides must be removed. The lanthanides have large neutron cross sections and hence they would poison a neutron driven nuclear reaction. To date the extraction system for the SANEX process has not been defined, but currently several different research groups are working towards a process. For instance the French CEA is working on a bis-triazinyl pyridine (BTP) based process. Other systems such as the dithiophosphinic acids are being worked on by some other workers.

UNEX

The *UNiversal* EXtraction process was developed in Russia and the Czech Republic; it is designed to completely remove the most troublesome radioisotopes (Sr, Cs and minor actinides) from the raffinate remaining after the extraction of uranium and plutonium from used nuclear fuel. The chemistry is based upon the interaction of caesium and strontium with polyethylene glycol) and a cobalt carborane anion (known as chlorinated cobalt dicarbollide). The actinides are extracted by CMPO, and the diluent is a polar aromatic such as nitrobenzene. Other dilents such as *meta*-nitrobenzotrifluoride and phenyl trifluoromethyl sulfone have been suggested as well.

Electrochemical Methods

An exotic method using electrochemistry and ion exchange in ammonium carbonate has been reported.

Obsolete Methods

Bismuth Phosphate

The bismuth phosphate process is an obsolete process that adds significant unnecessary material to the final radioactive waste. The bismuth phosphate process has been replaced by solvent extraction processes. The bismuth phosphate process was designed to extract plutonium from aluminium-clad nuclear fuel rods, containing uranium. The fuel was decladded by boiling it in caustic soda. After decladding, the uranium metal was dissolved in nitric acid.

The plutonium at this point is in the +4 oxidation state. It was then precipitated out of the solution by the addition of bismuth nitrate and phosphoric acid to form the bismuth phosphate. The plutonium was coprecipitated with this. The supernatant liquid (containing many of the fission products) was separated from the solid. The precipitate was then dissolved in nitric acid before the addition of an oxidant such as potassium permanganate which converted the plutonium to PuO_2^{2+} (Pu VI), then a dichromate salt was added to maintain the plutonium in the +6 oxidation state.

The bismuth phosphate was next re-precipitated leaving the plutonium in solution. Then an iron (II) salt such as ferrous sulfate was added, and the plutonium re-precipitated again using a bismuth phosphate carrier precipitate. Then lanthanum salts and fluoride were added to create solid lanthanum fluoride which acted as a carrier for the plutonium. This was converted to the oxide by the action of an alkali. The lanthanum plutonium oxide was next collected and extracted with nitric acid to form plutonium nitrate.

Hexone or Redox

This is a liquid-liquid extraction process which uses methyl isobutyl ketone as the extractant. The extraction is by a *solvation* mechanism. This process has the disadvantage of requiring the use of a salting-out reagent (aluminium nitrate) to increase the nitrate concentration in the aqueous phase to obtain a reasonable distribution ratio (D value). Also, hexone is degraded by concentrated nitric acid. This process has been replaced by the PUREX process.

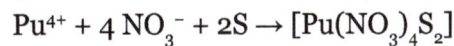

$$Pu^{4+} + 4\,NO_3^- + 2S \rightarrow [Pu(NO_3)_4S_2]$$

Butex, β,β'-dibutyoxydiethyl Ether

A process based on a solvation extraction process using the triether extractant named above. This process has the disadvantage of requiring the use of a salting-out reagent (aluminium nitrate) to increase the nitrate concentration in the aqueous phase to obtain a reasonable distribution ratio. This process was used at Windscale many years ago. This process has been replaced by PUREX.

Pyroprocessing

Pyroprocessing is a generic term for high-temperature methods. Solvents are molten salts (e.g. LiCl+KCl or LiF+CaF2) and molten metals (e.g. cadmium, bismuth, magnesium) rather than water and organic compounds. Electrorefining, distillation, and solvent-solvent extraction are common steps.

These processes are not currently in significant use worldwide, but they have been researched and developed at Argonne National Laboratory and elsewhere.

Advantages

- The principles behind them are well understood, and no significant technical barriers exist to their adoption.

- Readily applied to high-burnup spent fuel and requires little cooling time, since the operating temperatures are high already.

- Does not use solvents containing hydrogen and carbon, which are neutron moderators creating risk of criticality accidents and can absorb the fission product tritium and the activation product carbon-14 in dilute solutions that cannot be separated later.

o   Alternatively, voloxidation can remove 99% of the tritium from used fuel and recover it in the form of a strong solution suitable for use as a supply of tritium.

- More compact than aqueous methods, allowing on-site reprocessing at the reactor site, which avoids transportation of spent fuel and its security issues, instead storing a much smaller volume of fission products on site as high-level waste until decommissioning. For example, the Integral Fast Reactor and Molten Salt Reactor fuel cycles are based on on-site pyroprocessing.

- It can separate many or even all actinides at once and produce highly radioactive fuel which is harder to manipulate for theft or making nuclear weapons. (However, the difficulty has been questioned.) In contrast the PUREX process was designed to separate plutonium only for weapons, and it also leaves the minor actinides (americium and curium) behind, producing waste with more long-lived radioactivity.

- Most of the radioactivity in roughly 10^2 to 10^5 years after the use of the nuclear fuel is produced by the actinides, since there are no fission products with half-lives in this range. These actinides can fuel fast reactors, so extracting and reusing (fissioning) them reduces the long-term radioactivity of the wastes.

Disadvantages

- Reprocessing as a whole is not currently (2005) in favor, and places that do reprocess already have PUREX plants constructed. Consequently, there is little demand for new pyrometalurgical systems, although there could be if the Generation IV reactor programs become reality.

- The used salt from pyroprocessing is less suitable for conversion into glass than the waste materials produced by the PUREX process.

- If the goal is to reduce the longevity of spent nuclear fuel in burner reactors, then better recovery rates of the minor actinides need to be achieved.

Electrolysis

PYRO-A and -B for IFR

These processes were developed by Argonne National Laboratory and used in the Integral Fast Reactor project.

PYRO-A is a means of separating actinides (elements within the actinide family, generally heavier than U-235) from non-actinides. The spent fuel is placed in an anode basket which is immersed in a molten salt electrolyte. An electric current is applied, causing the uranium metal (or sometimes oxide, depending on the spent fuel) to plate out on a solid metal cathode while the other actinides (and the rare earths) can be absorbed into a liquid cadmium cathode. Many of the fission products (such as caesium, zirconium and strontium) remain in the salt. As alternatives to the molten cadmium electrode it is possible to use a molten bismuth cathode, or a solid aluminium cathode.

As an alternative to electrowinning, the wanted metal can be isolated by using a molten alloy of an electropositive metal and a less reactive metal.

Since the majority of the long term radioactivity, and volume, of spent fuel comes from actinides, removing the actinides produces waste that is more compact, and not nearly as dangerous over the long term. The radioactivity of this waste will then drop to the level of various naturally occurring minerals and ores within a few hundred, rather than thousands of, years.

The mixed actinides produced by pyrometallic processing can be used again as nuclear fuel, as they are virtually all either fissile, or fertile, though many of these materials would require a fast breeder reactor in order to be burned efficiently. In a thermal neutron spectrum, the concentrations of several heavy actinides (curium-242 and plutonium-240) can become quite high, creating fuel that is substantially different from the usual uranium or mixed uranium-plutonium oxides (MOX) that most current reactors were designed to use.

Another pyrochemical process, the PYRO-B process, has been developed for the processing and recycling of fuel from a transmuter reactor (a fast breeder reactor designed to convert transuranic nuclear waste into fission products). A typical transmuter fuel is free from uranium and contains recovered transuranics in an inert matrix such as metallic zirconium. In the PYRO-B processing of such fuel, an electrorefining step is used to separate the residual transuranic elements from the fission products and recycle the transuranics to the reactor for fissioning. Newly generated technetium and iodine are extracted for incorporation into transmutation targets, and the other fission products are sent to waste.

Voloxidation

Voloxidation (for *volumetric oxidation*) involves heating oxide fuel with oxygen, sometimes with alternating oxidation and reduction, or alternating oxidation by ozone to uranium trioxide with decomposition by heating back to triuranium octoxide. A major purpose is to capture tritium as tritiated water vapor before further processing where it would be difficult to retain the tritium. Other volatile elements leave the fuel and must be recovered, especially iodine, technetium, and carbon-14. Voloxidation also breaks up the fuel or increases its surface area to enhance penetration of reagents in following reprocessing steps.

Volatilization in Isolation

Simply heating spent oxide fuel in an inert atmosphere or vacuum at a temperature between 700 °C and 1000 °C as a first reprocessing step can remove several volatile elements, including caesium whose isotope caesium-137 emits about half of the heat produced by the spent fuel over the following 100 years of cooling (however, most of the other half is from strontium-90 which remains). The estimated overall mass balance for 20,000 grams of processed fuel with 2,000 grams of cladding is:

	Input	Residue	Zeolite filter	Carbon filter	Particle filters
Palladium	28	14	14		
Tellurium	10	5	5		
Molybdenum	70		70		
Caesium	46		46		

Rubidium	8		8		
Silver	2		2		
Iodine	4			4	
Cladding	2000	2000			
Uranium	19218	19218			?
Others	614	614			?
Total	22000	21851	145	4	0

Tritium is not mentioned in this paper.

Fluoride Volatility

In the fluoride volatility process, fluorine is reacted with the fuel. Fluorine is so much more reactive than even oxygen that small particles of ground oxide fuel will burst into flame when dropped into a chamber full of fluorine. This is known as flame fluorination; the heat produced helps the reaction proceed. Most of the uranium, which makes up the bulk of the fuel, is converted to uranium hexafluoride, the form of uranium used in uranium enrichment, which has a very low boiling point. Technetium, the main long-lived fission product, is also efficiently converted to its volatile hexafluoride. A few other elements also form similarly volatile hexafluorides, pentafluorides, or heptafluorides. The volatile fluorides can be separated from excess fluorine by condensation, then separated from each other by fractional distillation or selective reduction. Uranium hexafluoride and technetium hexafluoride have very similar boiling points and vapor pressures, which makes complete separation more difficult.

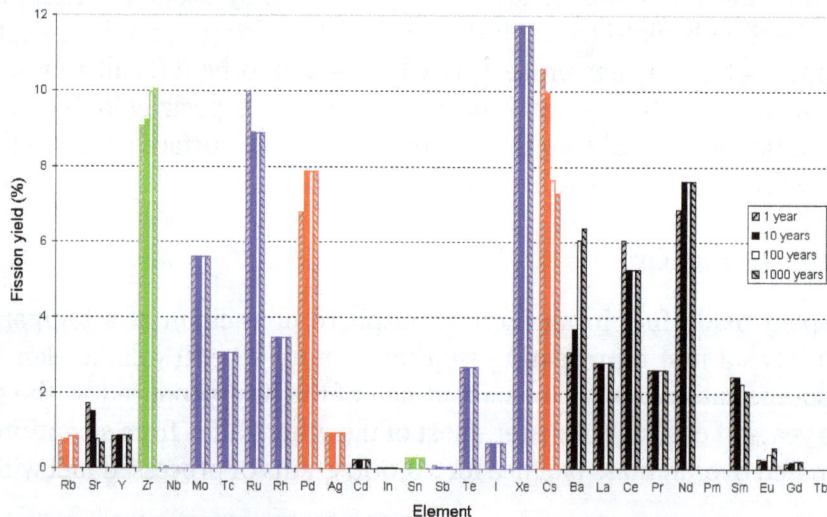

Blue elements have volatile fluorides or are already volatile; green elements do not but have volatile chlorides; red elements have neither, but the elements themselves or their oxides are volatile at very high temperatures. Yields at $10^{0,1,2,3}$ years after fission, not considering later neutron capture, fraction of 100% not 200%. Beta decay Kr-85→Rb, Sr-90→Zr, Ru-106→Pd, Sb-125→Te, Cs-137→Ba, Ce-144→Nd, Sm-151→Eu, Eu-155→Gd visible.

Many of the fission products volatilized are the same ones volatilized in non-fluorinated, higher-temperature volatilization, such as iodine, tellurium and molybdenum; notable differences are that technetium is volatilized, but caesium is not.

Some transuranium elements such as plutonium, neptunium and americium can form volatile fluorides, but these compounds are not stable when the fluorine partial pressure is decreased. Most of the plutonium and some of the uranium will initially remain in ash which drops to the bottom of the flame fluorinator. The plutonium-uranium ratio in the ash may even approximate the composition needed for fast neutron reactor fuel. Further fluorination of the ash can remove all the uranium, neptunium, and plutonium as volatile fluorides; however, some other minor actinides may not form volatile fluorides and instead remain with the alkaline fission products. Some noble metals may not form fluorides at all, but remain in metallic form; however ruthenium hexafluoride is relatively stable and volatile.

Distillation of the residue at higher temperatures can separate lower-boiling transition metal fluorides and alkali metal (Cs, Rb) fluorides from higher-boiling lanthanide and alkaline earth metal (Sr, Ba) and yttrium fluorides. The temperatures involved are much higher, but can be lowered somewhat by distilling in a vacuum. If a carrier salt like lithium fluoride or sodium fluoride is being used as a solvent, high-temperature distillation is a way to separate the carrier salt for reuse.

Molten salt reactor designs carry out fluoride volatility reprocessing continuously or at frequent intervals. The goal is to return actinides to the molten fuel mixture for eventual fission, while removing fission products that are neutron poisons, or that can be more securely stored outside the reactor core while awaiting eventual transfer to permanent storage.

Chloride Volatility and Solubility

Many of the elements that form volatile high-valence fluorides will also form volatile high-valence chlorides. Chlorination and distillation is another possible method for separation. The sequence of separation may differ usefully from the sequence for fluorides; for example, zirconium tetrachloride and tin tetrachloride have relatively low boiling points of 331 °C and 114.1 °C. Chlorination has even been proposed as a method for removing zirconium fuel cladding, instead of mechanical decladding.

Chlorides are likely to be easier than fluorides to later convert back to other compounds, such as oxides.

Chlorides remaining after volatilization may also be separated by solubility in water. Chlorides of alkaline elements like americium, curium, lanthanides, strontium, caesium are more soluble than those of uranium, neptunium, plutonium, and zirconium.

Radioanalytical Separations

In order to determine the distribution of radioactive metals for analytical purposes, Solvent Impregnated Resins (SIRs) can be used. SIRs are porous particles, which contain an extractant inside their pores. This approach avoids the liquid-liquid separation step required in conventional liquid-liquid extraction. For the preparation of SIRs for radioanalytical separations, organic Amberlite XAD-4 or XAD-7 can be used. Possible extractants are e.g. trihexyltetradecylphosphonium chloride(CYPHOS IL-101) or N,N0-dialkyl-N,N0-diphenylpyridine-2,6-dicarboxyamides (R-PDA; R = butyl, octy I, decyl, dodecyl).

Economics

The relative economics of reprocessing-waste disposal and interim storage-direct disposal has been the focus of much debate over the past ten years. Studies have modeled the total fuel cycle costs of a reprocessing-recycling system based on one-time recycling of plutonium in existing thermal reactors (as opposed to the proposed breeder reactor cycle) and compare this to the total costs of an open fuel cycle with direct disposal. The range of results produced by these studies is very wide, but all are agreed that under current (2005) economic conditions the reprocessing-recycle option is the more costly.

If reprocessing is undertaken only to reduce the radioactivity level of spent fuel it should be taken into account that spent nuclear fuel becomes less radioactive over time. After 40 years its radioactivity drops by 99.9%, though it still takes over a thousand years for the level of radioactivity to approach that of natural uranium. However the level of transuranic elements, including plutonium-239, remains high for over 100,000 years, so if not reused as nuclear fuel, then those elements need secure disposal because of nuclear proliferation reasons as well as radiation hazard.

On 25 October 2011 a commission of the Japanese Atomic Energy Commission revealed during a meeting calculations about the costs of recycling nuclear fuel for power generation. These costs could be twice the costs of direct geological disposal of spent fuel: the cost of extracting plutonium and handling spent fuel was estimated at 1.98 to 2.14 yen per kilowatt-hour of electricity generated. Discarding the spent fuel as waste would cost only 1 to 1.35 yen per kilowatt-hour.

In July 2004 Japanese newspapers reported that the Japanese Government had estimated the costs of disposing radioactive waste, contradicting claims four months earlier that no such estimates had been made. The cost of non-reprocessing options was estimated to be between a quarter and a third ($5.5–7.9 billion) of the cost of reprocessing ($24.7 billion). At the end of the year 2011 it became clear that Masaya Yasui, who had been director of the Nuclear Power Policy Planning Division in 2004, had instructed his subordinate in April 2004 to conceal the data. The fact that the data were deliberately concealed obliged the ministry to re-investigate the case and to reconsider whether to punish the officials involved.

Base Load

The base load on a grid is the minimum level of demand on an electrical grid over 24 hours. Base load power sources are power stations which can consistently generate the electrical power needed to satisfy this minimum demand.

Historically, large power grids have used base load power plants exclusively. However, there is no specific technical requirement for this to be so. The base load requirement can equally well be met by the appropriate mix of intermittent power sources, peaking power plants, hydroelectric power and other sources.

Description

Baseload plant, (also baseload power plant or base load power station) is an energy station devoted to the production of base load supply. Baseload plants are the production facilities used to meet some or all of a given region's continuous energy demand, and produce energy at a constant rate, usually at a low cost relative to other production facilities available to the system. Examples of baseload plants using nonrenewable fuels include nuclear and coal-fired plants. Baseload plants typically run at all times through the year except in the case of repairs or maintenance. These plants are often designed for relatively high efficiency, and may be combined cycle plants, but may take several days to start up and shut down.

A base load power station in Taiwan.

Each baseload power plant on a grid is allotted a specific amount of the baseload power demand to handle. The base load power is determined by the load duration curve of the system. For a typical power system, the rule of thumb is that the base load power is usually 35-40% of the maximum load during the year.

Peaks or spikes in customer power demand are frequently handled by smaller and more responsive, but perhaps somewhat less efficient types of power plants called peaking power plants, often powered with gas turbines.

While historically large power grids have had base load power plant to exclusively meet the base load, there is no specific technical requirement for this to be so. The base load can equally well be met by the appropriate quantity of intermittent power sources and peaking power plant.

Among the renewable energy sources, hydroelectric, geothermal, biogas, biomass, solar thermal with storage and ocean thermal energy conversion can provide base load power. A significant fraction of the average wind power production is available with 95% or greater probability, and so may be used for baseload power.

Hydroelectric power also has the desirable attribute of dispatchability, but conversely a hydroelectric plant may run low on its "fuel" (water at the reservoir elevation) if a long drought occurs over its drainage basin.

Economics

Power plants are designated *baseload* based on their low cost generation, efficiency and safety at rated output power levels. Baseload power plants do not change production to match power consumption demands since it is more economical to operate them at constant production levels. Use of higher cost combined-cycle plants or combustion turbines is thus minimized, and these plants can be cycled up and down to match more rapid fluctuations in consumption. Baseload generators, such as nuclear and coal, often have very high fixed costs, high plant load factor but very low marginal costs. On the other hand, peak load generators, such as natural gas, have low fixed costs, low plant load factor and high marginal costs. Typically baseload plants are large and provide a majority of the power used by a grid. Thus, they are more effective when used continuously to cover the power baseload required by the grid.

Base Load Power Plant Usage

Nuclear power plants may take many hours, if not days, to change their power output, although modern stations, and those in France, can and do operate as load following power plants and alter their output to meet varying demands. Because they require a long period of time to heat up to operating temperature, these plants typically handle large amounts of baseload demand. Different plants and technologies may have differing capacities to increase or decrease output on demand: nuclear plants are generally run at close to maximum output continuously (apart from maintenance, refueling and periodic refurbishment), while coal-fired plants may be cycled over the course of a day to meet demand. Plants with multiple generating units may be used as a group to improve the "fit" with demand, by operating each unit as close to peak efficiency as possible.

According to National Grid plc chief executive officer Steve Holliday in 2015, baseload is "outdated", as microgrids would become the primary means of production, and large powerplants relegated to supply the remainder.

In 2016, Ambrose Evans-Pritchard of the Daily Telegraph wrote that, with advances in energy storage, 'there ceases to be much point in building costly "baseload" power plants' and goes on to argue 'Nuclear reactors cannot be switched on and off as need demands - unlike gas plants. They are useless as a back-up for the decentralized grid of the future, when wind, solar, hydro, and other renewables will dominate the power supply'.

References

- Walker, J. Samuel (10 January 2006). Three Mile Island: A Nuclear Crisis in Historical Perspective. University of California Press. pp. 10–11. ISBN 978-0-520-24683-6.

- Giugni, Marco (2004). Social protest and policy change: ecology, antinuclear, and peace movements in comparative perspective. Rowman & Littlefield. pp. 44–. ISBN 978-0-7425-1827-8.

- Bernstein, Jeremy (2008). Nuclear Weapons: What You Need to Know. Cambridge University Press. p. 312. ISBN 978-0-521-88408-2. Retrieved 17 March 2011.

- Tsetkov, Pavel; Usman, Shoaib (2011). Krivit, Steven, ed. Nuclear Energy Encyclopedia: Science, Technology, and Applications. Hoboken, NJ: Wiley. pp. 48; 85. ISBN 978-0-470-89439-2.

- Kragh, Helge (1999). Quantum Generations: A History of Physics in the Twentieth Century. Princeton NJ: Princeton University Press. p. 286. ISBN 0-691-09552-3.

- National Research Council (U.S.). Committee on Future Nuclear Power, Nuclear power: technical and institutional options for the future National Academies Press, 1992, ISBN 0-309-04395-6.

- Helmreich, J.E. Gathering Rare Ores: The Diplomacy of Uranium Acquisition, 1943–1954, Princeton UP, 1986: ch. 10 ISBN 0-7837-9349-9

- Frank von Hippel; et al. (February 2010). Fast Breeder Reactor Programs: History and Status (PDF). International Panel on Fissile Materials. ISBN 978-0-9819275-6-5. Retrieved 28 April 2014.

- Wiser, Wendell H. (2000). Energy resources: occurrence, production, conversion, use. Birkhäuser. p. 190. ISBN 978-0-387-98744-6.

- Whitaker, Jerry C. (2006). AC power systems handbook. Boca Raton, FL: Taylor and Francis. p. 35. ISBN 978-0-8493-4034-5.

- Leyzerovich, Alexander (2005). Wet-steam Turbines for Nuclear Power Plants. Tulsa OK: PennWell Books. p. 111. ISBN 978-1-59370-032-4.

- William P. Sanders (ed), Turbine Steam Path Mechanical Design and Manufacture, Volume Iiia (PennWell Books, 2004) ISBN 1-59370-009-1 page 292

- Irvine, Maxwell (2011). Nuclear power : a very short introduction. Oxford: Oxford University Press. p. 55. ISBN 9780199584970.

- Neeb, Karl-Heinz (1997). "The radiochemistry of nuclear power plants with light water reactors". Walter de Gruyter. ISBN 3-11-013242-7.

- Karel Beckman (11 September 2015). "Steve Holliday CEO National Grid: baseload is outdated". EnergyPost. eu. Archived from the original on 10 September 2016. Retrieved 6 October 2016.

- Evans-Pritchard, Ambrose (10 August 2016). "Holy Grail of energy policy in sight as battery technology smashes the old order". The Telegraph.

- "Estimating China's Production of Plutonium for Weapons" (PDF). doi:10.1080/08929880390214133 (inactive 2015-02-27).

- "Запущен первый реактор на быстрых нейтронах БН-800, построенный в России". mining24.ru. Retrieved 22 December 2015.

- S. R. Pillai, M. V. Ramana (2014). "Breeder reactors: A possible connection between metal corrosion and sodium leaks". Bulletin of the Atomic Scientists. 70 (3). doi:10.1177/0096340214531178. Retrieved 15 February 2015.

Nuclear Power Reactor: A Design Classification

With the passage of time, nuclear reactors have evolved and have incorporated cutting-edge technology to improve thermal efficiency as well as reduce maintenance and optimize fuel technology. Based on their evolutionary improvements, the reactors have been arranged into generations. In this chapter, the reader is informed about Generations II, III and IV with differentiating characteristics and designs of each generation.

Generation II Reactor

A generation II reactor is a design classification for a nuclear reactor, and refers to the class of commercial reactors built up to the end of the 1990s. Prototypical generation II reactors include the PWR, CANDU, BWR, AGR, and VVER.

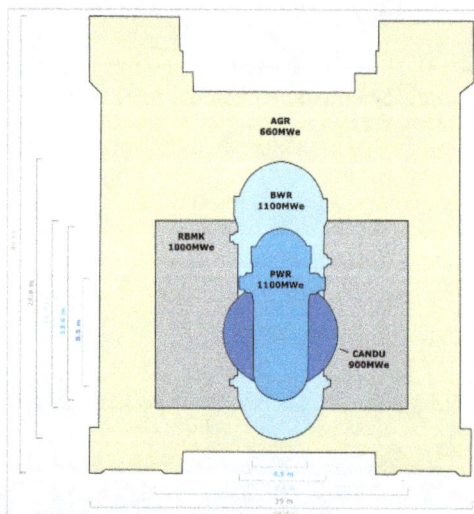

Generation II reactor vessels size comparison.

These are contrasted to generation I reactors, which refer to the early prototype of power reactors, such as Shippingport, Magnox/UNGG, Fermi 1, and Dresden. The nomenclature for reactor designs, describing four 'generations', was proposed by the US Department of Energy when it introduced the concept of generation IV reactors.

The designation *generation II+ reactor* is sometimes used for modernized generation II designs built post-2000, such as the Chinese CPR-1000, in competition with more expensive generation III reactor designs. Typically, the modernization includes improved safety systems and a 60-year design life.

Generation II reactor designs generally had an original design life of 30 or 40 years. This date was set as the period over which loans taken out for the plant would be paid off. However, many generation II reactor are being life-extended to 50 or 60 years, and a second life-extension to 80 years may also be economic in many cases. By 2013 about 75% of still operating U.S. reactors had been granted life extension licenses to 60 years.

Fukushima Daiichi's three destroyed reactors are Mark I Boiling water reactors (BWR) designed by General Electric. In 2015, unit 2 at the Watts Bar Nuclear Generating Station is likely to be the last generation II reactor to come online.

Generation III Reactor

A generation III reactor is a development of generation II nuclear reactor designs incorporating evolutionary improvements in design developed during the lifetime of the generation II reactor designs. These include improved fuel technology, superior thermal efficiency, passive nuclear safety systems and standardised design for reduced maintenance and capital costs. The first Generation III reactor to begin operation was Kashiwazaki 6 (an ABWR) in 1996.

Model of Toshiba ABWR. The first Generation III reactor to come online in 1996.

Due to the lack of reactor construction in the Western world, very few third generation reactors have been built in developed nations. In general, Generation IV designs are still in development, and might come online in the 2030s.

Overview

Generation III+ AP1000 reactor.

Though the distinction is arbitrary, the improvements in reactor technology in third generation reactors are intended to result in a longer operational life (60 years of operation, extendable to

120+ years of operation prior to complete overhaul and reactor pressure vessel replacement) compared with currently used generation II reactors (designed for 40 years of operation, extendable to 80+ years of operation prior to complete overhaul and pressure vessel replacement).

The core damage frequencies for these reactors are designed to be lower than for Generation II reactors – 60 core damage events for the EPR and 3 core damage events for the ESBWR per 100 million reactor-years are significantly lower than the 1,000 core damage events per 100 million reactor-years for BWR/4 generation II reactors.

The third generation EPR reactor was also designed to use uranium more efficiently than older Generation II reactors, using approximately 17% less uranium per unit of electricity generated than these older reactor technologies.

Response and Criticism

Proponents of nuclear power and some who have historically been critical have acknowledged that third generation reactors as a whole are safer than older reactors. However, while there are some strong proponents of the American third generation designs that claim they are much safer than existing reactors in the US, other engineers, although not outright saying that they are not safer, are more conservative and have some specific concerns.

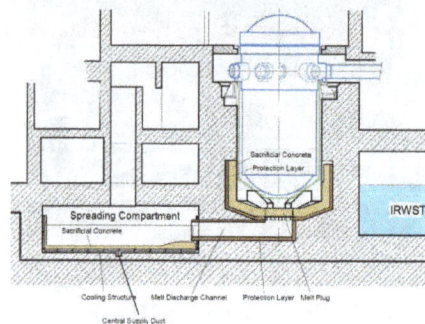

EPR core catching room designed to catch the corium in case of a meltdown.

Edwin Lyman, a senior staff scientist at the Union of Concerned Scientists, has challenged specific cost-saving design choices made for two generation III reactors, both the AP1000 and ESBWR. Lyman, John Ma (a senior structural engineer at the NRC), and Arnold Gundersen (an anti-nuclear consultant) are concerned about what they perceive as weaknesses in the steel containment vessel and the concrete shield building around the AP1000 in that its containment vessel does not have sufficient safety margins in the event of a direct airplane strike. Other engineers do not agree with these concerns, and claim the containment building is more than sufficient in safety margins and factors of safety.

The Union of Concerned Scientists in 2008 referred to the EPR as the only new reactor design under consideration in the United States that "...appears to have the potential to be significantly safer and more secure against attack than today's reactors."

There have also been issues in fabricating the precision parts necessary to maintain safe operation of these reactors, with cost overruns, broken parts, and extremely fine steel tolerances causing issues with new reactors under construction in France.

Existing and Future Reactors

The first generation III reactors were built in Japan, in the form of Advanced Boiling Water Reactors, while several others are in construction in Europe, including the EPR at Flamanville. The next third generation reactor predicted to come on line is a Westinghouse AP1000 reactor, the Sanmen Nuclear Power Station in China, which was scheduled to become operational in 2015. Its completion has since been delayed until 2017.

In the USA, reactor designs are certified by the Nuclear Regulatory Commission (NRC). As of October 2014 the commission has approved five designs, and is considering another five designs as well.

Generation III Reactors

Generation III reactors operational or under construction.				
Developer	**Reactor name**	**Type**	**MWe (gross)**	**Notes**
General Electric/Toshiba/ Hitachi	ABWR	BWR	1356	In operation at Kashiwazari, Japan in 1996. NRC certified in 1997.
KEPCO	APR-1400	PWR	1450	In operation at Kori, Korea since Jan 2016.
CGNPG/CNNC	Hualong One			Development of the CPR-1000 and ACP1000 Chinese reactors.
OKB Gidropress	VVER-1200/392M		1200	Prototype unit expected to be operating by 2017 at Novovoronezh II, Russia.
	VVER-1200/491			Sold as MIR-1200. Prototype unit expected to be operating by 2018 at Leningrad.
	VVER-1200/513			First unit expected to be completed by 2022 at Akkuyu, Turkey.
OKBM	BN-800	FBR	880	Demonstration fast reactor in commercial operation since Aug 2016 at Beloyarsk.

Generation III Designs Not Adopted or Built Yet

Developer	**Reactor name**	**Type**	**MWe (gross)**	**Notes**
Mitsubishi	APWR	PWR	1700	Two units planned at Tsuruga cancelled in 2011.
Westinghouse	AP600		600	NRC certified in 1999. Evolved into the larger AP1000 design.
Combustion Engineering	System 80+			NRC certified in 1997. Provided a basis for the Korean APR-1400.
Candu Energy Inc.	EC6	PHWR	750	
BARC	Advanced Heavy Water Reactor			In development by India to utilize thorium.

Generation III+ Reactors

Generation III+ designs offer significant improvements in safety and economics over Generation III advanced reactor designs.

Generation III+ reactors under construction.				
Developer	**Reactor name**	**Type**	**MWe (gross)**	**Notes**
Westinghouse/Toshiba	AP1000	PWR	1200	NRC certified Dec 2005. First unit expected to be completed by 2017 at Sanmen, China.
Areva	EPR		1750	First unit expected to be completed by 2017 at Taishan, China.
Areva/Mitsubishi	ATMEA1		1150	First unit expected to be completed by 2023 at Sinop, Turkey.

Generation III+ Designs Not Adopted or Built Yet

Developer	**Reactor name**	**Type**	**MWe (gross)**	**Notes**
General Electric/Hitachi	ESBWR	BWR	1600	Based on the ABWR.
KEPCO	APR+	PWR	1550	APR-1400 with increased output and safety features.
OKB Gidropress	VVER-1300/510		1255	Also referred to as VVER-TOI, based on V392M.
Babcock & Wilcox/Bechtel	B&W mPower		160	Small modular reactor.
Candu Energy Inc.	ACR-1000	PHWR	1200	

Generation IV Reactor

Nuclear Energy Systems Deployable no later than 2030 and offering significant advances in sustainability, safety and reliability, and economics

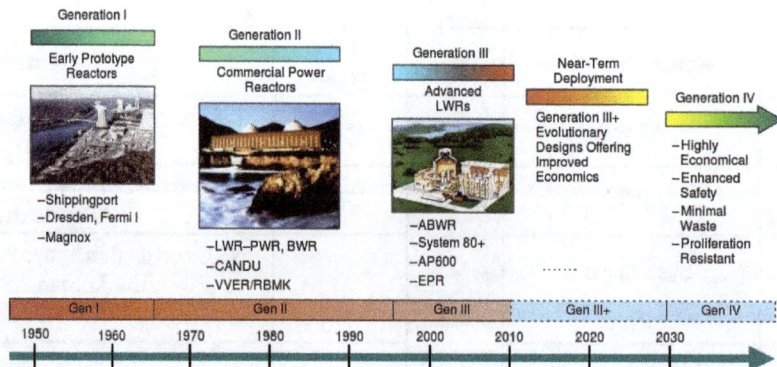

Generation IV: Nuclear Energy Systems Deployable no later than 2030 and offering significant advances in sustainability, safety and reliability, and economics

Generation IV reactors (Gen IV) are a set of nuclear reactor designs currently being researched for commercial applications by the Generation IV International Forum, with Technology readiness levels varying between the level requiring a demonstration, to economical competitive implementation. They are motivated by a variety of goals including improved safety, sustainability, efficiency, and cost.

Most of these designs are generally not expected to be available for commercial construction before 2030–40. Presently the majority of reactors in operation around the world are considered second generation reactor systems, as the vast majority of the first-generation systems were retired some time ago, and there are only a dozen or so Generation III reactors in operation (2014). Generation V reactors refer to reactors that are purely theoretical and are therefore not yet considered feasible in the short term, resulting in limited R&D funding.

History

The Generation IV International Forum (GIF) is "a co-operative international endeavour which was set up to carry out the research and development needed to establish the feasibility and performance capabilities of the next generation nuclear energy systems." It was founded in 2001. There are currently ten active members of the Generation IV International Forum (GIF): Canada, China, the European Atomic Energy Community (Euratom), France, Japan, Russia, South Africa, South Korea, Switzerland, and the United States. The non-active members are Argentina, Brazil, and the United Kingdom. Switzerland joined in 2002, Euratom in 2003, and China and Russia in 2006. The remaining countries (excluding Australia) were founding members.

The 36th GIF meeting in Brussels was held in November 2013. The *Technology Roadmap Update for Generation IV Nuclear Energy Systems* was published in January 2014 which details R&D objectives for the next decade. A breakdown of the reactor designs being researched by each forum member has been made available.

Australia joined the forum in 2016.

Reactor Types

Many reactor types were considered initially; however, the list was downsized to focus on the most promising technologies and those that could most likely meet the goals of the Gen IV initiative. Three systems are nominally thermal reactors and three are fast reactors. The Very High Temperature Reactor (VHTR) is also being researched for potentially providing high quality process heat for hydrogen production. The fast reactors offer the possibility of burning actinides to further reduce waste and of being able to "breed more fuel" than they consume. These systems offer significant advances in sustainability, safety and reliability, economics, proliferation resistance (depending on perspective) and physical protection.

System	Neutron Spectrum	Coolant	Temperature (°C)	Fuel Cycle	Size (MW)	Example developers
VHTR	Thermal	Helium	900–1000	Open	250–300	JAEA (HTTR), Tsinghua University (HTR-10), X-energy
SFR	Fast	Sodium	550	Closed	30–150, 300–1500, 1000–2000	TerraPower (TWR), Toshiba (4S), GE Hitachi Nuclear Energy (PRISM)

SCWR	Thermal/ fast	Water	510–625	Open/closed	300–700, 1000–1500	
GFR	Fast	Helium	850	Closed	1200	Energy Multiplier Module
LFR	Fast	Lead	480–800	Closed	20–180, 300–1200, 600–1000	
MSR	Fast/ thermal	Fluoride/Chloride salts	700–800	Closed	250, 1000	Flibe Energy (LFTR), Transatomic Power, Thorium Tech Solution (FUJI MSR), Terrestrial Energy (IMSR), Southern Company

Thermal Reactors

A thermal reactor is a nuclear reactor that uses slow or thermal neutrons. A neutron moderator is used to slow the neutrons emitted by fission to make them more likely to be captured by the fuel.

Very-high-temperature Reactor (VHTR)

The very high temperature reactor concept uses a graphite-moderated core with a once-through uranium fuel cycle, using helium or molten salt as the coolant. This reactor design envisions an outlet temperature of 1,000 °C. The reactor core can be either a prismatic-block or a pebble bed reactor design. The high temperatures enable applications such as process heat or hydrogen production via the thermochemical iodine-sulfur process. It would also be passively safe.

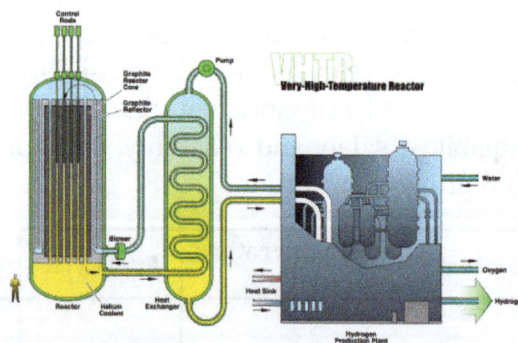

Very-High-Temperature Reactor (VHTR)

The planned construction of the first VHTR, the South African PBMR (pebble bed modular reactor), lost government funding in February, 2010. A pronounced increase of costs and concerns about possible unexpected technical problems had discouraged potential investors and customers.

The Peoples Republic of China began construction of a 200-MW High Temperature Pebble bed reactor in 2012 as a successor to its HTR-10.

Also in 2012, as part of the Next Generation Nuclear Plant competition, Idaho National Laboratory approved a design similar to Areva's prismatic block Antares reactor as the chosen HTGR to be deployed as a prototype by 2021. It was in competition with General Atomics' Gas turbine modular helium reactor and Westinghouse's Pebble Bed Modular Reactor.

Molten-salt Reactor (MSR)

A molten salt reactor is a type of nuclear reactor where the primary coolant, or even the fuel itself is a molten salt mixture. There have been many designs put forward for this type of reactor and a few prototypes built. The early concepts and many current ones rely on nuclear fuel dissolved in the molten fluoride salt as uranium tetrafluoride (UF_4) or thorium tetrafluoride (ThF_4). The fluid would reach criticality by flowing into a graphite core which would also serve as the moderator. Many current concepts rely on fuel that is dispersed in a graphite matrix with the molten salt providing low pressure, high temperature cooling.

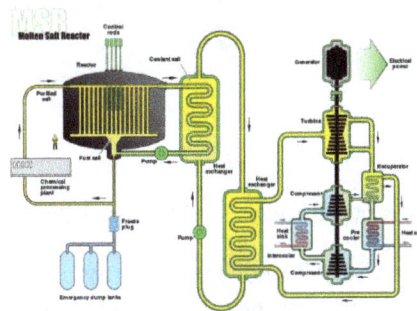

Molten Salt Reactor (MSR)

The Gen IV MSR is more accurately termed an epithermal reactor than a thermal reactor due to the average speed of the neutrons that would cause the fission events within its fuel being faster than thermal neutrons.

The principle of a MSR can be used for thermal, epithermal and fast reactors. Since 2005 the focus has moved towards a fast spectrum MSR (MSFR).

While most MSR designs being pursued are largely derived from the 1960s Molten-Salt Reactor Experiment(MSRE), variants of molten salt technology include the conceptual *Dual fluid reactor* which is being designed with lead as a cooling medium but molten salt fuel, commonly as the metal chloride e.g Plutonium(III) chloride, to aid in greater "nuclear waste" closed-fuel cycle capabilities. Other notable approaches differing substantially from those with MSRE pedigree include the *Stable Salt Reactor*(SSR) concept promoted by MOLTEX, which encases the molten salt in hundreds of the common solid fuel rods that are already well established in the nuclear industry. This latter British design was found to be the most competitive for Small modular reactor development by a British based consultancy firm *Energy Process Development* in 2015.

Supercritical-water-cooled Reactor (SCWR)

The supercritical water reactor (SCWR) is a reduced moderation water reactor concept that, due to the average speed of the neutrons that would cause the fission events within the fuel being faster than thermal neutrons, it is more accurately termed an epithermal reactor than a thermal reactor.

It uses supercritical water as the working fluid. SCWRs are basically light water reactors (LWR) operating at higher pressure and temperatures with a direct, once-through heat exchange cycle. As most commonly envisioned, it would operate on a direct cycle, much like a boiling water reactor (BWR), but since it uses supercritical water as the working fluid, it would have only one water phase present, which makes the supercritical heat exchange method more similar to a pressurized water reactor (PWR). It could operate at much higher temperatures than both current PWRs and BWRs.

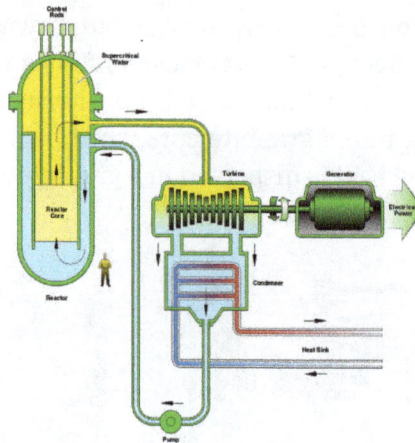

Supercritical-Water-Cooled Reactor (SCWR)

Supercritical water-cooled reactors (SCWRs) are promising advanced nuclear systems because of their high thermal efficiency (i.e., about 45% vs. about 33% efficiency for current LWRs) and considerable plant simplification.

The main mission of the SCWR is generation of low-cost electricity. It is built upon two proven technologies, LWRs, which are the most commonly deployed power generating reactors in the world, and superheated fossil fuel fired boilers, a large number of which are also in use around the world. The SCWR concept is being investigated by 32 organizations in 13 countries.

A SCWR Design under development is the VVER-1700/393 (VVER-SCWR or VVER-SKD) — a Russian Supercritical-water-cooled reactor with double-inlet-core and a breeding ratio of 0.95.

Fast Reactors

A fast reactor directly uses the fast neutrons emitted by fission, without moderation. Unlike thermal neutron reactors, fast neutron reactors can be configured to "burn", or fission, all actinides, and given enough time, therefore drastically reduce the actinides fraction in spent nuclear fuel produced by the present world fleet of thermal neutron light water reactors, thus closing the nuclear fuel cycle. Alternatively, if configured differently, they can also breed more actinide fuel than they consume.

Gas-cooled Fast Reactor (GFR)

The gas-cooled fast reactor (GFR) system features a fast-neutron spectrum and closed fuel cycle for efficient conversion of fertile uranium and management of actinides. The reactor is helium-cooled and with an outlet temperature of 850 °C it is an evolution of the very-high-temperature reactor (VHTR) to a more sustainable fuel cycle. It will use a direct Brayton cycle gas turbine for high thermal efficiency. Several fuel forms are being considered for their potential to operate at very high temperatures and to ensure an excellent retention of fission products: composite ceramic fuel, advanced fuel particles, or ceramic clad elements of actinide compounds. Core configurations are being considered based on pin- or plate-based fuel assemblies or prismatic blocks.

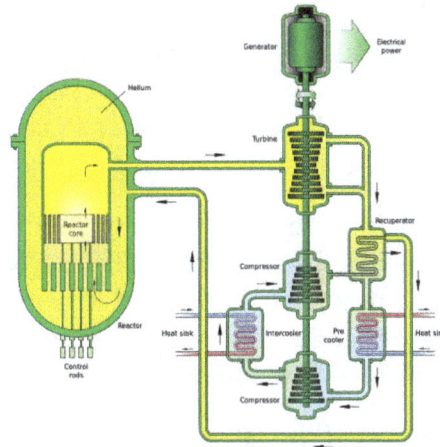

Gas-Cooled Fast Reactor (GFR)

The European Sustainable Nuclear Industrial Initiative is funding three Generation IV reactor systems, one of which is a gas-cooled fast reactor, called *Allegro*, 100 MW(t), which will be built in a central or eastern European country with construction expected to begin in 2018. The central European Visegrád Group are committed to pursuing the technology. In 2013 German, British, and French institutes finished a 3-year collaboration study on the follow on industrial scale design, known as *GoFastR*. They were funded by the EU's 7th FWP framework programme, with the goal of making a sustainable VHTR.

Sodium-cooled Fast Reactor (SFR)

The SFR is a project that builds on two closely related existing projects, the liquid metal fast breeder reactor and the integral fast reactor.

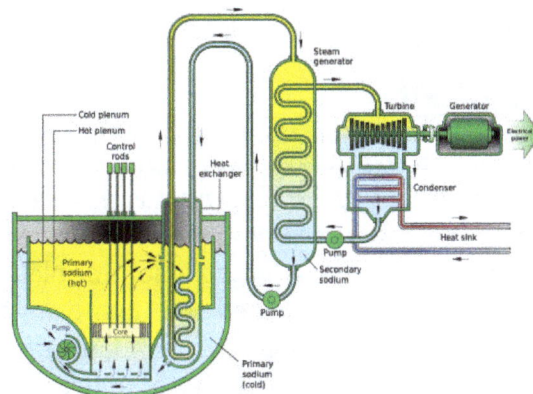

Pool design Sodium-Cooled Fast Reactor (SFR)

The goals are to increase the efficiency of uranium usage by breeding plutonium and eliminating the need for transuranic isotopes ever to leave the site. The reactor design uses an unmoderated core running on fast neutrons, designed to allow any transuranic isotope to be consumed (and in some cases used as fuel). In addition to the benefits of removing the long half-life transuranics from the waste cycle, the SFR fuel expands when the reactor overheats, and the chain reaction automatically slows down. In this manner, it is passively safe.

The SFR reactor concept is cooled by liquid sodium and fueled by a metallic alloy of uranium and plutonium or spent nuclear fuel, the "nuclear waste" of light water reactors. The SFR fuel is contained in steel cladding with liquid sodium filling in the space between the clad elements which make up the fuel assembly. One of the design challenges of an SFR is the risks of handling sodium, which reacts explosively if it comes into contact with water. However, the use of liquid metal instead of water as coolant allows the system to work at atmospheric pressure, reducing the risk of leakage.

The European Sustainable Nuclear Industrial Initiative is funding three Generation IV reactor systems, one of which is a sodium-cooled fast reactor, called *ASTRID*, Advanced Sodium Technical Reactor for Industrial Demonstration, Areva, CEA and EDF are leading the design with British collaboration. Astrid will be rated about 600 MWe and is proposed to be built in France, near to the Phénix reactor. A final decision in construction is to be made in 2019

The PRC's first commercial-scale, 800 MWe, fast neutron reactor, to be situated near Sanming in Fujian province will be a SFR. In 2009 an agreement was signed that would entail the Russian BN-800 reactor design to be sold to the PRC once it is completed, this would be the first time commercial-scale fast neutron reactors have ever been exported. The BN-800 reactor became operational in 2014.

In India, the Prototype Fast Breeder Reactor, a 500 MWe Sodium cooled fast reactor is being built at a cost of INR 5,677 crores (~US$900 million) and is expected to be commissioned by September 2016. The PFBR will be followed by six more Commercial Fast Breeder Reactors (CFBRs) of 500 MWe each.

The 400 MWe Fast Flux Test Facility operated successfully for ten years at the Hanford site in Washington State.

The 20 MWe EBR II operated successfully for over thirty years at the Idaho National Laboratory, until it was shut down in 1994.

Lead-cooled Fast Reactor (LFR)

The lead-cooled fast reactor features a fast-neutron-spectrum lead or lead/bismuth eutectic (LBE) liquid-metal-cooled reactor with a closed fuel cycle. Options include a range of plant ratings, including a "battery" of 50 to 150 MW of electricity that features a very long refueling interval, a modular system rated at 300 to 400 MW, and a large monolithic plant option at 1,200 MW. (The term *battery* refers to the long-life, factory-fabricated core, not to any provision for electrochemical energy conversion.) The fuel is metal or nitride-based containing fertile uranium and transuranics. The LFR is cooled by natural convection with a reactor outlet coolant temperature of 550 °C, possibly ranging up to 800 °C with advanced materials. The higher temperature enables the production of hydrogen by thermochemical processes.

The European Sustainable Nuclear Industrial Initiative is funding three Generation IV reactor systems, one of which is a lead-cooled fast reactor that is also an accelerator-driven sub-critical reactor, called *MYRRHA*, 100 MW(t), which will be built in Belgium with construction expected to begin after 2014 and the industrial scale version, known as *Alfred*, slated to be constructed sometime after 2017. A reduced-power model of Myrrha called *Guinevere* was started up at Mol in March 2009. In 2012 the research team reported that Guinevere was operational.

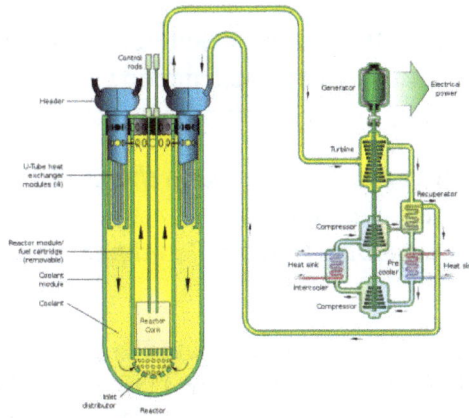

Lead-Cooled Fast Reactor (LFR)

Two other lead-cooled fast reactors under development are the SVBR-100, a modular 100 MWe lead-bismuth cooled fast neutron reactor concept designed by OKB Gidropress in Russia and the BREST-OD-300 (Lead-cooled fast reactor) 300 MWe, to be developed after the SVBR-100, and built over 2016-20, it will dispense with the fertile blanket around the core and will supersede the sodium cooled BN-600 reactor design, to purportedly give enhanced proliferation resistance.

Advantages and Disadvantages

Relative to current nuclear power plant technology, the claimed benefits for 4th generation reactors include:

- Nuclear waste that remains radioactive for a few centuries instead of millennia

- 100-300 times more energy yield from the same amount of nuclear fuel

- Broader range of fuels, and even unencapsulated raw fuels (non-pebble MSR, LFTR).

- In some reactors, the ability to consume existing nuclear waste in the production of electricity, that is, a Closed nuclear fuel cycle. This strengthens the argument to deem nuclear power as renewable energy.

- Improved operating safety features, such as (depending on design) avoidance of pressurized operation, automatic passive (unpowered, uncommanded) reactor shutdown, avoidance of water cooling and the associated risks of loss of water (leaks or boiling) and hydrogen generation/explosion and contamination of coolant water.

Nuclear reactors do not emit CO_2 during operation, although like all low carbon power sources, the mining and construction phase can result in CO_2 emissions, if energy sources which are not carbon neutral (such as fossil fuels), or CO_2 emitting cements are used during the construction process. A 2012 Yale University review published in the Journal of Industrial Ecology analyzing CO_2 life cycle assessment (LCA) emissions from nuclear power determined that:

"The collective LCA literature indicates that life cycle GHG [greenhouse gas] emissions from nuclear power are only a fraction of traditional fossil sources and comparable to renewable technologies."

Although the paper primarily dealt with data from Generation II reactors, and did not analyze the CO_2 emissions by 2050 of the presently under construction Generation III reactors, it did summarize the Life Cycle Assessment findings of in development reactor technologies.

FBRs [Fast Breeder Reactors] have been evaluated in the LCA literature. The limited literature that evaluates this potential future technology reports median life cycle GHG emissions... similar to or lower than LWRs[Gen II light water reactors] and purports to consume little or no uranium ore.

A specific risk of the sodium-cooled fast reactor is related to using metallic sodium as a coolant. In case of a breach, sodium explosively reacts with water. Fixing breaches may also prove dangerous, as the cheapest noble gas argon is also used to prevent sodium oxidation. Argon, like helium, can displace oxygen in the air and can pose hypoxia concerns, so workers may be exposed to this additional risk. This is a pertinent problem as can be testified by the events at the loop type Prototype Fast Breeder Reactor Monju at Tsuruga, Japan. Using lead or molten salts mitigates this problem by making the coolant less reactive and allowing a high freezing temperature and low pressure in case of a leak. Disadvantages of lead compared to sodium are much higher viscosity, much higher density, lower heat capacity, and more radioactive neutron activation products.

In many cases, there is already a large amount of experience built up with numerous proof of concept Gen IV designs. For example, the reactors at Fort St. Vrain Generating Station and HTR-10 are similar to the proposed Gen IV VHTR designs, and the pool type EBR-II, Phénix and BN-600 reactor are similar to the proposed pool type Gen IV Sodium Cooled Fast reactors being designed.

Types of Generation IV Reactors

Very-high-temperature Reactor

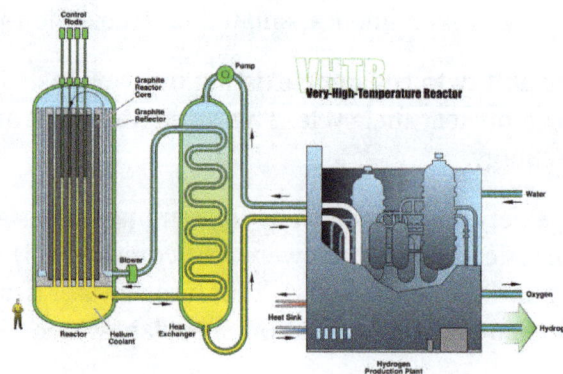

Very-high-temperature reactor scheme.

The very-high-temperature reactor (VHTR), or high-temperature gas-cooled reactor (HTGR), is a Generation IV reactor concept that uses a graphite-moderated nuclear reactor with a once-through uranium fuel cycle. The VHTR is a type of high-temperature reactor (HTR) that can conceptually have an outlet temperature of 1000 °C. The reactor core can be either a "prismatic block" or a "pebble-bed" core. The high temperatures enable applications such as process heat or hydrogen production via the thermochemical sulfur–iodine cycle.

Overview

The VHTR is a type of high-temperature reactor that conceptually can reach higher outlet temperatures (up to 1000 °C); however, in practice the term "VHTR" is usually thought of as a gas-cooled reactor, and commonly used interchangeably with "HTGR" (high-temperature gas-cooled reactor).

AVR in Germany.

There are two main types of HTGRs: pebble bed reactors (PBR) and prismatic block reactors (PMR).The prismatic block reactor refers to a prismatic block core configuration, in which hexagonal graphite blocks are stacked to fit in a cylindrical pressure vessel. The pebble bed reactor (PBR) design consists of fuel in the form of pebbles, stacked together in a cylindrical pressure vessel, like a gum-ball machine. Both reactors may have the fuel stacked in an annulus region with a graphite center spire, depending on the design and desired reactor power.

The Russian VHTR is also a HTGR.

History

The HTGR design was first proposed by the staff of the Power Pile Division of the Clinton Laboratories (known now as Oak Ridge National Laboratory) in 1947. Professor Dr. Rudolf Schulten in Germany also played a role in development during the 1950s. The Peach Bottom reactor in the United States was the first HTGR to produce electricity, and did so very successfully, with operation from 1966 through 1974 as a technology demonstrator. Fort St. Vrain Generating Station was one example of this design that operated as an HTGR from 1979 to 1989; though the reactor was beset by some problems which led to its decommissioning due to economic factors, it served as proof of the HTGR concept in the United States (though no new commercial HTGRs have been developed there since). HTGRs have also existed in the United Kingdom (the Dragon reactor) and Germany (AVR reactor and THTR-300), and currently exist in Japan (the HTTR using prismatic fuel with 30 MW_{th} of capacity) and China (the HTR-10, a pebble-bed design with 10 MW_e of generation). Two full-scale pebble-bed HTGRs HTR-PM, each with 100 – 195 MW_e of electrical production capacity are under construction in China as of November 2009, and are promoted in several countries by reactor designers.

Nuclear Reactor Design

Neutron Moderator

The neutron moderator is graphite, although whether the reactor core is configured in graphite prismatic blocks or in graphite pebbles depends on the HTGR design.

Nuclear Fuel

The fuel used in HTGRs is coated fuel particles, such as TRISO fuel particles. Coated fuel particles have fuel kernels, usually made of uranium dioxide, however, uranium carbide or uranium oxycarbide are also possibilities. Uranium oxycarbide combines uranium carbide with the uranium dioxide to reduce the oxygen stoichiometry. Less oxygen may lower the internal pressure in the TRISO particles caused by the formation of carbon monoxide, due to the oxidization of the porous carbon layer in the particle. The TRISO particles are either dispersed in a pebble for the pebble bed design or molded into compacts/rods that are then inserted into the hexagonal graphite blocks. The QUADRISO fuel concept conceived at Argonne National Laboratory has been used to better manage the excess of reactivity.

Coolant

Helium

Helium has been the coolant used in most HTGRs to date, and the peak temperature and power depend on the reactor design. Helium is an inert gas, so it will generally not chemically react with any material. Additionally, exposing helium to neutron radiation does not make it radioactive, unlike most other possible coolants.

Molten Salt

The molten salt cooled variant, the LS-VHTR, similar to the advanced high-temperature reactor (AHTR) design, uses a liquid fluoride salt for cooling in a pebble core. It shares many features with a standard VHTR design, but uses molten salt as a coolant instead of helium. The pebble fuel floats in the salt, and thus pebbles are injected into the coolant flow to be carried to the bottom of the pebble bed, and are removed from the top of the bed for recirculation. The LS-VHTR has many attractive features, including: the ability to work at high temperatures (the boiling point of most molten salts being considered are > 1,400 °C), low-pressure operation, high power density, better electric conversion efficiency than a helium-cooled VHTR operating at similar conditions, passive safety systems, and better retention of fission products in case an accident occurred.

Control

In the prismatic designs, control rods are inserted in holes cut in the graphite blocks that make up the core. The VHTR will be controlled like current PBMR designs if it utilizes a pebble bed core, the control rods will be inserted in the surrounding graphite reflector. Control can also be attained by adding pebbles containing neutron absorbers.

Materials Challenges

The high-temperature, high-neutron dose, and, if using a molten salt coolant, the corrosive environment,(p46) of the VHTR require materials that exceed the limitations of current nuclear reactors. In a study of Generation IV reactors in general (of which there are numerous designs, including the VHTR), Murty and Charit suggest that materials that have high dimensional stability, either with or without stress, maintain their tensile strength, ductility, creep resistance, etc. after aging, and are corrosion resistant are primary candidates for use in VHTRs. Some materials suggested include nickel-base superalloys, silicon carbide, specific grades of graphite, high-chromium steels, and refractory alloys. Further research is being conducted at US national laboratories as to which specific issues must be addressed in the Generation IV VHTR prior to construction.

Safety Features and Other Benefits

The design takes advantage of the inherent safety characteristics of a helium-cooled, graphite-moderated core with specific design optimizations. The graphite has large thermal inertia and the helium coolant is single phase, inert, and has no reactivity effects. The core is composed of graphite, has a high heat capacity and structural stability even at high temperatures. The fuel is coated uranium-oxycarbide which permits high burn-up (approaching 200 GWd/t) and retains fission products. The high average core-exit temperature of the VHTR (1,000 °C) permits emissions-free production of process heat. Reactor is designed for 60 years of service.

Sodium-cooled Fast Reactor

The sodium-cooled fast reactor (SFR) is a Generation IV reactor project to design an advanced fast neutron reactor.

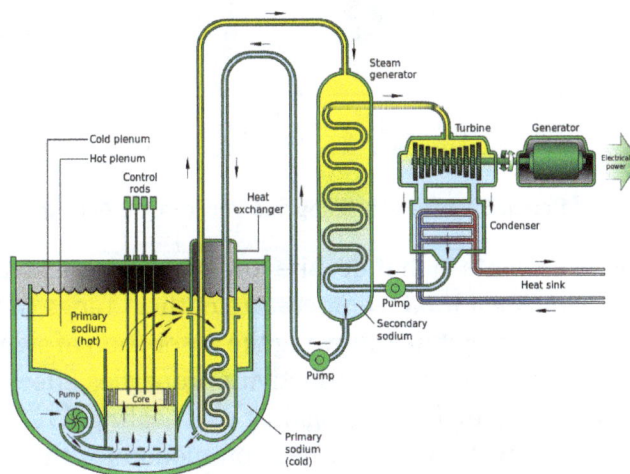

Pool type sodium-cooled fast reactor (SFR)

It builds on two closely related existing projects, the LMFBR and the Integral Fast Reactor, with the objective of producing a fast-spectrum, sodium-cooled reactor.

The reactors are intended for use in nuclear power plants to produce nuclear power from nuclear fuel.

Fuel Cycle

The nuclear fuel cycle employs a full actinide recycle with two major options: One is an intermediate-size (150–600 MWe) sodium-cooled reactor with uranium-plutonium-minor-actinide-zirconium metal alloy fuel, supported by a fuel cycle based on pyrometallurgical reprocessing in facilities integrated with the reactor. The second is a medium to large (500–1,500 MWe) sodium-cooled reactor with mixed uranium-plutonium oxide fuel, supported by a fuel cycle based upon advanced aqueous processing at a central location serving a number of reactors. The outlet temperature is approximately 510–550 degrees Celsius for both.

Sodium as A Coolant

Liquid metallic sodium may be used as the sole coolant, carrying heat from the core. Sodium has only one stable isotope, sodium-23. Sodium-23 is a very weak absorber of neutrons. When it does absorb a neutron it produces sodium-24, which has a half-life of 15 hours and decays in to magnesium-24.

Advantages

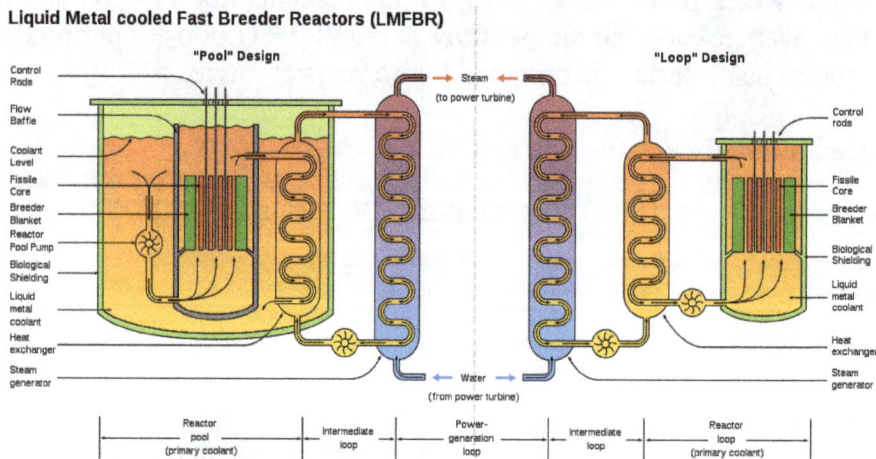

Schematic diagram showing the difference between the Loop and Pool designs of a liquid metal fast breeder reactor

An advantage of liquid metal coolants is that despite low specific heat, sodium melts at 371K and boils / vaporizes at 1156K, allowing a total "temperature outlier" range of 785K of heat variation between solid / frozen and gas / vapor states allowing the absorption of significant heat, less safety margins, in liquid phase. The high thermal conductivity properties effectively create a reservoir of heat capacity which provides thermal inertia against overheating. Water is difficult to use as a coolant for a fast reactor because water acts as a neutron moderator that slows the fast neutrons into thermal neutrons. Unlike liquid sodium, water has a higher specific heat, with a smaller liquid range of just 100K between ice and gas at normal, sea-level atmospheric pressure conditions. While it may be possible to use supercritical water as a coolant in a fast reactor, this would require a very high pressure. In contrast, sodium atoms are much heavier than both the oxygen and hydrogen atoms found in water, and therefore the neutrons lose less energy in collisions with sodium atoms. Sodium also need not be pressurized since its boiling point is much higher than the reactor's operating temperature, and sodium does not corrode steel reactor parts. The high temperatures reached by the coolant (up to 1156K for pure molten sodium, less all Generation IV margins of alarm call safety) permit a higher thermodynamic efficiency than in water cooled reactors. The molten sodium, being electrically conductive, can be pumped by electromagnetic pumps.

Disadvantages

A disadvantage of sodium is its chemical reactivity, which requires special precautions to prevent and suppress fires. If sodium comes into contact with water it explodes, and it burns when in contact with air. This was the case at the Monju Nuclear Power Plant in a 1995 accident. In addition, neutrons cause it to become radioactive; however, activated sodium has a half-life of only 15 hours.

Design Goals

The operating temperature should not exceed the melting temperature of the fuel. Fuel-to-cladding chemical interaction (FCCI) has to be designed against. FCCI is eutectic melting between the fuel and the cladding; uranium, plutonium, and lanthanum (a fission product) inter-diffuse with the iron of the cladding. The alloy that forms has a low eutectic melting temperature. FCCI causes the cladding to reduce in strength and could eventually rupture. The amount of transuranic transmutation is limited by the production of plutonium from uranium. A design work-around has been proposed to have an inert matrix. Magnesium oxide has been proposed as the inert matrix. Magnesium oxide has an entire order of magnitude smaller probability of interacting with neutrons (thermal and fast) than elements like iron.

Actinides and fission products by half-life								
Actinides by decay chain				Half-life range (y)		Fission products of ^{235}U by yield		
$4n$	$4n+1$	$4n+2$	$4n+3$			4.5–7%	0.04–1.25%	<0.001%
^{228}Ra$^{№}$				4–6	†		^{155}Eu^{b}	
^{244}Cm^{f}	^{241}Pu^{f}	^{250}Cf	^{227}Ac$^{№}$	10–29		^{90}Sr	^{85}Kr	^{113m}Cd^{b}
^{232}U^{f}		^{238}Pu$^{f№}$	^{243}Cm^{f}	29–97		^{137}Cs	^{151}Sm^{b}	^{121m}Sn
^{248}Bk	^{249}Cf^{f}	^{242m}Am^{f}		141–351				
	^{241}Am^{f}		^{251}Cf^{f}	430–900				
		^{226}Ra$^{№}$	^{247}Bk	1.3 k – 1.6 k		No fission products have a half-life in the range of 100–210 k years …		
^{240}Pu$^{f№}$	^{229}Th$^{№}$	^{246}Cm^{f}	^{243}Am^{f}	4.7 k – 7.4 k				
	^{245}Cm^{f}	^{250}Cm		8.3 k – 8.5 k				
			^{239}Pu$^{f№}$	24.1 k				
		^{230}Th$^{№}$	^{231}Pa$^{№}$	32 k – 76 k				
^{236}Np^{f}	^{233}U$^{f№}$	^{234}U$^{№}$		150 k – 250 k	‡	^{99}Tc$^{□}$	^{126}Sn	
^{248}Cm		^{242}Pu^{f}		327 k – 375 k			^{79}Se$^{□}$	
				1.53 M		^{93}Zr		
	^{237}Np$^{f№}$			2.1 M – 6.5 M		^{135}Cs$^{□}$	^{107}Pd	
^{236}U$^{№}$			^{247}Cm^{f}	15 M – 24 M			^{129}I$^{□}$	
^{244}Pu$^{№}$				80 M		… nor beyond 15.7 M years		
^{232}Th$^{№}$		^{238}U$^{№}$	^{235}U$^{f№}$	0.7 G – 14.1 G				

Legend for superscript symbols
⊠ has thermal neutron capture cross section in the range of 8–50 barns
f fissile
m metastable isomer
№ naturally occurring radioactive material (NORM)
þ neutron poison (thermal neutron capture cross section greater than 3k barns)
† range 4–97 y: Medium-lived fission product
‡ over 200,000 y: Long-lived fission product

The SFR is designed for management of high-level wastes and, in particular, management of plutonium and other actinides. Important safety features of the system include a long thermal response time, a large margin to coolant boiling, a primary system that operates near atmospheric pressure, and intermediate sodium system between the radioactive sodium in the primary system and the water and steam in the power plant. With innovations to reduce capital cost, such as making a modular design, removing a primary loop, integrating the pump and intermediate heat exchanger, or simply find better materials for construction, the SFR can be a viable technology for electricity generation.

The SFR's fast spectrum also makes it possible to use available fissile and fertile materials (including depleted uranium) considerably more efficiently than thermal spectrum reactors with once-through fuel cycles.

Reactors

Sodium-cooled reactors have included:

- BN-350 reactor, Russia
- BN-600 reactor, Russia (operational for electricity production as of 2015, 560MW). Pool design.
- BN-800 reactor, Russia (first criticality achieved 27.06.2014, grid connection 10.12.2015)
- China Experimental Fast Reactor (operational as of 2013)
- Clinch River Breeder Reactor Project, United States
- Dounreay Prototype Fast Reactor, United Kingdom
- Fermi 1, United States
- Experimental Breeder Reactor I, United States
- Experimental Breeder Reactor II, United States. Pool design.
- Fast Breeder Test Reactor, India
- Jōyō, Japan (MK-III (140-150 MWt) 2003–2007)
- Monju Nuclear Power Plant, Japan (as of March 2015 awaiting permission to restart). Loop design.
- Phénix, France. Pool design.
- Prototype Fast Breeder Reactor, India (under construction, reported criticality in 2014)
- S1G and S2G, United States Navy
- SNR-300, Germany
- Sodium Reactor Experiment, United States
- Superphénix, France
- Rapsodie, France

Most of these were experimental plants, which are no longer operational

Related:

- Fast Flux Test Facility, United States, a sodium-cooled fast neutron reactor

Supercritical Water Reactor

The supercritical water reactor (SCWR) is a concept Generation IV reactor, mostly designed as light water reactor (LWR) that operates at supercritical pressure (i.e. greater than 22.1 MPa). The term *critical* in this context refers to the critical point of water, and must not be confused with the concept of criticality of the nuclear reactor.

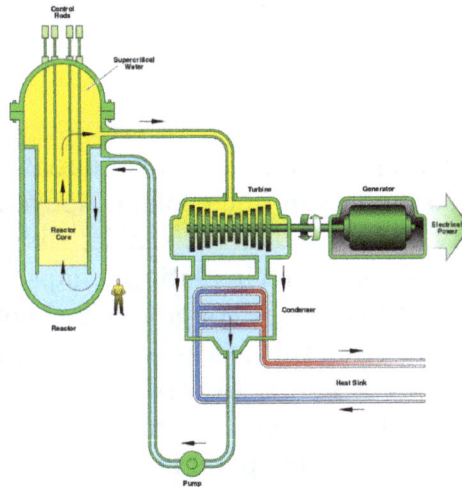

Supercritical water reactor scheme.

The water heated in the reactor core becomes a supercritical fluid above the critical temperature of 374 °C, transitioning from a fluid more resembling liquid water to a fluid more resembling saturated steam (which can be used in a steam turbine), without going through the distinct phase transition of boiling.

In contrast, the well-established pressurized water reactors (PWR) have a primary cooling loop of liquid water at a subcritical pressure, transporting heat from the reactor core to a secondary cooling loop, where the steam for driving the turbines is produced in a boiler (called the steam generator). Boiling water reactors (BWR) operate at even lower pressures, with the boiling process to generate the steam happening in the reactor core.

The supercritical steam generator is a proven technology. The development of SCWR systems is considered a promising advancement for nuclear power plants because of its high thermal efficiency (~45 % vs. ~33 % for current LWRs) and simpler design. As of 2012 the concept was being investigated by 32 organizations in 13 countries.

History

The super-heated steam cooled reactors operating at subcritical-pressure were experimented with in both Soviet Union and in the United States as early as the 1950s and 1960s such as Beloyarsk Nuclear Power Station, Pathfinder and Bonus of GE's Operation Sunrise program. These are not SCWRs. SCWRs were developed from the 1990s onwards. Both a LWR-type SCWR with a reactor pressure vessel and a CANDU-type SCWR with pressure tubes are being developed.

A 2010 book includes conceptual design and analysis methods such as core design, plant system, plant dynamics and control, plant startup and stability, safety, fast reactor design etc.

A 2013 document saw the completion of a prototypical fueled loop test in 2015. A Fuel Qualification Test was completed in 2014.

A 2014 book saw reactor conceptual design of a thermal spectrum reactor (Super LWR) and a fast reactor (Super FR) and experimental results of thermal hydraulics, materials and material-coolant interactions.

Design

Moderator-coolant

The SCWR operates at supercritical pressure. The reactor outlet coolant is supercritical water. Light water is used as a neutron moderator and coolant. Above the critical point, steam and liquid become the same density and are indistinguishable, eliminating the need for pressurizers and steam generators (PWR), or jet/recirculation pumps, steam separators and dryers (BWR). Also by avoiding boiling, SCWR does not generate chaotic voids (bubbles) with less density and moderating effect. In a LWR this can affect heat transfer and water flow, and the feedback can make the reactor power harder to predict and control. SCWR's simplification should reduce construction costs and improve reliability and safety. The neutron spectrum will be only partly moderated, perhaps to the point of being a fast neutron reactor. This is because the supercritical water has a lower density and moderating effect than liquid water, but is better at heat transfer, so less is needed. In some designs with a faster neutron spectrum the water is a reflector outside the core, or else only part of the core is moderated. A fast neutron spectrum has three main advantages:

- A higher power density, generating more power for the same size of reactor

- A conversion ratio of greater than 1, which makes breeder reactors possible. This allows for the efficient use of Uranium-238 (which makes up over 99% of natural uranium).

- The fast neutrons split actinides, while long-lived fission products can be transmuted with excess neutrons

Fuel

The fuel will resemble traditional LWR fuel, likely with channelized fuel assemblies like the BWR to reduce the risk of hotspots caused by local pressure/temperature variations. The enrichment of the fuel will have to be higher to compensate for the neutron absorption by the cladding, which can't be made from the zirconium customary in LWRs, as zirconium would corrode rapidly. Stainless steel or nickel alloys may be used. The fuel rods must withstand the corrosive supercritical environment, as well as a power surge in case of an accident. There are four failure modes considered during an accident: brittle failure, buckling collapse, overpressure damage and creep failure. To reduce corrosion, hydrogen can be added to the water.

At least one concept uses high temperature gas cooled reactor fuel particles, BISO.

This uses corrosion resistant silicon carbide coatings on uranium fuel particles, solving the challenge of the cladding using an innovative yet proven fuel.

Control

SCWRs would likely have control rods inserted through the top, as is done in PWRs.

Material

The conditions inside an SCWR are harsher than those in LWRs, LMFBRs and supercritical fossil fuel plants (with which much experience has been gained, though this does not include the combination of harsh environment and intense neutron radiation). SCWRs need a higher standard of core materials (especially fuel cladding) than either of these. In addition, some elements become very radioactive from absorbing neutrons, e.g. cobalt-59 captures neutrons to become cobalt-60, a strong gamma emitter, so cobalt-containing alloys are unsuitable for reactors. R&D focuses on:

- The chemistry of supercritical water under radiation (preventing stress corrosion cracking, and maintaining corrosion resistance under neutron radiation and high temperatures)

- Dimensional and microstructural stability (preventing embrittlement, retaining strength and creep resistance also under radiation and high temperatures)

- Materials that both resist the harsh conditions and do not absorb too many neutrons, which affects fuel economy

Advantages

- Supercritical water has excellent heat transfer properties allowing a high power density, a small core, and a small containment structure.

- The use of a supercritical Rankine cycle with its typically higher temperatures improves efficiency (would be ~45 % versus ~33 % of current PWR/BWRs).

- This higher efficiency would lead to better fuel economy and a lighter fuel load, lessening residual (decay) heat.

- SCWR is typically designed as a direct-cycle, whereby steam or hot supercritical water from the core is used directly in a steam turbine. This makes the design simple. As a BWR is simpler than a PWR, a SCWR is a lot simpler and more compact than a less-efficient BWR having the same electrical output. There are no steam separators, steam dryers, internal recirculation pumps, or recirculation flow inside the pressure vessel. The design is a once-through, direct-cycle, the simplest type of cycle possible. The stored thermal and radiologic energy in the smaller core and its (primary) cooling circuit would also be less than that of either a BWR's or a PWR's.

- Water is liquid at room temperature, cheap, non-toxic and transparent, simplifying inspection and repair (compared to liquid metal cooled reactors).

- A fast SCWR could be a breeder reactor, like the proposed Clean And Environmentally Safe Advanced Reactor, and could burn the long-lived actinide isotopes.

- A heavy-water SCWR could breed fuel from thorium (4x more abundant than uranium), with increased proliferation resistance over plutonium breeders.

Disadvantages

- Lower water inventory (due to compact primary loop) means less heat capacity to buffer transients and accidents (e.g. loss of feedwater flow or large break loss-of-coolant accident) resulting in accident and transient temperatures that are too high for conventional metallic cladding.

- Higher pressure combined with higher temperature and also a higher temperature rise across the core (compared to PWR/BWRs) result in increased mechanical and thermal stresses on vessel materials that are difficult to solve. A pressure-tube design, where the core is divided up into smaller tubes for each fuel channel, has potentially fewer issues here, as smaller diameter tubing can be much thinner than massive single pressure vessels, and the tube can be insulated on the inside with inert ceramic insulation so it can operate at low (calandria water) temperature.

The coolant greatly reduces its density at the end of the core, resulting in a need to place extra moderator there. Most designs use an internal calandria where part of the feedwater flow is guided through top tubes through the core, that provide the added moderation (feedwater) in that region. This has the added advantage of being able to cool the entire vessel wall with feedwater, but results in a complex and materially demanding (high temperature, high temperature differences, high radiation) internal calandria and plena arrangement. Again a pressure-tube design has potentially fewer issues, as most of the moderator is in the calandria at low temperature and pressure, reducing the coolant density effect on moderation, and the actual pressure tube can be kept cool by the calandria water.

- Extensive material development and research on supercritical water chemistry under radiation is needed

- Special start-up procedures needed to avoid instability before the water reaches supercritical conditions

- A fast SCWR needs a relatively complex reactor core to have a negative void coefficient

Gas-cooled Fast Reactor

The gas-cooled fast reactor (GFR) system is a nuclear reactor design which is currently in development. Classed as a Generation IV reactor, it features a fast-neutron spectrum and closed fuel cycle for efficient conversion of fertile uranium and management of actinides. The reference reactor design is a helium-cooled system operating with an outlet temperature of 850 °C using a direct Brayton closed-cycle gas turbine for high thermal efficiency. Several fuel forms are being considered for their potential to operate at very high temperatures and to ensure an excellent retention of fission products: composite ceramic fuel, advanced fuel particles, or ceramic clad elements of actinide compounds. Core configurations are being considered based on pin- or plate-based fuel assemblies or prismatic blocks, which allows for better coolant circulation than traditional fuel assemblies.

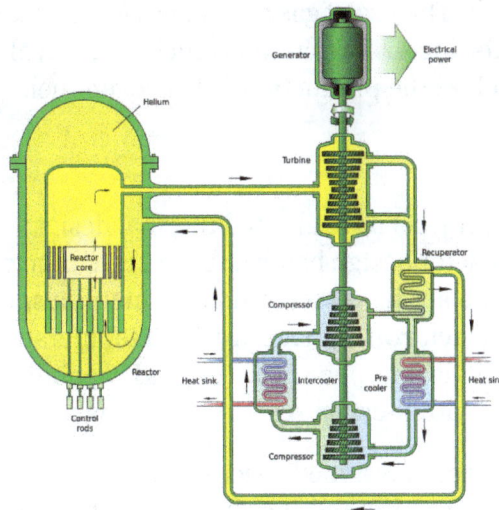

Gas-cooled Fast Reactor scheme.

The reactors are intended for use in nuclear power plants to produce electricity, while at the same time producing (breeding) new nuclear fuel.

Nuclear Reactor Design

Fast reactors were originally designed to be primarily breeder reactors. This was because of a view at the time of their conception that there was an imminent shortage of uranium fuel for existing reactors. The projected increase in uranium price did not materialize, but if uranium demand increases in the future, then there may be renewed interest in fast reactors.

The GFR base design is a fast reactor, but in other ways similar to a high temperature gas-cooled reactor. It differs from the HTGR design in that the core has a higher fissile fuel content as well as a non-fissile, fertile, breeding component, and of course there is no neutron moderator. Due to the higher fissile fuel content, the design has a higher power density than the HTGR.

Fuel

In a GFR reactor design, the unit operates on fast neutrons, no moderator is needed to slow neutrons down. This means that, apart from nuclear fuel such as uranium, other fuels can be used. The most common is thorium, which absorbs a fast neutron and decays into Uranium 233. This means GFR designs have breeding properties—they can use fuel that is unsuitable in light water reactor designs and breed fuel. Because of these properties, once the initial loading of fuel has been applied into the reactor, the unit can go years without needing fuel. If these reactors are used for breeding, it is economical to remove the fuel and separate the generated fuel for future use.

Coolant

The gas used can be many different types, including carbon dioxide or helium. It must be composed of elements with low neutron capture cross sections to prevent positive void coefficient and induced radioactivity. The use of gas also removes the possibility of phase transition—induced explosions, such as when the water in a water-cooled reactor (PWR or BWR) flashes to steam upon overheating or depressurization. The use of gas also allows for higher operating temperatures than are possible with other coolants, increasing thermal efficiency, and allowing other non-mechanical applications of the energy, such as the production of hydrogen fuel.

Research History

Past pilot and demonstration projects have all used thermal designs with graphite moderators. As such, no true gas-cooled fast reactor design has ever been brought to criticality. The main challenges that have yet to be overcome are in-vessel structural materials, both in-core and out-of-core, that will have to withstand fast-neutron damage and high temperatures, (up to 1600 °C). Another problem is the low thermal inertia and poor heat removal capability at low helium pressures, although these issues are shared with thermal reactors which have been constructed.

Gas-cooled projects include decommissioned reactors such as the Dragon reactor, built and operated in the United Kingdom, the AVR and the THTR-300, built and operated in Germany, and Peach Bottom and Fort St. Vrain, built and operated in the United States. Ongoing demonstrations include the HTTR in Japan, which reached full power (30 MWth) using fuel compacts inserted in prismatic blocks in 1999, and the HTR-10 in China, which may reach 10 MWth in 2002 using pebble fuel. A 400 MWth pebble bed modular reactor demonstration plant was designed by PBMR Pty for deployment in South Africa but withdrawn in 2010, and a consortium of Russian institutes is designing a 600 MWth GT-MHR (prismatic block reactor) in cooperation with General Atomics. In 2010, General Atomics announced the Energy Multiplier Module reactor design, an advanced version of the GT-MHR.

The Chinese are, as of 2015, building a commercial helium gas-cooled pebble-bed reactor: HTR-PM.

Lead-cooled Fast Reactor

Molten lead or lead-bismuth eutectic can be used as the primary coolant in a nuclear reactor, because lead and bismuth have low neutron absorption and relatively low melting points. Neutrons are slowed less by interaction with these heavy nuclei, (thus not being neutron moderators) and therefore help make this type of reactor a fast-neutron reactor. The coolant does however serve as a neutron reflector, returning some escaping neutrons to the core.

Few have been constructed, including some Soviet nuclear submarine reactors in the 1970s, but a number of proposed new nuclear reactor designs are lead-cooled. Some designs are claimed to be able to circulate the primary coolant via convection without requiring pumps, at least in emergency shutdown conditions.

Generation IV Reactor Design

The Gen IV lead-cooled fast reactor is a nuclear reactor that features a fast neutron spectrum, molten lead or lead-bismuth eutectic coolant. Options include a range of plant ratings, including a number of 50 to 150 MWe (megawatts electric) units featuring long-life, pre-manufactured cores. Plans include modular arrangements rated at 300 to 400 MWe, and a large monolithic plant rated at 1,200 MWe. The fuel is metal or nitride-based containing fertile uranium and transuranics. A smaller capacity LFR such as SSTAR can be cooled by natural convection, larger proposals such as ELSY use forced circulation in normal power operation, but with natural circulation emergency cooling. The reactor outlet coolant temperature is typically in the range of 500 to 600 °C, possibly ranging over 800 °C with advanced materials for later designs. Temperatures higher than 800 °C are high enough to support thermochemical production of hydrogen.

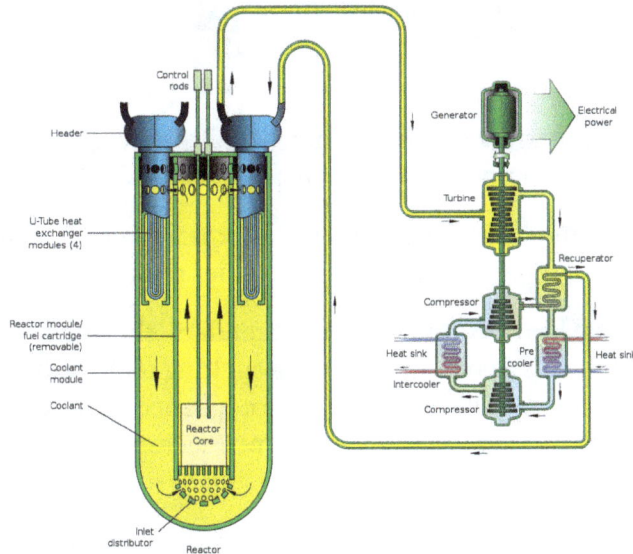

Lead cooled fast reactor scheme.

Modular Nuclear Reactors

The LFR battery is a small turnkey-type power plant using cassette cores running on a closed fuel cycle with 15 to 20 years' refuelling interval, or entirely replaceable reactor modules. It is designed for generation of electricity on small grids (and other resources, including hydrogen and potable water).

Advantages

- Instead of refueling, the whole core can be replaced after many years of operation. Such a reactor is suitable for countries that do not plan to build their own nuclear infrastructure.

- As no electricity is required for the cooling after shutdown, this design has the potential to be safer than a water-cooled reactor.

- Liquid lead-bismuth systems can't cause an explosion and quickly solidify in case of a leak, further improving safety.

- Lead is very dense, and therefore a good shield against gamma rays.

- Lead's nuclear properties allow it to prevent a positive void coefficient, which is difficult to prevent in large sodium fast reactor cores.

- The operating pressure is very low and lead has an extremely high boiling point of 1750 degrees Celsius, which is over 1100 degrees Celsius higher than the peak coolant operating temperature. This makes significant reactor pressurization by overheating virtually impossible.

- Lead does not react significantly with water or air, unlike sodium which burns readily in air and can explode in contact with water. This allows easier, cheaper and safer containment and heat exchanger/steam generator design.

Disadvantages

- Lead and lead-bismuth are very dense, increasing the weight of the system therefore requiring more structural support and seismic protection which increases building cost.

- While lead is cheap and abundant, bismuth is expensive and quite rare. A lead-bismuth reactor will require hundreds to thousands of tonnes of bismuth depending on reactor size.

- Solidification of the lead-bismuth solution renders the reactor inoperable. However, lead-bismuth eutectic has a comparatively low melting temperature of 123.5 °C (254.3 °F), making desolidification a relatively easily accomplished task. Lead has a higher melting point of 327.5 °C, but is often used as a pool type reactor where the large bulk of lead does not easily freeze.

- By leaking and solidifying, the coolant may damage the equipment.

- Lead-bismuth produces a considerable amount of polonium, a highly radioactive and quite mobile element. This can complicate maintenance and pose a plant contamination problem. Lead produces orders of magnitudes less polonium, and so has an advantage over lead-bismuth in this regard.

Implementation

Belgium

The MYRRHA project (for Multi-purpose hYbrid Research Reactor for High-tech Applications) is a first-of-a-kind design of a nuclear reactor coupled to a proton accelerator (so-called Accelerator-driven system (ADS)). This will be a 'Lead-cooled fast reactor' with two possible configurations: sub-critical or critical. The project is managed by SCK•CEN, the Belgium center for nuclear energy. It will be built based on a first successful demonstrator: GUINEVERE. The project entered a new phase of development in 2013 when a contract for the front-end engineering design was awarded to a consortium led by Areva. MYRRHA enjoys international recognition and was listed in December 2010 by the European Commission as one of 50 projects for maintaining European leadership in high-tech research in the next 20 years.

Russia/USSR

Two types of LFR reactor were used in Soviet Alfa class submarines of the 1970s. The OK-550 and BM-40A designs were both capable of producing 155MWt. They were significantly lighter than typical water-cooled reactors and had an advantage of being capable to quickly switch between maximum power and minimum noise operation modes.

A joint venture called AKME Engineering was announced on 25 December 2009 between Rosatom and En+ Group, to develop a commercial lead-bismuth reactor. The SVBR-100 ('Svintsovo-Vismutovyi Bystryi Reaktor' - lead-bismuth fast reactor) is based on the Alfa designs and will produce 100MWe electricity from gross thermal power of 280MWt, about twice that of the submarine reactors. They can also be used in groups of up to 16 if more power is required. The coolant increases from 345 °C (653 °F) to 495 °C (923 °F) as it goes through the core. Uranium oxide enriched to 16.5% U-235 could be used as fuel, and refuelling would be required every 7–8 years. A prototype is planned for 2017.

Another two lead cooled reactors are developed by Russians: BREST-300 and BREST-1200 The BREST-300 design was completed in September 2014.

WNA mentions Russia role on boosting other countries interest in this field:

In 1998 Russia declassified a lot of research information derived from its experience with submarine reactors, and US interest in using Pb or Pb-Bi for small reactors has increased subsequently.

United States

According to Nuclear Engineering International, the initial design of the Hyperion Power Module will be of this type, using uranium nitride fuel encased in HT-9 tubes, using a quartz reflector, and lead-bismuth eutectic as coolant.

Germany

The *dual fluid reactor* (DFR) is a German project combining the advantages of the molten salt reactor with the ones of the liquid metal cooled reactor. As a breeder reactor the DFR can burn both natural uranium and thorium, as well as recycle nuclear waste. Due to the high thermal conductivity of the molten metal, the DFR is a inherently safe reactor (the decay heat can be removed passively).

Molten Salt Reactor

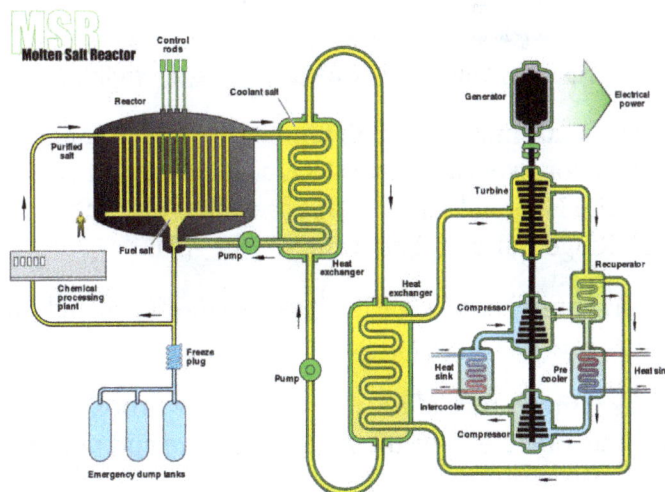

Example of a molten salt reactor scheme

A molten salt reactor (MSR) is a class of generation IV nuclear fission reactor in which the primary nuclear reactor coolant, or even the fuel itself, is a molten salt mixture. MSRs can run at higher temperatures than water-cooled reactors for a higher thermodynamic efficiency, while staying at low vapour pressure.

The nuclear fuel may be solid or dissolved in the coolant. In many designs the nuclear fuel dissolved in the coolant is uranium tetrafluoride (UF_4). The fluid becomes critical in a graphite core that serves as the moderator. Some solid-fuel designs propose ceramic fuel dispersed in a graphite matrix, with the molten salt providing low pressure, high temperature cooling. The salts are much more efficient than compressed helium (another potential coolant in Generation IV reactor designs) at removing heat from the core, reducing the need for pumping and piping and reducing the core size.

The concept was established in the 1950s. The early Aircraft Reactor Experiment (1954) was primarily motivated by the small size that the design could provide, while the Molten-Salt Reactor Experiment (1965–1969) was a prototype for a thorium fuel cycle breeder reactor nuclear power plant. The increased research into Generation IV reactor designs included a renewed interest in the technology.

History

Extensive research into molten salt reactors started with the U.S. aircraft reactor experiment (ARE) in support of the U.S. Aircraft Nuclear Propulsion program. The ARE was a 2.5 MW_{th} nuclear reactor experiment designed to attain a high energy density for use as an engine in a nuclear-powered bomber.

Aircraft Reactor Experiment building at ORNL. It was later retrofitted for the MSRE.

The project included experiments, including high temperature reactor and engine tests collectively called the Heat Transfer Reactor Experiments: HTRE-1, HTRE-2 and HTRE-3 at the National Reactor Test Station (now Idaho National Laboratory) as well as an experimental high-temperature molten salt reactor at Oak Ridge National Laboratory – the ARE.

The ARE used molten fluoride salt NaF-ZrF_4-UF_4 (53-41-6 mol%) as fuel, moderated by beryllium oxide (BeO). Liquid sodium was a secondary coolant.

The experiment had a peak temperature of 860 °C. It produced 100 MWh over nine days in 1954. This experiment used Inconel 600 alloy for the metal structure and piping.

After ARE, another reactor was operated at the Critical Experiments Facility of the Oak Ridge National Laboratory in 1957. It was part of the circulating-fuel reactor program of the Pratt &

Whitney Aircraft Company (PWAC). This was called the PWAR-1, the Pratt and Whitney Aircraft Reactor-1. The experiment was run for only a few weeks and at essentially zero nuclear power, but it reached criticality. The operating temperature was held constant at approximately 675 °C (1,250 °F). The PWAR-1 used NaF-ZrF_4-UF_4 as the primary fuel and coolant, making it one of the three critical molten salt reactors ever built.

Molten-salt Reactor Experiment

Oak Ridge National Laboratory (ORNL) took the lead in researching the MSR through the 1960s. Much of their work culminated with the Molten-Salt Reactor Experiment (MSRE). The MSRE was a 7.4 MW_{th} test reactor simulating the neutronic "kernel" of a type of epithermal thorium molten salt breeder reactor called the liquid fluoride thorium reactor. The large (expensive) breeding blanket of thorium salt was omitted in favor of neutron measurements.

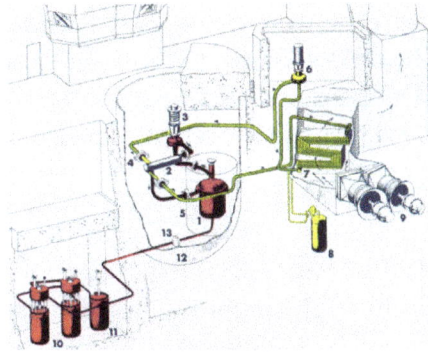

MSRE plant diagram

The MSRE was located at ORNL. Its piping, core vat and structural components were made from Hastelloy-N, moderated by pyrolytic graphite. It went critical in 1965 and ran for four years. The fuel for the MSRE was LiF-BeF_2-ZrF_4-UF_4 (65-29-5-1). The graphite core moderated it. Its secondary coolant was FLiBe ($2LiF$-BeF_2). It reached temperatures as high as 650 °C and operated for the equivalent of about 1.5 years of full power operation.

Oak Ridge National Laboratory Molten Salt Breeder Reactor

The culmination of the Oak Ridge National Laboratory research during the 1970–1976 timeframe resulted in a proposed molten salt breeder reactor (MSBR) design which would use LiF-BeF_2-ThF_4-UF_4 (72-16-12-0.4) as fuel. It was to be moderated by graphite with a 4-year replacement schedule. The secondary coolant was to be NaF-$NaBF_4$. Its peak operating temperature was to be 705 °C. Despite the success, the MSR program closed down in the early 1970s in favor of the liquid metal fast-breeder reactor (LMFBR), after which research stagnated in the United States. As of 2011, the ARE and the MSRE remained the only molten-salt reactors ever operated.

The MSBR project received funding until 1976. Inflation-adjusted to 1991 dollars, the project received $38.9 million from 1968 to 1976.

Officially, the program was cancelled because:

- The political and technical support for the program in the United States was too thin geographically. Within the United States, only in Oak Ridge, Tennessee, was the technology well understood.

- The MSR program was in competition with the fast breeder program at the time, which got an early start and had copious government development funds allocated to many parts of the United States. When the MSR development program had progressed far enough to justify an expanded program leading to commercial development, the AEC could not justify the diversion of substantial funds from the LMFBR to a competing program.

Oak Ridge National Laboratory Denatured Molten Salt Reactor (DMSR)

In 1980, the engineering technology division at Oak Ridge National Laboratory published a paper entitled "Conceptual Design Characteristics of a Denatured Molten-Salt Reactor with Once-Through Fueling." In it, the authors "examine the conceptual feasibility of a molten-salt power reactor fueled with denatured uranium-235 (i.e. with low-enriched uranium) and operated with a minimum of chemical processing." The main priority behind the design characteristics is proliferation resistance. Lessons learned from past projects and research at ORNL were considered. Although the DMSR can theoretically be fueled partially by thorium or plutonium, fueling solely with low enriched uranium (LEU) helps maximize proliferation resistance.

Another important goal of the DMSR was to minimize R&D and to maximize feasibility. The Generation IV international Forum (GIF) includes "salt processing" as a technology gap for molten salt reactors. The DMSR requires minimal chemical processing because it is a burner rather than a breeder. Both reactors built at ORNL were burner designs. In addition, the choices to use graphite for neutron moderation and enhanced Hastelloy-N for piping simplify the design and reduce R&D.

United Kingdom

The UK's Atomic Energy Research Establishment (AERE) were developing an alternative MSR design across its National Laboratories at Harwell, Culham, Risley and Winfrith. AERE opted to focus on a lead-cooled 2.5 GWe Molten Salt Fast Reactor (MSFR) concept using a chloride. They also researched the option of helium gas as an alternative coolant.

The UK MSFR would be fuelled by plutonium, a fuel considered to be 'free' by the program's research scientists, because of the UK's plutonium stockpile.

Despite their different designs, ORNL and AERE maintained contact during this period with information exchange and expert visits. Theoretical work on the concept was conducted between 1964 and 1966, while experimental work was ongoing between 1968 and 1973. The program received annual government funding of around £100,000-£200,000 (equivalent to £2m-£3m in 2005). This funding came to an end in 1974, partly due to the success of the Prototype Fast Reactor at Dounreay which was considered a priority for funding as it went critical in the same year.

AERE reports and findings from its MSR Program conducted in the 60's and 70's are available for public viewing at the UK National Archives in Kew, London.

Soviet Union

In the USSR, a molten-salt reactor research program was started in the second half of the 1970s at the Kurchatov Institute. It included theoretical and experimental studies, particularly the investigation of mechanical, corrosion and radiation properties of the molten salt container materials. The main findings supported the conclusion that there were no physical nor technological obstacles to the practical implementation of MSRs. A reduction in activity occurred after 1986 due to the Chernobyl accident, along with a general stagnation of nuclear power and the nuclear industry.

Twenty-first Century

Canada

Terrestrial Energy Inc. (TEI), a Canadian based company, is developing a DMSR design called the *Integral Molten Salt Reactor* (IMSR). The IMSR is designed to be deployable as a small modular reactor (SMR) and will be constructed in three configurations ranging from 80 to 600 MW. With high operating temperatures, the IMSR has applications in industrial heat markets as well as traditional power markets. The main design features include neutron moderation from graphite, fueling with low-enriched uranium and a compact and replaceable Core-unit. The latter feature permits the operational simplicity necessary for industrial deployment.

China

Under Jiang Mianheng's direction, China initiated a thorium molten-salt reactor research project. It was formally announced at the Chinese Academy of Sciences (CAS) annual conference in January 2011. A 100-MW demonstrator of the solid fuel version (*TMSR-SF*), based on pebble bed technology, was to be ready by 2024. A 10-MW pilot and a larger demonstrator of the liquid fuel (*TMSR-LF*) variant are targeted for 2024 and 2035 respectively.

Denmark

Seaborg Technologies, a company based in Denmark, is developing the core for a *Molten Salt Waste-burner* (MSW). The MSW is a high temperature, single salt, thermal MSR designed to go critical on a combination of thorium and nuclear waste from conventional nuclear reactors. The MSW design is modular. The reactor core is estimated to be replaced every 6–10 years. However, the fuel will not be replaced and will burn for the entire power plant lifetime. The first version of the Seaborg core is planned to produce 50 MW_{th} power and could consume approximately 1 ton (not considering natural decays) of transuranic waste over its 60 years power plant lifetime. After 60 years the ^{233}U concentration in the fuel salt is high enough to initiate a closed thorium fuel cycle in the next generation power plant.

France

The CNRS project *EVOL* (Evaluation and viability of liquid fuel fast reactor system) project, with the objective of proposing a design of the MSFR (Molten Salt Fast Reactor), released its final report in 2014. The various molten salt reactor projects like FHR, MOSART, MSFR, and TMSR have common themes in basic R&D areas, according to a 2014 paper giving an overview of the MSR in a GenV context. Another paper gives an overview of the MSFR. More resources are available in the MSFR bibliography.

India

Ratan Kumar Sinha, Chairman of Atomic Energy Commission of India, stated in 2013: "India is also investigating Molten Salt Reactor (MSR) technology. We have molten salt loops operational at BARC."

Japan

The Fuji Molten Salt Reactor is a 100 to 200 MW_e LFTR, using technology similar to the Oak Ridge project. A consortium including members from Japan, the U.S. and Russia are developing the project. The project would likely take 20 years to develop a full size reactor, but the project seems to lack funding.

United Kingdom

The Alvin Weinberg Foundation is a British non-profit organization founded in 2011, dedicated to raising awareness about the potential of thorium energy and LFTR. It was formally launched at the House of Lords on 8 September 2011. It is named after American nuclear physicist Alvin M. Weinberg, who pioneered thorium molten salt reactor research.

A study on MSRs completed in July 2015 by Energy Process Developments, funded by Innovate UK, summarizes MSR activity internationally. It looks at the feasibility of developing a pilot scale demonstration MSR in the UK. A review of potential UK sites is given along with an insight into the UK regulatory process for innovative reactor technology. The technical review of six MSR designs led to the selection of the Stable Salt Reactor, designed by Moltex Energy, as most suitable for UK implementation.

United States

Idaho National Laboratory designed a molten-salt-cooled, molten-salt-fuelled reactor with a prospective output of 1000 MW_e.

Kirk Sorensen, former NASA scientist and chief nuclear technologist at Teledyne Brown Engineering, is a long-time promoter of the thorium fuel cycle, coining the term liquid fluoride thorium reactor. In 2011, Sorensen founded Flibe Energy, a company aimed at developing 20–50 MW LFTR reactor designs to power military bases. (It is easier to approve novel military designs than civilian power station designs in today's US nuclear regulatory environment).

Transatomic Power was created by Ph.D. students from MIT including CEO Leslie Dewan and Mark Massie, and Russ Wilcox of E Ink. They are pursuing what they term a *Waste-Annihilating Molten Salt Reactor* (acronym WAMSR), intending to consume existing spent nuclear fuel. Transatomic received venture capital funding in early 2015.

In January 2016, the United States Department of Energy announced a $80m award fund to develop Generation IV reactor designs. One of the two beneficiaries, Southern Company will use the funding to develop a *Molten Chloride Fast Reactor* (MCFR), a type of MSR developed earlier by British scientists.

Variants

Liquid-salt Very-high-temperature Reactor

As of September 2010, research was continuing for reactors that utilize molten salts for coolant. Both the traditional molten-salt reactor and the very high temperature reactor (VHTR) were selected as potential designs for study under the Generation Four Initiative (GEN-IV). One version of the VHTR under study was the *liquid-salt very-high-temperature reactor* (LS-VHTR), also commonly called the *advanced high-temperature reactor* (AHTR).

It is essentially a standard VHTR design that uses liquid salt as a coolant in the primary loop, rather than a single helium loop. It relies on "TRISO" fuel dispersed in graphite. Early AHTR research focused on graphite would be in the form of graphite rods that would be inserted in hexagonal moderating graphite blocks, but current studies focus primarily on pebble-type fuel. The LS-VHTR has many attractive features, including the ability to work at very high temperatures (the boiling point of most molten salt candidates is >1400 °C); low-pressure cooling that can be used to more easily match hydrogen production facility conditions (most thermochemical cycles require temperatures in excess of 750 °C); better electric conversion efficiency than a helium-cooled VHTR operating at similar conditions; passive safety systems and better retention of fission products in the event of an accident. This concept is now referred to as "fluoride salt-cooled high-temperature reactor" (FHR).

Liquid Fluoride Thorium Reactor (LFTR)

Reactors containing molten thorium salt, called liquid fluoride thorium reactors (LFTR), would tap the abundant energy source of the thorium fuel cycle. Private companies from Japan, Russia, Australia and the United States, and the Chinese government, have expressed interest in developing this technology.

Advocates estimate that five hundred metric tons of thorium could supply all U.S. energy needs for one year. The U.S. Geological Survey estimates that the largest known U.S. thorium deposit, the Lemhi Pass district on the Montana-Idaho border, contains thorium reserves of 64,000 metric tons.

Molten-salt Fueling Options

The LFTR design was strongly supported by Alvin Weinberg, who patented the light-water reactor and was a director of the U.S.'s Oak Ridge National Laboratory. In 2016 Nobel prize winning physicist Carlo Rubbia, former Director General of CERN, claimed that one of the main reasons why research was cut is that thorium is difficult to turn into a nuclear weapon.

Thorium is not for tomorrow but unless you do any development, it will not get there. —*Dr Carlo Rubbia, Nobel Laureate and former Director General of CERN, January 2016*

Alternatives to thorium include enriched uranium-235 or fissile material from dismantled nuclear weapons.

Molten-salt-cooled Reactors

Molten-salt-cooled solid-fuel reactors are quite different from molten-salt-fueled reactors. They are called "molten salt reactor system" in the Generation IV proposal, also called *Molten Salt Converter Reactor* (MSCR). These reactors were additionally referred to as *advanced high-temperature reactors* (AHTRs), but since about 2010 the preferred DOE designation is *fluoride high-temperature reactors* (FHR).

The FHR concept cannot reprocess fuel easily and has fuel rods that need to be fabricated and validated, delaying deployment by up to twenty years from project inception. However, since it uses fabricated fuel, reactor manufacturers can still profit by selling fuel assemblies.

The FHR retains the safety and cost advantages of a low-pressure, high-temperature coolant, also shared by liquid metal cooled reactors. Notably, steam is not created in the core (as is present in BWRs), and no large, expensive steel pressure vessel (as required for PWRs). Since it can operate at high temperatures, the conversion of the heat to electricity can use an efficient, lightweight Brayton cycle gas turbine.

Much of the current research on FHRs is focused on small, compact heat exchangers that reduce molten salt volumes and associated costs.

Molten salts can be highly corrosive and corrosivity increases with temperature. For the primary cooling loop, a material is needed that can withstand corrosion at high temperatures and intense radiation. Experiments show that Hastelloy-N and similar alloys are suited to these tasks at operating temperatures up to about 700 °C. However, operating experience is limited. Still higher operating temperatures are desirable – at 850 °C thermochemical production of hydrogen becomes possible. Materials for this temperature range have not been validated, though carbon composites, molybdenum alloys (e.g. TZM), carbides, and refractory metal based or ODS alloys might be feasible.

Fused Salt Selection

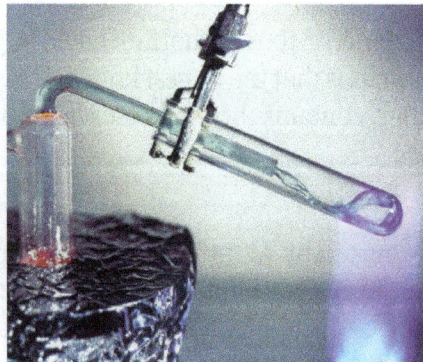

Molten FLiBe

The salt mixtures are chosen to make the reactor safer and more practical. Fluoride salts are favored, because fluorine has only one stable isotope (F-19), and does not easily become radioactive under neutron bombardment. Both of these make fluorine better than chlorine, which has two stable isotopes (Cl-35 and Cl-37), as well as a slow-decaying isotope between them which facilitates neutron absorption by Cl-35. Compared to chlorine and other halides, fluorine also absorbs fewer neutrons and slows ("moderates") neutrons better. Low-valence fluorides boil at high temperatures, though many pentafluorides and hexafluorides boil at low temperatures. They also must be very hot before they break down into their constituent elements. Such molten salts are "chemically stable" when maintained well below their boiling points.

On the other hand, some salts are so useful that isotope separation of the halide is worthwhile. Chlorides permit fast breeder reactors to be constructed using molten salts. Much less research has been done on reactor designs using chloride salts. Chlorine, unlike fluorine, must be purified to isolate the heavier stable isotope, chlorine-37, thus reducing production of sulfur tetrafluoride that occurs when chlorine-35 absorbs a neutron to become chlorine-36, then degrades by beta decay to sulfur-36.

Similarly, any lithium present in a salt mixture must be in the form of purified lithium-7, because lithium-6 effectively captures neutrons and produces tritium. Even if pure ^{7}Li is used, salts containing lithium will cause significant tritium production, comparable with heavy water reactors.

Reactor salts are usually close to eutectic mixtures to reduce their melting point. A low melting point simplifies melting the salt at startup and reduces the risk of the salt freezing as it is cooled in the heat exchanger.

Due to the high "redox window" of fused fluoride salts, the redox potential of the fused salt system can be changed. Fluorine-Lithium-Beryllium ("FLiBe") can be used with beryllium additions to lower the redox potential and almost eliminate corrosion. However, since beryllium is extremely toxic, special precautions must be engineered into the design to prevent its release into the environment. Many other salts can cause plumbing corrosion, especially if the reactor is hot enough to make highly reactive hydrogen.

To date, most research has focused on FLiBe, because lithium and beryllium are reasonably effective moderators and form a eutectic salt mixture with a lower melting point than each of the constituent salts. Beryllium also performs neutron doubling, improving the neutron economy. This process occurs when the beryllium nucleus re-emits two neutrons after absorbing a single neutron. For the fuel carrying salts, generally 1% or 2% (by mole) of UF_4 is added. Thorium and plutonium fluorides have also been used.

Comparison of the neutron capture and moderating efficiency of several materials. Red are Be-bearing, blue are ZrF_4-bearing and green are LiF-bearing salts.		
Material	**Total neutron capture relative to graphite (per unit volume)**	**Moderating ratio (Avg. 0.1 to 10 eV)**
Heavy water	0.2	11449
ZrH	~0.2	~0 if <0.14 eV, ~11449 if >0.14 eV
Light water	75	246
Graphite	1	863
Sodium	47	2
UCO	285	2

UO_2	3583	0.1
$2LiF–BeF_2$	8	60
$LiF–BeF_2–ZrF_4$ (64.5–30.5–5)	8	54
$NaF–BeF_2$ (57–43)	28	15
$LiF–NaF–BeF_2$ (31–31–38)	20	22
$LiF–ZrF_4$ (51–49)	9	29
$NaF–ZrF_4$ (59.5–40.5)	24	10
$LiF-NaF–ZrF_4$ (26–37–37)	20	13
$KF–ZrF_4$ (58–42)	67	3
$RbF–ZrF_4$ (58–42)	14	13
$LiF–KF$ (50–50)	97	2
$LiF–RbF$ (44–56)	19	9
$LiF–NaF–KF$ (46.5–11.5–42)	90	2
$LiF–NaF–RbF$ (42–6–52)	20	8

Fused Salt Purification

Techniques for preparing and handling molten salt were first developed at Oak Ridge National Lab. The purpose of salt purification was to eliminate oxides, sulfur and metal impurities. Oxides could result in the deposition of solid particles in reactor operation. Sulfur had to be removed because of its corrosive attack on nickel-based alloys at operational temperature. Structural metal such as chromium, nickel, and iron had to be removed for corrosion control.

A water content reduction purification stage using HF and helium sweep gas was specified to run at 400 °C. Oxide and sulfur contamination in the salt mixtures were removed using gas sparging of $HF – H_2$ mixture, with the salt heated to 600 °C.[p8] Structural metal contamination in the salt mixtures were removed using hydrogen gas sparging, at 700 °C.[p26] Solid ammonium hydrofluoride was proposed as a safer alternative for oxide removal.

Fused Salt Processing

The possibility of online processing can be an MSR advantage. Continuous processing would reduce the inventory of fission products, control corrosion and improve neutron economy by removing fission products with high neutron absorption cross-section, especially xenon. This makes the MSR particularly suited to the neutron-poor thorium fuel cycle. Online fuel processing can introduce risks of fuel processing accidents, which can trigger release of radio isotopes.

In some thorium breeding scenarios, the intermediate product protactinium-233 would be removed from the reactor and allowed to decay into highly pure uranium-233, an attractive bomb-making material. More modern designs propose to use a lower specific power or a separate large thorium breeding blanket. This dilutes the protactinium to such an extent that few protactinium atoms absorb a second neutron or, via a (n, 2n) reaction (in which an incident neutron is not absorbed but instead knocks a neutron out of the nucleus), generate uranium-232. Because U-232 has a short half-life and its decay chain contains hard gamma emitters, it makes the isotopic mix of uranium less attractive for bomb-making. This benefit would come with the added expense of a larger fissile inventory or a 2-fluid design with a large quantity of blanket salt.

The necessary fuel salt reprocessing technology has been demonstrated, but only at laboratory scale. A prerequisite to full-scale commercial reactor design is the R&D to engineer an economically competitive fuel salt cleaning system.

Fissile Fuel Reprocessing Issues

Reprocessing refers to the chemical separation of fissionable uranium and plutonium from spent nuclear fuel. The recovery of uranium or plutonium could increase the risk of nuclear proliferation. In the United States the regulatory regime has varied dramatically in different administrations.

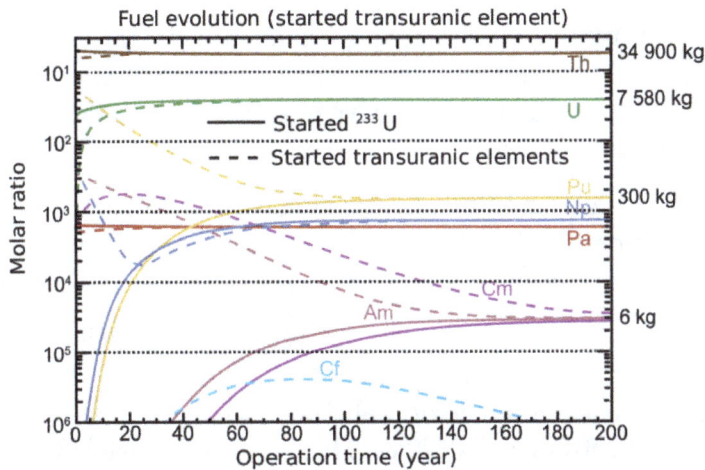

Changes in the composition of a MSR fast neutron (kg/GW)

In the original 1971 Molten Salt Breeder Reactor proposal, uranium reprocessing was scheduled every ten days as part of reactor operation.[p181] Subsequently a once-through fueling design was proposed that limited uranium reprocessing to every thirty years at the end of useful salt life.[p98] A mixture of uranium-238 was called for to make sure recovered uranium would not be weapons grade. This design is referred to as denatured molten salt reactor. If reprocessing were to be prohibited then the uranium would be disposed with other fission products.

Comparison to Light Water Reactors

MSRs, especially those with the fuel dissolved in the salt differ considerably from conventional reactors. Reactor core pressure can be low and the temperature much higher. In this respect an MSR is more similar to a liquid metal cooled reactor than to a conventional light water cooled reactor. MSRs are often planned as breeding reactors with a closed fuel cycle – as opposed to the once-through fuel currently used in U.S. nuclear reactors.

Safety concepts rely on a negative temperature coefficient of reactivity and a large possible temperature rise to limit reactivity excursions. As an additional method for shutdown, a separate, passively cooled container below the reactor can be included. In case of problems and for regular maintenance the fuel is drained from the reactor. This stops the nuclear reaction and acts as another second cooling system. Neutron-producing accelerators have been proposed for some super-safe subcritical experimental designs.

Cost estimates from the 1970s were slightly lower than for conventional light-water reactors.

The temperatures of some proposed designs are high enough to produce process heat for hydrogen production or other chemical reactions. Because of this, they are included in the GEN-IV roadmap for further study.

Advantages

MSR offers many potential advantages over current light water reactors:

- Inherently safe design (safety by passive components and the strong negative temperature coefficient of reactivity of some designs). In some designs, the fuel and the coolant are the same fluid, so a loss of coolant removes the reactor's fuel. Unlike steam, fluoride salts dissolve poorly in water, and do not form burnable hydrogen. Unlike steel and solid uranium oxide, molten salts are not damaged by the core's neutron bombardment.

- A low-pressure MSR lacks a LWR's high-pressure radioactive steam and therefore do not experience leaks of radioactive steam and cooling water, and the expensive containment, steel core vessel, piping and safety equipment needed to contain radioactive steam.

- MSRs make closed nuclear fuel cycles cheaper and more practical. If fully implemented, a closed nuclear fuel cycle reduces environmental impacts: The chemical separation makes long-lived actinides back into reactor fuel. The discharged wastes are mostly fission products (nuclear ashes) with short half-lives. This reduces the needed geologic containment to 300 years rather than the tens of thousands of years needed by a light-water reactor's spent nuclear fuel. It also permits society to use more-abundant nuclear fuels.

- The fuel's liquid phase might be pyroprocessed to separate fission products (nuclear ashes) from actinide fuels. This may have advantages over conventional reprocessing, though much development is still needed.

- Fuel rods are not required.

- In new solid-fueled reactor designs, the longest-lead item is the safety testing of fuel element designs. Fuel tests usually must cover several three-year refueling cycles. However, several molten salt fuels have already been validated.

- Some designs can "burn" problematic transuranic elements from traditional solid-fuel nuclear reactors.

- An MSR can react to load changes in less than 60 seconds (unlike "traditional" solid-fuel nuclear power plants that suffer from xenon poisoning).

- Molten salt reactors can run at high temperatures, yielding high production efficiency. This reduces the size, expense and environmental impacts of a power plant.

- MSRs can offer a high "specific power," that is high power at a low mass as demonstrated by the ARE. Simplified MSR power plants may be suitable for ships.

- A possibly good neutron economy makes the MSR attractive for the neutron poor thorium fuel cycle.

- LWR's (and most other solid-fuel reactors) have no fundamental "off switch", but once the initial criticality is overcome, an MSR is comparatively easy and fast turn to off by letting the freeze plug melt.

Disadvantages

- Little development compared to most Gen IV designs .

- Required onsite chemical plant to manage core mixture and remove fission products.

- Required regulatory changes to deal with radically different design features.

- MSR designs rely on nickel-based alloys to hold the molten salt. Alloys based on nickel and iron are prone to embrittlement under high neutron flux.

- Corrosion risk.

- As a breeder reactor, a modified MSR might be able to produce weapons-grade nuclear material.

- The MSRE and aircraft nuclear reactors used enrichment levels so high that they approach the levels of nuclear weapons. These levels would be illegal in most modern regulatory regimes for power plants. Some modern designs avoid this issue.

- Neutron damage to solid moderator materials can limit the core lifetime of an MSR that makes moderately fast neutrons. For example, the MSRE was designed so that its graphite moderator sticks had very low tolerances, so neutron damage could change their size without damage. "Two fluid" MSR designs are unable to use graphite piping because graphite changes size when it is bombarded with neutrons, and graphite pipes would crack and leak.

References

- Jamasb, Tooraj; William J. Nuttall; Michael G. Pollitt (2006). Future electricity technologies and systems (illustrated ed.). Cambridge University Press. p. 203. ISBN 978-0-521-86049-9.

- Oka, Yoshiaki; Koshizuka, Seiichi; Ishiwatari, Yuki; Yamaji, Akifumi (2010). Super Light Water Rectors and Super Fast Reactors. Springer. ISBN 978-1-4419-6034-4.

- Yoshiaki Oka; Hideo Mori, eds. (2014). Supercritical-Pressure Light Water Cooled Reactors. Springer. ISBN 978-4-431-55024-2.

- Cohen, Linda R.; Noll, Roger G. (1991). The Technology pork barrel. Brookings Institution. p. 234. ISBN 0-8157-1508-0. Retrieved 28 February 2012.

- Weinberg, Alvin (1997). The First Nuclear Era: The Life and Times of a Technological Fixer. Springer. ISBN 978-1-56396-358-2.

- "Energy Department Announces New Investments in Advanced Nuclear Power Reactors...". US Department of Energy. Retrieved 16 January 2016.

- Halper, Mark. "China eyes thorium MSRs for industrial heat, hydrogen; revises timeline". Weinberg Next Nuclear. The Alvin Weinberg Foundation. Retrieved 9 June 2016.

- Griffiths, Trevor; Tomlinson, Jasper; O'Sullivan, Rory. "MSR Review - Feasibility of Developing a Pilot Scale Molten Salt Reactor in the UK" (PDF). Energy Process Developments. Retrieved 14 January 2016.

- "Energy Department Announces New Investments in Advanced Nuclear Power Reactors...". US Department of Energy. Retrieved 16 January 2016.

- Baron, Matthias; Böck, Helmuth; Villa, Mario. "TRIGA Reactor Characteristics". IAEA Education and Training. IAEA. Retrieved 2 June 2016.

- Gylfe, J.D. "US Patent 3,145,150, Aug. 18, 1954, Fuel Moderator Element for a Nuclear Reactor, and Method of Making". U.S. Patent Office. U.S. Government. Retrieved 2 June 2016.

- Massie, Mark; Dewan, Leslie C. "US 20130083878 A1, April 4, 2013, NUCLEAR REACTORS AND RELATED METHODS AND APPARATUS". U.S. Patent Office. U.S. Government. Retrieved 2 June 2016.

- Williams, Stephen (16 January 2015). "Molten Salt Reactors: The Future of Green Energy?". ZME Science. Retrieved 18 February 2015.

- Hellemans, Alexander (12 January 2012). "Reactor-Accelerator Hybrid Achieves Successful Test Run". Science Insider. Retrieved 29 December 2014.

- "Design features of BREST reactors and experimental work to advance the concept of BREST reactors" (PDF). US DoE, Small Modular Reactor Program. Retrieved 2013-05-16.

- Jha, Saurav (16 September 2013). "The Thorium Question – An interview with India's nuclear czar". Retrieved 19 September 2013.

4

Nuclear Reaction: An Integrated Study

Nuclear reactions are those that involve the collision of two nuclei or the bombardment of a nucleus with subatomic particles like neutrons, protons or high energy electrons to produce one or more lighter nuclei. Nuclear reactions occur naturally when matter interacts with cosmic rays and can also be artificially induced in nuclear reactors. This chapter comprehensively studies nuclear reactions and the topics of energy conversion and nuclear binding energy.

Nuclear Reaction

In nuclear physics and nuclear chemistry, a nuclear reaction is semantically considered to be the process in which two nuclei, or else a nucleus of an atom and a subatomic particle (such as a proton, neutron, or high energy electron) from outside the atom, collide to produce one or more nuclides that are different from the nuclide(s) that began the process. Thus, a nuclear reaction must cause a transformation of at least one nuclide to another. If a nucleus interacts with another nucleus or particle and they then separate without changing the nature of any nuclide, the process is simply referred to as a type of nuclear scattering, rather than a nuclear reaction.

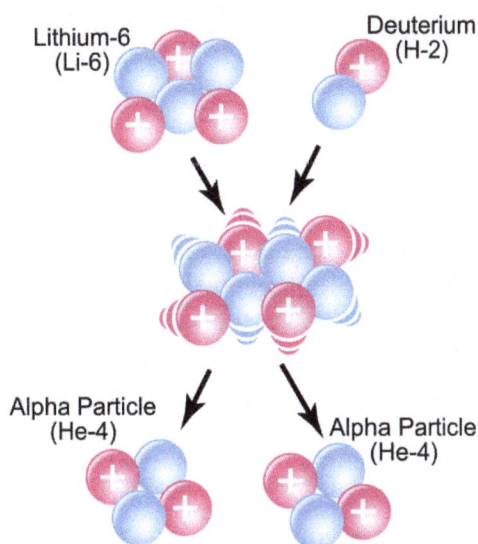

Lithium-6 – Deuterium Reaction

In this symbolic representing of a nuclear reaction, lithium-6 ($^6_3$Li) and deuterium ($^2_1$H) react to form the highly excited intermediate nucleus $^8_4$Be which then decays immediately into two alpha particles of helium-4 ($^4_2$He). Protons are symbolically represented by red spheres, and neutrons by blue spheres.

In principle, a reaction can involve more than two particles colliding, but because the probability of three or more nuclei to meet at the same time at the same place is much less than for two nuclei, such an event is exceptionally rare. "Nuclear reaction" is a term implying an induced change in a nuclide, and thus it does not apply to any type of radioactive decay (which by definition is a spontaneous process).

Natural nuclear reactions occur in the interaction between cosmic rays and matter, and nuclear reactions can be employed artificially to obtain nuclear energy, at an adjustable rate, on demand. Perhaps the most notable nuclear reactions are the nuclear chain reactions in fissionable materials that produce induced nuclear fission, and the various nuclear fusion reactions of light elements that power the energy production of the Sun and stars. Both of these types of reactions are employed in nuclear weapons.

Notation

Nuclear reactions may be shown in a form similar to chemical equations, for which invariant mass must balance for each side of the equation, and in which transformations of particles must follow certain conservation laws, such as conservation of charge and baryon number (total atomic mass number). An example of this notation follows:

$$\ _3^6\text{Li} \ + \ \ _1^2\text{H} \ \rightarrow \ \ _2^4\text{He} \ + \ ?.$$

To balance the equation above for mass, charge and mass number, the second nucleus to the right must have atomic number 2 and mass number 4; it is therefore also helium-4. The complete equation therefore reads:

$$\ _3^6\text{Li} \ + \ \ _1^2\text{H} \ \rightarrow \ \ _2^4\text{He} \ + \ \ _2^4\text{He}$$

or more simply:

$$\ _3^6\text{Li} \ + \ \ _1^2\text{H} \ \rightarrow \ 2\,_2^4\text{He}.$$

Instead of using the full equations in the style above, in many situations a compact notation is used to describe nuclear reactions. This style of the form A(b,c)D is equivalent to A + b producing c + D. Common light particles are often abbreviated in this shorthand, typically p for proton, n for neutron, d for deuteron, α representing an alpha particle or helium-4, β for beta particle or electron, γ for gamma photon, etc. The reaction above would be written as Li-6(d,α)α.

History

In 1917, Ernest Rutherford was able to accomplish transmutation of nitrogen into oxygen at the University of Manchester, using alpha particles directed at nitrogen $^{14}\text{N} + \alpha \rightarrow \ ^{17}\text{O} + \text{p}$. This was the first observation of an induced nuclear reaction, that is, a reaction in which particles from one decay are used to transform another atomic nucleus. Eventually, in 1932 at Cambridge University, a fully artificial nuclear reaction and nuclear transmutation was achieved by Rutherford's colleagues John Cockcroft and Ernest Walton, who used artificially accelerated protons against lithium-7, to split the nucleus into two alpha particles. The feat was popularly known as "splitting the atom", although it was not the modern nuclear fission reaction later discovered in heavy elements, in 1938 by the German scientists Otto Hahn and Fritz Straßmann.

Energy Conservation

Kinetic energy may be released during the course of a reaction (exothermic reaction) or kinetic energy may have to be supplied for the reaction to take place (endothermic reaction). This can be calculated by reference to a table of very accurate particle rest masses, as follows: according to the reference tables, the $^6_3$Li nucleus has a relative atomic mass of 6.015 atomic mass units (abbreviated u), the deuterium has 2.014 u, and the helium-4 nucleus has 4.0026 u. Thus:

- total rest mass on left side = 6.015 + 2.014 = 8.029 u;

- total rest mass on right side = 2 × 4.0026 = 8.0052 u;

- missing rest mass = 8.029 − 8.0052 = 0.0238 atomic mass units.

In a nuclear reaction, the total (relativistic) energy is conserved. The "missing" rest mass must therefore reappear as kinetic energy released in the reaction; its source is the nuclear binding energy. Using Einstein's mass-energy equivalence formula $E = mc^2$, the amount of energy released can be determined. We first need the energy equivalent of one atomic mass unit:

$$1\,u\,c^2 = (1.66054 \times 10^{-27}\,kg) \times (2.99792 \times 10^8\,m/s)^2$$

$$= 1.49242 \times 10^{-10}\,kg\,(m/s)^2 = 1.49242 \times 10^{-10}\,J\,(joule) \times (1\,MeV\,/\,1.60218 \times 10^{-13}\,J)$$

$$= 931.49\,MeV,$$

so $1\,u\,c^2 = 931.49$ MeV.

Hence, the energy released is 0.0238 × 931 MeV = 22.2 MeV.

Expressed differently: the mass is reduced by 0.3%, corresponding to 0.3% of 90 PJ/kg is 270 TJ/kg.

This is a large amount of energy for a nuclear reaction; the amount is so high because the binding energy per nucleon of the helium-4 nucleus is unusually high, because the He-4 nucleus is "doubly magic". (The He-4 nucleus is unusually stable and tightly bound for the same reason that the helium atom is inert: each pair of protons and neutrons in He-4 occupies a filled 1s nuclear orbital in the same way that the pair of electrons in the helium atom occupy a filled 1s electron orbital). Consequently, alpha particles appear frequently on the right hand side of nuclear reactions.

The energy released in a nuclear reaction can appear mainly in one of three ways:

- kinetic energy of the product particles;

- emission of very high energy photons, called gamma rays;

- some energy may remain in the nucleus, as a metastable energy level.

When the product nucleus is metastable, this is indicated by placing an asterisk ("*") next to its atomic number. This energy is eventually released through nuclear decay.

A small amount of energy may also emerge in the form of X-rays. Generally, the product nucleus has a different atomic number, and thus the configuration of its electron shells is wrong. As the electrons rearrange themselves and drop to lower energy levels, internal transition X-rays (X-rays with precisely defined emission lines) may be emitted.

Q-value and Energy Balance

In writing down the reaction equation, in a way analogous to a chemical equation, one may in addition give the reaction energy on the right side:

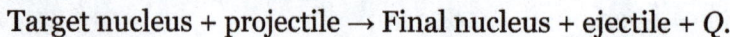

Target nucleus + projectile → Final nucleus + ejectile + Q.

For the particular case discussed above, the reaction energy has already been calculated as Q = 22.2 MeV. Hence:

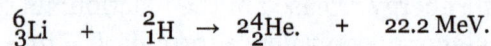

$$\,^{6}_{3}\text{Li} \;+\; \,^{2}_{1}\text{H} \;\rightarrow\; 2\,^{4}_{2}\text{He.} \;+\; 22.2\,\text{MeV.}$$

The reaction energy (the "Q-value") is positive for exothermal reactions and negative for endothermal reactions. On the one hand, it is the difference between the sums of kinetic energies on the final side and on the initial side. But on the other hand, it is also the difference between the nuclear rest masses on the initial side and on the final side (in this way, we have calculated the Q-value above).

Reaction Rates

If the reaction equation is balanced, that does not mean that the reaction really occurs. The rate at which reactions occur depends on the particle energy, the particle flux and the reaction cross section. An example of a large repository of reaction rates is the REACLIB database, as maintained by the Joint Institute for Nuclear Astrophysics.

Neutrons Vs. Ions

In the initial collision which begins the reaction, the particles must approach closely enough so that the short range strong force can affect them. As most common nuclear particles are positively charged, this means they must overcome considerable electrostatic repulsion before the reaction can begin. Even if the target nucleus is part of a neutral atom, the other particle must penetrate well beyond the electron cloud and closely approach the nucleus, which is positively charged. Thus, such particles must be first accelerated to high energy, for example by:

- particle accelerators;

- nuclear decay (alpha particles are the main type of interest here, since beta and gamma rays are rarely involved in nuclear reactions);

- very high temperatures, on the order of millions of degrees, producing thermonuclear reactions;

- cosmic rays.

Also, since the force of repulsion is proportional to the product of the two charges, reactions between heavy nuclei are rarer, and require higher initiating energy, than those between a heavy and light nucleus; while reactions between two light nuclei are the most common ones.

Neutrons, on the other hand, have no electric charge to cause repulsion, and are able to initiate a nuclear reaction at very low energies. In fact, at extremely low particle energies (corresponding, say, to thermal equilibrium at room temperature), the neutron's de Broglie wavelength is greatly increased, possibly greatly increasing its capture cross section, at energies close to resonances of the nuclei involved. Thus low energy neutrons *may* be even more reactive than high energy neutrons.

Notable Types

While the number of possible nuclear reactions is immense, there are several types which are more common, or otherwise notable. Some examples include:

- Fusion reactions — two light nuclei join to form a heavier one, with additional particles (usually protons or neutrons) thrown off to conserve momentum.

- Spallation — a nucleus is hit by a particle with sufficient energy and momentum to knock out several small fragments or smash it into many fragments.

- Induced gamma emission belongs to a class in which only photons were involved in creating and destroying states of nuclear excitation.

- Alpha decay — Though driven by the same underlying forces as spontaneous fission, α decay is usually considered to be separate from the latter. The often-quoted idea that "nuclear reactions" are confined to induced processes is incorrect. "Radioactive decays" are a subgroup of "nuclear reactions" that are spontaneous rather than induced. For example, so-called "hot alpha particles" with unusually high energies may actually be produced in induced ternary fission, which is an induced nuclear reaction (contrasting with spontaneous fission). Such alphas occur from spontaneous ternary fission as well.

- Fission reactions — a very heavy nucleus, after absorbing additional light particles (usually neutrons), splits into two or sometimes three pieces. This is an induced nuclear reaction. Spontaneous fission, which occurs without assistance of the neutron, is usually not considered a nuclear reaction. At most, it is not an *induced* nuclear reaction.

Direct Reactions

An intermediate energy projectile transfers energy or picks up or loses nucleons to the nucleus in a single quick (10^{-21} second) event. Energy and momentum transfer are relatively small. These are particularly useful in experimental nuclear physics, because the reaction mechanisms are often simple enough to calculate with sufficient accuracy to probe the structure of the target nucleus.

Inelastic Scattering

Only energy and momentum are transferred.

- (p,p') tests differences between nuclear states.

- (α,α') measures nuclear surface shapes and sizes. Since α particles that hit the nucleus react more violently, elastic and shallow inelastic α scattering are sensitive to the shapes and sizes of the targets, like light scattered from a small black object.

- (e,e') is useful for probing the interior structure. Since electrons interact less strongly than do protons and neutrons, they reach to the centers of the targets and their wave functions are less distorted by passing through the nucleus.

Transfer Reactions

Usually at moderately low energy, one or more nucleons are transferred between the projectile and target. These are useful in studying outer shell structure of nuclei.

- (α,n) and (α,p) reactions. Some of the earliest nuclear reactions studied involved an alpha particle produced by alpha decay, knocking a nucleon from a target nucleus.

- (d,n) and (d,p) reactions. A deuteron beam impinges on a target; the target nuclei absorb either the neutron or proton from the deuteron. The deuteron is so loosely bound that this is almost the same as proton or neutron capture. A compound nucleus may be formed, leading to additional neutrons being emitted more slowly. (d,n) reactions are used to generate energetic neutrons.

- The strangeness exchange reaction (K, π) has been used to study hypernuclei.

- The reaction $^{14}N(\alpha,p)^{17}O$ performed by Rutherford in 1917 (reported 1919), is generally regarded as the first nuclear transmutation experiment.

Reactions with Neutrons

	→ T	→ ^{7}Li	→ ^{14}C		
(n,α)	^{6}Li + n → T + α	^{10}B + n → ^{7}Li + α	^{17}O + n → ^{14}C + α	^{21}Ne + n → ^{18}O + α	^{37}Ar + n → ^{34}S + α
(n,p)	^{3}He + n → T + p	^{7}Be + n → ^{7}Li + p	^{14}N + n → ^{14}C + p	^{22}Na + n → ^{22}Ne + p	

Reactions with neutrons are important in nuclear reactors and nuclear weapons. While the best known neutron reactions are neutron scattering, neutron capture, and nuclear fission, for some light nuclei (especially odd-odd nuclei) the most probable reaction with a thermal neutron is a transfer reaction:

Some reactions are only possible with fast neutrons:

- (n,2n) reactions produce small amounts of protactinium-231 and uranium-232 in the thorium cycle which is otherwise relatively free of highly radioactive actinide products.

- ^{9}Be + n → 2α + 2n can contribute some additional neutrons in the beryllium neutron reflector of a nuclear weapon.

- $^7Li + n \rightarrow T + \alpha + n$ unexpectedly contributed additional yield in Castle Bravo, Castle Romeo, and Castle Yankee, the three highest-yield nuclear tests conducted by the U.S.

Compound Nuclear Reactions

Either a low energy projectile is absorbed or a higher energy particle transfers energy to the nucleus, leaving it with too much energy to be fully bound together. On a time scale of about 10^{-19} seconds, particles, usually neutrons, are "boiled" off. That is, it remains together until enough energy happens to be concentrated in one neutron to escape the mutual attraction. Charged particles rarely boil off because of the coulomb barrier. The excited quasi-bound nucleus is called a compound nucleus.

- Low energy (e, e' xn), (γ, xn) (the xn indicating one or more neutrons), where the gamma or virtual gamma energy is near the giant dipole resonance. These increase the need for radiation shielding around electron accelerators.

Nuclear Fusion

The Sun is a main-sequence star, and thus generates its energy by nuclear fusion of hydrogen nuclei into helium. In its core, the Sun fuses 620 million metric tons of hydrogen each second.

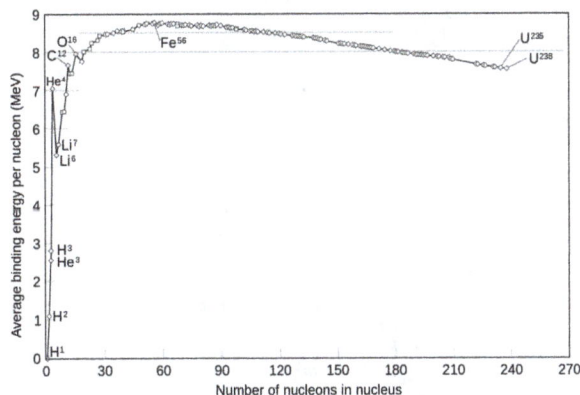

The nuclear binding energy curve. The formation of nuclei with masses up to Iron-56 releases energy, while forming those that are heavier requires energy input. This is because the nuclei below Iron-56 have high binding energies, while the heavier ones have lower binding energies, as illustrated above.

In nuclear physics, nuclear fusion is a nuclear reaction in which two or more atomic nuclei come close enough to form one or more different atomic nuclei and subatomic particles (neutrons and/or protons). The difference in mass between the products and reactants is manifested as the release of large amounts of energy. This difference in mass arises due to the difference in atomic "binding energy" between the atomic nuclei before and after the reaction. Fusion is the process that powers active or "main sequence" stars, or other high magnitude stars.

The fusion process that produces a nucleus lighter than iron-56 or nickel-62 will generally yield a net energy release. These elements have the smallest mass per nucleon and the largest binding

energy per nucleon, respectively. Fusion of light elements toward these releases energy (an exothermic process), while a fusion producing nuclei heavier than these elements, will result in energy
retained by the resulting nucleons, and the resulting reaction is endothermic. The opposite is true
for the reverse process, nuclear fission. This means that the lighter elements, such as hydrogen and
helium, are in general more fusable; while the heavier elements, such as uranium and plutonium,
are more fissionable. The extreme astrophysical event of a supernova can produce enough energy
to fuse nuclei into elements heavier than iron.

Following the discovery of quantum tunneling by physicist Friedrich Hund, in 1929 Robert Atkin-
son and Fritz Houtermans used the measured masses of light elements to predict that large amounts
of energy could be released by fusing small nuclei. Building upon the nuclear transmutation experiments by Ernest Rutherford, carried out several years earlier, the laboratory fusion of hydrogen
isotopes was first accomplished by Mark Oliphant in 1932. During the remainder of that decade
the steps of the main cycle of nuclear fusion in stars were worked out by Hans Bethe. Research into
fusion for military purposes began in the early 1940s as part of the Manhattan Project. Fusion was
accomplished in 1951 with the Greenhouse Item nuclear test. Nuclear fusion on a large scale in an
explosion was first carried out on November 1, 1952, in the Ivy Mike hydrogen bomb test.

Research into developing controlled thermonuclear fusion for civil purposes also began in earnest
in the 1950s, and it continues to this day.

Process

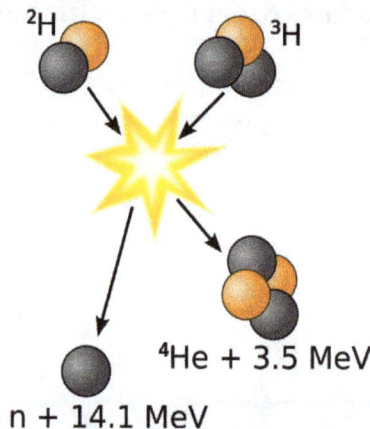

^{2}H

^{3}H

^{4}He + 3.5 MeV

n + 14.1 MeV

$\Delta mc^2|$, where Δm is the decrease in the total rest mass of particles.

The origin of the energy released in fusion of light elements is due to interplay of two opposing
forces, the nuclear force which combines together protons and neutrons, and the Coulomb force,
which causes protons to repel each other. The protons are positively charged and repel each other
but they nonetheless stick together, demonstrating the existence of another force referred to as
nuclear attraction. This force, called the strong nuclear force, overcomes electric repulsion at very
close range. The effect of this force is not observed outside the nucleus, hence the force is called a
short-range force. The same force also pulls the nucleons (neutrons and protons) together allowing ordinary matter to exist. Light nuclei (or nuclei smaller than iron and nickel), are sufficiently
small and proton-poor allowing the nuclear force to overcome the repulsive Coulomb force. This
is because the nucleus is sufficiently small that all nucleons feel the short-range attractive force at

least as strongly as they feel the infinite-range Coulomb repulsion. Building up these nuclei from lighter nuclei by fusion thus releases the extra energy from the net attraction of these particles. For larger nuclei, however, no energy is released, since the nuclear force is short-range and cannot continue to act across still larger atomic nuclei. Thus, energy is no longer released when such nuclei are made by fusion; instead, energy is required as input to such processes.

Fusion reactions create the light elements that power the stars and produce virtually all elements in a process called nucleosynthesis. The fusion of lighter elements in stars releases energy and the mass that always accompanies it. For example, in the fusion of two hydrogen nuclei to form helium, 0.7% of the mass is carried away from the system in the form of kinetic energy of an alpha particle or other forms of energy, such as electromagnetic radiation.

Research into controlled fusion, with the aim of producing fusion power for the production of electricity, has been conducted for over 60 years. It has been accompanied by extreme scientific and technological difficulties, but has resulted in progress. At present, controlled fusion reactions have been unable to produce break-even (self-sustaining) controlled fusion. Workable designs for a reactor that theoretically will deliver ten times more fusion energy than the amount needed to heat plasma to the required temperatures are in development. The ITER facility is expected to finish its construction phase in 2019. It will start commissioning the reactor that same year and initiate plasma experiments in 2020, but is not expected to begin full deuterium-tritium fusion until 2027.

It takes considerable energy to force nuclei to fuse, even those of the lightest element, hydrogen. This is because all nuclei have a positive charge due to their protons, and as like charges repel, nuclei strongly resist being pushed close together. Accelerated to high speeds, they can overcome this electrostatic repulsion and be forced close enough such that the attractive nuclear force is stronger than the repulsive force. As the strong force grows very rapidly once beyond that critical distance, the fusing nucleons "fall" into one another and result is fusion and net energy produced. The fusion of lighter nuclei, which creates a heavier nucleus and often a free neutron or proton, generally releases more energy than it takes to force the nuclei together; this is an exothermic process that can produce self-sustaining reactions. The US National Ignition Facility, which uses laser-driven inertial confinement fusion, was designed with a goal of break-even fusion.

The first large-scale laser target experiments were performed in June 2009 and ignition experiments began in early 2011.

Energy released in most nuclear reactions are much larger than in chemical reactions, because the binding energy that holds a nucleus together is far greater than the energy that holds electrons to a nucleus. For example, the ionization energy gained by adding an electron to a hydrogen nucleus is 13.6 eV—less than one-millionth of the 17.6 MeV released in the deuterium–tritium (D–T) reaction shown in the diagram to the right. The complete conversion of one gram of matter would release 9×10^{13} joules of energy. Fusion reactions have an energy density many times greater than nuclear fission; the reactions produce far greater energy per unit of mass even though *individual* fission reactions are generally much more energetic than *individual* fusion ones, which are themselves millions of times more energetic than chemical reactions. Only direct conversion of mass into energy, such as that caused by the annihilatory collision of matter and antimatter, is more energetic per unit of mass than nuclear fusion.

Nuclear Fusion in Stars

The most important fusion process in nature is the one that powers stars, stellar nucleosynthesis. In the 20th century, it was realized that the energy released from nuclear fusion reactions accounted for the longevity of the Sun and other stars as a source of heat and light. The fusion of nuclei in a star, starting from its initial hydrogen and helium abundance, provides that energy and synthesizes new nuclei as a byproduct of the fusion process. The prime energy producer in the Sun is the fusion of hydrogen to form helium, which occurs at a solar-core temperature of 14 million kelvin. The net result is the fusion of four protons into one alpha particle, with the release of two positrons, two neutrinos (which changes two of the protons into neutrons), and energy. Different reaction chains are involved, depending on the mass of the star. For stars the size of the sun or smaller, the proton-proton chain dominates. In heavier stars, the CNO cycle is more important.

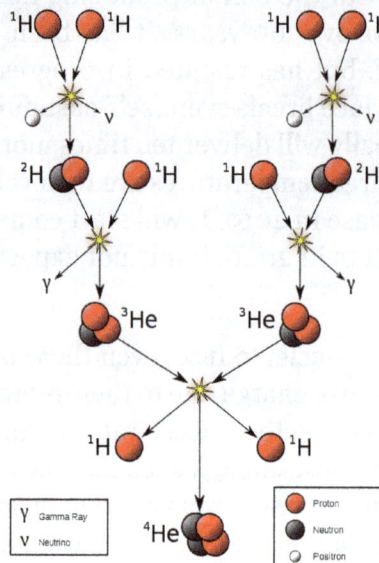

The proton-proton chain dominates in stars the size of the Sun or smaller.

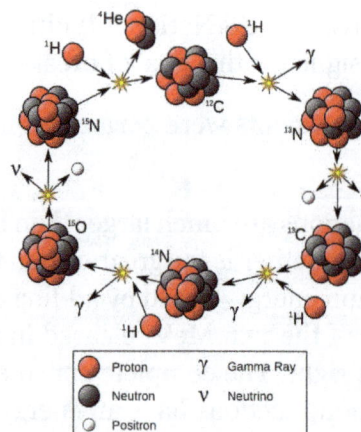

The CNO cycle dominates in stars heavier than the Sun.

As a star uses up a substantial fraction of its hydrogen, it begins to synthesize heavier elements. However the heaviest elements are synthesized by fusion that occurs as a more massive star undergoes a violent supernova at the end of its life, a process known as supernova nucleosynthesis.

Requirements

Details and supporting references on the material in this section can be found in textbooks on nuclear physics or nuclear fusion.

A substantial energy barrier of electrostatic forces must be overcome before fusion can occur. At large distances, two naked nuclei repel one another because of the repulsive electrostatic force between their positively charged protons. If two nuclei can be brought close enough together, however, the electrostatic repulsion can be overcome by the quantum effect in which nuclei can tunnel through columb forces.

When a nucleon such as a proton or neutron is added to a nucleus, the nuclear force attracts it to all the other nucleons of the nucleus (if the atom is small enough), but primarily to its immediate neighbours due to the short range of the force. The nucleons in the interior of a nucleus have more neighboring nucleons than those on the surface. Since smaller nuclei have a larger surface area-to-volume ratio, the binding energy per nucleon due to the nuclear force generally increases with the size of the nucleus but approaches a limiting value corresponding to that of a nucleus with a diameter of about four nucleons. It is important to keep in mind that nucleons are quantum objects. So, for example, since two neutrons in a nucleus are identical to each other, the goal of distinguishing one from the other, such as which one is in the interior and which is on the surface, is in fact meaningless, and the inclusion of quantum mechanics is therefore necessary for proper calculations.

The electrostatic force, on the other hand, is an inverse-square force, so a proton added to a nucleus will feel an electrostatic repulsion from *all* the other protons in the nucleus. The electrostatic energy per nucleon due to the electrostatic force thus increases without limit as nuclei atomic number grows.

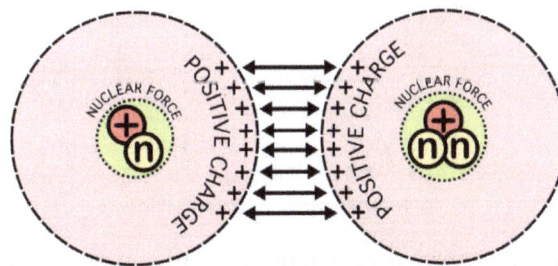

The electrostatic force between the positively charged nuclei is repulsive, but when the separation is small enough, the quantum effect will tunnel through the wall. Therefore, the prerequisite for fusion is that the two nuclei be brought close enough together for a long enough time for quantum tunnelling to act.

The net result of the opposing electrostatic and strong nuclear forces is that the binding energy per nucleon generally increases with increasing size, up to the elements iron and nickel, and then decreases for heavier nuclei. Eventually, the binding energy becomes negative and very heavy nuclei (all with more than 208 nucleons, corresponding to a diameter of about 6 nucleons) are not stable. The four most tightly bound nuclei, in decreasing order of binding energy per nucleon, are 62Ni, 58Fe, 56Fe, and 60Ni. Even though the nickel isotope, 62Ni, is more stable, the iron isotope 56Fe is an order of magnitude more common. This is due to the fact that there is no easy way for stars to create 62Ni through the alpha process.

An exception to this general trend is the helium-4 nucleus, whose binding energy is higher than that of lithium, the next heaviest element. This is because protons and neutrons are fermions, which according to the Pauli exclusion principle cannot exist in the same nucleus in exactly the same state. Each proton or neutron energy state in a nucleus can accommodate both a spin up particle and a spin down particle. Helium-4 has an anomalously large binding energy because its nucleus consists of two protons and two neutrons, so all four of its nucleons can be in the ground state. Any additional nucleons would have to go into higher energy states. Indeed, the helium-4 nucleus is so tightly bound that it is commonly treated as a single particle in nuclear physics, namely, the alpha particle.

The situation is similar if two nuclei are brought together. As they approach each other, all the protons in one nucleus repel all the protons in the other. Not until the two nuclei actually come close enough for long enough can the strong nuclear force take over (by way of tunneling). Consequently, even when the final energy state is lower, there is a large energy barrier that must first be overcome. It is called the Coulomb barrier.

The Coulomb barrier is smallest for isotopes of hydrogen, as their nuclei contain only a single positive charge. A diproton is not stable, so neutrons must also be involved, ideally in such a way that a helium nucleus, with its extremely tight binding, is one of the products.

Using deuterium-tritium fuel, the resulting energy barrier is about 0.1 MeV. In comparison, the energy needed to remove an electron from hydrogen is 13.6 eV, about 7500 times less energy. The (intermediate) result of the fusion is an unstable ^{5}He nucleus, which immediately ejects a neutron with 14.1 MeV. The recoil energy of the remaining ^{4}He nucleus is 3.5 MeV, so the total energy liberated is 17.6 MeV. This is many times more than what was needed to overcome the energy barrier.

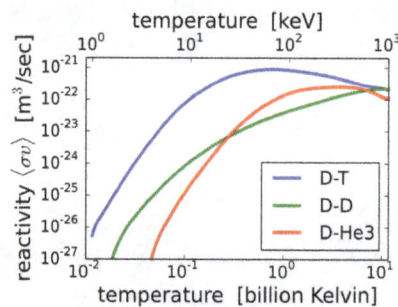

The fusion reaction rate increases rapidly with temperature until it maximizes and then gradually drops off. The DT rate peaks at a lower temperature (about 70 keV, or 800 million kelvin) and at a higher value than other reactions commonly considered for fusion energy.

The reaction cross section σ is a measure of the probability of a fusion reaction as a function of the relative velocity of the two reactant nuclei. If the reactants have a distribution of velocities, e.g. a thermal distribution, then it is useful to perform an average over the distributions of the product of cross section and velocity. This average is called the 'reactivity', denoted $<\sigma v>$. The reaction rate (fusions per volume per time) is $<\sigma v>$ times the product of the reactant number densities:

$$f = n_1 n_2 \langle \sigma v \rangle.$$

If a species of nuclei is reacting with itself, such as the DD reaction, then the product $n_1 n_2$ must be replaced by $(1 >$.

$\langle \sigma v \rangle$ increases from virtually zero at room temperatures up to meaningful magnitudes at temperatures of 10–100 keV. At these temperatures, well above typical ionization energies (13.6 eV in the hydrogen case), the fusion reactants exist in a plasma state.

The significance of σv as a function of temperature in a device with a particular energy confinement time is found by considering the Lawson criterion. This is an extremely challenging barrier to overcome on Earth, which explains why fusion research has taken many years to reach the current high state of technical prowess.

Methods for Achieving Fusion

Thermonuclear Fusion

If the matter is sufficiently heated (hence being plasma), the fusion reaction may occur due to collisions with extreme thermal kinetic energies of the particles. In the form of thermonuclear weapons, thermonuclear fusion is the only fusion technique so far to yield undeniably large amounts of useful fusion energy. Usable amounts of thermonuclear fusion energy released in a controlled manner have yet to be achieved. In nature, this is what produces energy in stars through stellar nucleosynthesis.

Inertial Confinement Fusion

Inertial confinement fusion (ICF) is a type of fusion energy research that attempts to initiate nuclear fusion reactions by heating and compressing a fuel target, typically in the form of a pellet that most often contains a mixture of deuterium and tritium.

Inertial Electrostatic Confinement

Inertial electrostatic confinement is a set of devices that use an electric field to heat ions to fusion conditions. The most well known is the fusor. Starting in 1999, a number of amateurs have been able to do amateur fusion using these homemade devices. Other IEC devices include: the Polywell, MIX POPS and Marble concepts.

Beam-beam or Beam-target Fusion

If the energy to initiate the reaction comes from accelerating one of the nuclei, the process is called *beam-target* fusion; if both nuclei are accelerated, it is *beam-beam* fusion.

Accelerator-based light-ion fusion is a technique using particle accelerators to achieve particle kinetic energies sufficient to induce light-ion fusion reactions. Accelerating light ions is relatively easy, and can be done in an efficient manner—requiring only a vacuum tube, a pair of electrodes, and a high-voltage transformer; fusion can be observed with as little as 10 kV between the electrodes. The key problem with accelerator-based fusion (and with cold targets in general) is that fusion cross sections are many orders of magnitude lower than Coulomb interaction cross sections. Therefore, the vast majority of ions expend their energy emitting bremsstrahlung radiation and the ionization of atoms of the target. Devices referred to as sealed-tube neutron generators are particularly relevant to this discussion. These small devices are miniature particle accelerators filled with deuterium and tritium gas in an arrangement that allows ions of those nuclei to be accelerated against hydride targets, also containing deuterium and tritium, where fusion takes place, releasing a flux of neutrons. Hundreds of neutron generators are produced annually for use in the petroleum industry where they are used in measurement equipment for locating and mapping oil reserves.

Muon-catalyzed Fusion

Muon-catalyzed fusion is a well-established and reproducible fusion process that occurs at ordinary temperatures. It was studied in detail by Steven Jones in the early 1980s. Net energy production from this reaction cannot occur because of the high energy required to create muons, their short 2.2 µs half-life, and the high chance that a muon will bind to the new alpha particle and thus stop catalyzing fusion.

Other Principles

Antimatter-initialized fusion uses small amounts of antimatter to trigger a tiny fusion explosion. This has been studied primarily in the context of making nuclear pulse propulsion, and pure fusion bombs feasible. This is not near becoming a practical power source, due to the cost of manufacturing antimatter alone.

The *Tokamak à configuration variable*, research fusion reactor, at the École Polytechnique Fédérale de Lausanne (Switzerland).

Some other confinement principles have been investigated.

Pyroelectric fusion was reported in April 2005 by a team at UCLA. The scientists used a pyroelectric crystal heated from −34 to 7 °C (−29 to 45 °F), combined with a tungsten needle to produce an electric field of about 25 gigavolts per meter to ionize and accelerate deuterium nuclei into an erbium deuteride target. At the estimated energy levels, the D-D fusion reaction may occur, producing helium-3 and a 2.45 MeV neutron. Although it makes a useful neutron generator, the apparatus is not intended for power generation since it requires far more energy than it produces.

Hybrid nuclear fusion-fission (hybrid nuclear power) is a proposed means of generating power by use of a combination of nuclear fusion and fission processes. The concept dates to the 1950s, and was briefly advocated by Hans Bethe during the 1970s, but largely remained unexplored until a revival of interest in 2009, due to the delays in the realization of pure fusion. Project PACER, carried out at Los Alamos National Laboratory (LANL) in the mid-1970s, explored the possibility of a fusion power system that would involve exploding small hydrogen bombs (fusion bombs) inside an underground cavity. As an energy source, the system is the only fusion power system that could be demonstrated to work using existing technology. However it would also require a large, continuous supply of nuclear bombs, making the economics of such a system rather questionable.

Important Reactions

Astrophysical Reaction Chains

At the temperatures and densities in stellar cores the rates of fusion reactions are notoriously slow. For example, at solar core temperature ($T \approx 15$ MK) and density (160 g/cm³), the energy release rate is only 276 µW/cm³—about a quarter of the volumetric rate at which a resting human body generates heat. Thus, reproduction of stellar core conditions in a lab for nuclear fusion power production is completely impractical. Because nuclear reaction rates depend on density as well as temperature and most fusion scemes operate at relatively low densities, those methods are strongly dependent on higher temperatures. The fusion rate as a function of temperature ($\exp(-E/kT)$), leads to the need to achieve temperatures in terrestrial reactors 10–100 times higher temperatures and in stellar interiors: $T \approx 0.1$–1.0×10^9 K.

Criteria and Candidates for Terrestrial Reactions

In artificial fusion, the primary fuel is not constrained to be protons and higher temperatures can be used, so reactions with larger cross-sections are chosen. Another concern is the production of neutrons, which activate the reactor structure radiologically, but also have the advantages of allowing volumetric extraction of the fusion energy and tritium breeding. Reactions that release no neutrons are referred to as *aneutronic*.

To be a useful energy source, a fusion reaction must satisfy several criteria. It must:

Be exothermic

> This limits the reactants to the low Z (number of protons) side of thc curve of binding energy. It also makes helium 4He the most common product because of its extraordinarily tight binding, although 3He and 3H also show up.

Involve low atomic number (Z) nuclei

> This is because the electrostatic repulsion must be overcome before the nuclei are close enough to fuse.

Have two reactants

> At anything less than stellar densities, three body collisions are too improbable. In inertial confinement, both stellar densities and temperatures are exceeded to compensate for the shortcomings of the third parameter of the Lawson criterion, ICF's very short confinement time.

Have two or more products

> This allows simultaneous conservation of energy and momentum without relying on the electromagnetic force.

Conserve both protons and neutrons

The cross sections for the weak interaction are too small.

Few reactions meet these criteria. The following are those with the largest cross sections:

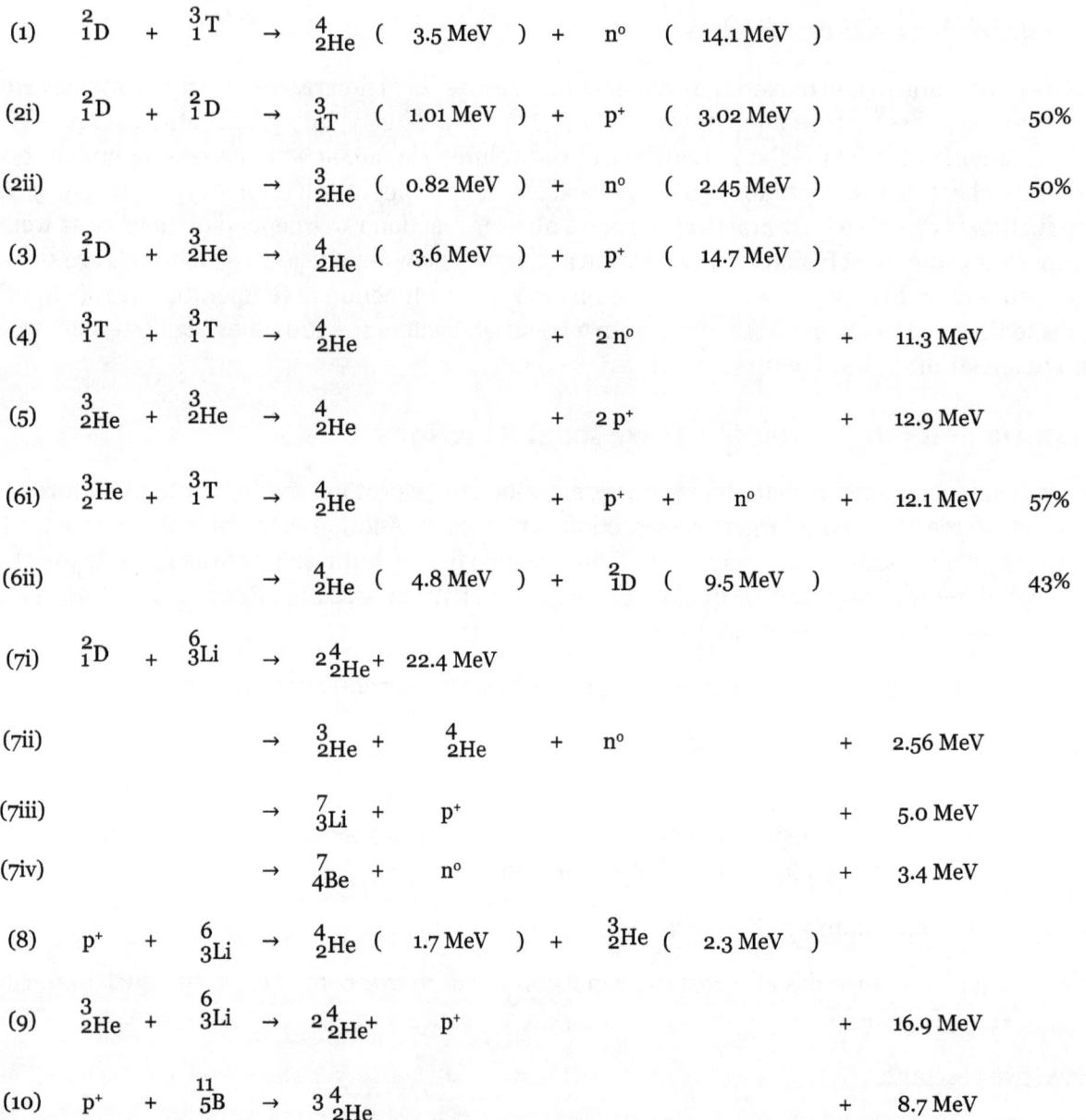

(1) $\quad {}^{2}_{1}D \ + \ {}^{3}_{1}T \ \rightarrow \ {}^{4}_{2}He \ (\ 3.5\,MeV \ ) \ + \ n^{o} \ (\ 14.1\,MeV \ )$

(2i) $\quad {}^{2}_{1}D \ + \ {}^{2}_{1}D \ \rightarrow \ {}^{3}_{1}T \ (\ 1.01\,MeV \ ) \ + \ p^{+} \ (\ 3.02\,MeV \ ) \qquad\qquad 50\%$

(2ii) $\qquad\qquad\qquad \rightarrow \ {}^{3}_{2}He \ (\ 0.82\,MeV \ ) \ + \ n^{o} \ (\ 2.45\,MeV \ ) \qquad\qquad 50\%$

(3) $\quad {}^{2}_{1}D \ + \ {}^{3}_{2}He \ \rightarrow \ {}^{4}_{2}He \ (\ 3.6\,MeV \ ) \ + \ p^{+} \ (\ 14.7\,MeV \ )$

(4) $\quad {}^{3}_{1}T \ + \ {}^{3}_{1}T \ \rightarrow \ {}^{4}_{2}He \qquad\qquad + \ 2\,n^{o} \qquad\qquad + \ 11.3\,MeV$

(5) $\quad {}^{3}_{2}He \ + \ {}^{3}_{2}He \ \rightarrow \ {}^{4}_{2}He \qquad\qquad + \ 2\,p^{+} \qquad\qquad + \ 12.9\,MeV$

(6i) $\quad {}^{3}_{2}He \ + \ {}^{3}_{1}T \ \rightarrow \ {}^{4}_{2}He \qquad\qquad + \ p^{+} \ + \ n^{o} \qquad + \ 12.1\,MeV \qquad 57\%$

(6ii) $\qquad\qquad\qquad \rightarrow \ {}^{4}_{2}He \ (\ 4.8\,MeV \ ) \ + \ {}^{2}_{1}D \ (\ 9.5\,MeV \ ) \qquad\qquad 43\%$

(7i) $\quad {}^{2}_{1}D \ + \ {}^{6}_{3}Li \ \rightarrow \ 2\,{}^{4}_{2}He + \ 22.4\,MeV$

(7ii) $\qquad\qquad\qquad \rightarrow \ {}^{3}_{2}He \ + \ {}^{4}_{2}He \quad + \ n^{o} \qquad\qquad + \ 2.56\,MeV$

(7iii) $\qquad\qquad\qquad \rightarrow \ {}^{7}_{3}Li \ + \ p^{+} \qquad\qquad\qquad + \ 5.0\,MeV$

(7iv) $\qquad\qquad\qquad \rightarrow \ {}^{7}_{4}Be \ + \ n^{o} \qquad\qquad\qquad + \ 3.4\,MeV$

(8) $\quad p^{+} \ + \ {}^{6}_{3}Li \ \rightarrow \ {}^{4}_{2}He \ (\ 1.7\,MeV \ ) \ + \ {}^{3}_{2}He \ (\ 2.3\,MeV \ )$

(9) $\quad {}^{3}_{2}He \ + \ {}^{6}_{3}Li \ \rightarrow \ 2\,{}^{4}_{2}He + \ p^{+} \qquad\qquad\qquad + \ 16.9\,MeV$

(10) $\quad p^{+} \ + \ {}^{11}_{5}B \ \rightarrow \ 3\,{}^{4}_{2}He \qquad\qquad\qquad + \ 8.7\,MeV$

For reactions with two products, the energy is divided between them in inverse proportion to their masses, as shown. In most reactions with three products, the distribution of energy varies. For reactions that can result in more than one set of products, the branching ratios are given.

Some reaction candidates can be eliminated at once. The D-^{6}Li reaction has no advantage compared to p$^+$ - $^{11}_{5}$B because it is roughly as difficult to burn but produces substantially more neutrons through $^{2}_{1}$D- $^{2}_{1}$D side reactions. There is also a p$^+$- $^{7}_{3}$Li reaction, but the cross section is far too low, except possibly when $T_i > 1$ MeV, but at such high temperatures an endothermic, direct neutron-producing reaction also becomes very significant. Finally there is also a p^+-$^9_4$Be reaction, which is not only difficult to burn, but $^9_4$Be can be easily induced to split into two alpha particles and a neutron.

In addition to the fusion reactions, the following reactions with neutrons are important in order to "breed" tritium in "dry" fusion bombs and some proposed fusion reactors:

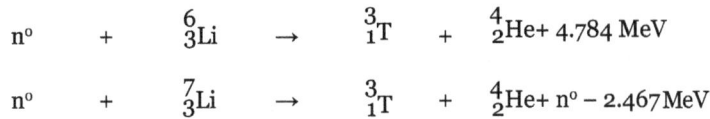

$$n^0 \quad + \quad ^6_3\text{Li} \quad \rightarrow \quad ^3_1\text{T} \quad + \quad ^4_2\text{He} + 4.784 \text{ MeV}$$

$$n^0 \quad + \quad ^7_3\text{Li} \quad \rightarrow \quad ^3_1\text{T} \quad + \quad ^4_2\text{He} + n^0 - 2.467 \text{MeV}$$

The latter of the two equations was unknown when the U.S. conducted the Castle Bravo fusion bomb test in 1954. Being just the second fusion bomb ever tested (and the first to use lithium), the designers of the Castle Bravo "Shrimp" had understood the usefulness of Lithium-6 in tritium production, but had failed to recognize that Lithium-7 fission would greatly increase the yield of the bomb. While Li-7 has a small neutron cross-section for low neutron energies, it has a higher cross section above 5 MeV. Li-7 also undergoes a chain reaction due to its release of a neutron after fissioning. The 15 Mt yield was 150% greater than the predicted 6 Mt and caused casualties from the fallout generated.

To evaluate the usefulness of these reactions, in addition to the reactants, the products, and the energy released, one needs to know something about the cross section. Any given fusion device has a maximum plasma pressure it can sustain, and an economical device would always operate near this maximum. Given this pressure, the largest fusion output is obtained when the temperature is chosen so that $<\sigma v>/T^2$ is a maximum. This is also the temperature at which the value of the triple product $nT\tau$ required for ignition is a minimum, since that required val-ue is inversely proportional to $<\sigma v>/T^2$. (A plasma is "ignited" if the fusion reactions produce enough power to maintain the temperature without external heating.) This optimum temperature and the value of $<\sigma v>/T^2$ at that temperature is given for a few of these reactions in the following table.

fuel	T [keV]	$<\sigma v>/T^2$ [m^3/s/keV^2]
$^2_1$D-$^3_1$T	13.6	1.24×10^{-24}
$^2_1$D-$^2_1$D	15	1.28×10^{-26}
$^2_1$D-$^3_2$He	58	2.24×10^{-26}
p^+-$^6_3$Li	66	1.46×10^{-27}
p^+-$^{11}_5$B	123	3.01×10^{-27}

Note that many of the reactions form chains. For instance, a reactor fueled with $^3_1$T and $^3_2$He creates some $^2_1$D, which is then possible to use in the $^2_1$D-$^3_2$He reaction if the energies are "right". An elegant idea is to combine the reactions (8) and (9). The $^3_2$He from reaction (8) can react with $^6_3$Li in reaction (9) before completely thermalizing. This produces an energetic proton, which in turn undergoes reaction (8) before thermalizing. Detailed analysis shows that this idea would not work well, but it is a good example of a case where the usual assumption of a Maxwellian plasma is not appropriate.

Neutronicity, Confinement Requirement, and Power Density

Any of the reactions above can in principle be the basis of fusion power production. In addition to the temperature and cross section discussed above, we must consider the total energy of the fusion products E_{fus}, the energy of the charged fusion products E_{ch}, and the atomic number Z of the non-hydrogenic reactant.

The only man-made fusion device to achieve ignition to date is the hydrogen bomb. The detonation of the first device, codenamed Ivy Mike, occurred in 1952 and is shown here.

Specification of the $^2_1$D-$^2_1$D reaction entails some difficulties, though. To begin with, one must average over the two branches (2i) and (2ii). More difficult is to decide how to treat the $^3_1$T and $^3_2$He products. $^3_1$T burns so well in a deuterium plasma that it is almost impossible to extract from the plasma. The $^2_1$D-$^3_2$He reaction is optimized at a much higher temperature, so the burnup at the optimum $^2_1$D-$^2_1$D temperature may be low, so it seems reasonable to assume the $^3_1$T but not the $^3_2$He gets burned up and adds its energy to the net reaction. Thus the total reaction would be the sum of (2i), (2ii), and (1):

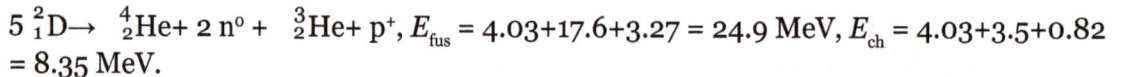

$$5\,^2_1\text{D} \rightarrow\ ^4_2\text{He} + 2\,\text{n}^0 +\ ^3_2\text{He} + \text{p}^+,\ E_{fus} = 4.03 + 17.6 + 3.27 = 24.9 \text{ MeV},\ E_{ch} = 4.03 + 3.5 + 0.82$$
$$= 8.35 \text{ MeV}.$$

We count the $^2_1$D-$^2_1$D fusion energy *per D-D reaction* (not per pair of deuterium atoms) as E_{fus} = (4.03 MeV + 17.6 MeV)×50% + (3.27 MeV)×50% = 12.5 MeV and the energy in charged particles as E_{ch} = (4.03 MeV + 3.5 MeV)×50% + (0.82 MeV)×50% = 4.2 MeV. (Note: if the tritium ion reacts with a deuteron while it still has a large kinetic energy, then the kinetic energy of the helium-4 produced may be quite different from 3.5 MeV, so this calculation of energy in charged particles is only approximate.)

Another unique aspect of the $^2_1$D-$^2_1$D reaction is that there is only one reactant, which must be taken into account when calculating the reaction rate.

With this choice, we tabulate parameters for four of the most important reactions

fuel	Z	E_{fus} [MeV]	E_{ch} [MeV]	neutronicity
$^2_1$D-$^3_1$T	1	17.6	3.5	0.80
$^2_1$D-$^2_1$D	1	12.5	4.2	0.66
$^2_1$D-$^3_2$He	2	18.3	18.3	~0.05
p$^+$-$^{11}_5$B	5	8.7	8.7	~0.001

The last column is the neutronicity of the reaction, the fraction of the fusion energy released as neutrons. This is an important indicator of the magnitude of the problems associated with neutrons like radiation damage, biological shielding, remote handling, and safety. For the first two reactions it is calculated as $(E_{fus}-E_{ch})/E_{fus}$. For the last two reactions, where this calculation would give zero, the values quoted are rough estimates based on side reactions that produce neutrons in a plasma in thermal equilibrium.

Of course, the reactants should also be mixed in the optimal proportions. This is the case when each reactant ion plus its associated electrons accounts for half the pressure. Assuming that the total pressure is fixed, this means that density of the non-hydrogenic ion is smaller than that of the hydrogenic ion by a factor $2/(Z+1)$. Therefore, the rate for these reactions is reduced by the same factor, on top of any differences in the values of $<\sigma v>/T^2$. On the other hand, because the $^2_1 D$-$^2_1 D$ reaction has only one reactant, its rate is twice as high as when the fuel is divided between two different hydrogenic species, thus creating a more efficient reaction.

Thus there is a "penalty" of $(2/(Z+1))$ for non-hydrogenic fuels arising from the fact that they require more electrons, which take up pressure without participating in the fusion reaction. (It is usually a good assumption that the electron temperature will be nearly equal to the ion temperature. Some authors, however discuss the possibility that the electrons could be maintained substantially colder than the ions. In such a case, known as a "hot ion mode", the "penalty" would not apply.) There is at the same time a "bonus" of a factor 2 for $^2_1 D$-$^2_1 D$ because each ion can react with any of the other ions, not just a fraction of them.

We can now compare these reactions in the following table.

fuel	$<\sigma v>/T^2$	penalty/bonus	reactivity	Lawson criterion	power density (W/m³/kPa²)	relation of power density
$^2_1 D$-$^3_1 T$	1.24×10^{-24}	1	1	1	34	1
$^2_1 D$-$^2_1 D$	1.28×10^{-26}	2	48	30	0.5	68
$^2_1 D$-$^3_2 He$	2.24×10^{-26}	2/3	83	16	0.43	80
p^+-$^6_3 Li$	1.46×10^{-27}	1/2	1700		0.005	6800
p^+-$^{11}_5 B$	3.01×10^{-27}	1/3	1240	500	0.014	2500

The maximum value of $<\sigma v>/T^2$ is taken from a previous table. The "penalty/bonus" factor is that related to a non-hydrogenic reactant or a single-species reaction. The values in the column "reactivity" are found by dividing 1.24×10^{-24} by the product of the second and third columns. It indicates the factor by which the other reactions occur more slowly than the $^2_1 D$-$^3_1 T$ reaction under comparable conditions. The column "Lawson criterion" weights these results with E_{ch} and gives an indication of how much more difficult it is to achieve ignition with these reactions, relative to the difficulty for the $^2_1 D$-$^3_1 T$ reaction. The last column is labeled "power density" and weights the practical reactivity with E_{fus}. It indicates how much lower the fusion power density of the other reactions is compared to the $^2_1 D$-$^3_1 T$ reaction and can be considered a measure of the economic potential.

Bremsstrahlung Losses in Quasineutral, Isotropic Plasmas

The ions undergoing fusion in many systems will essentially never occur alone but will be mixed with electrons that in aggregate neutralize the ions' bulk electrical charge and form a plasma. The electrons will generally have a temperature comparable to or greater than that of the ions, so they will collide with the ions and emit x-ray radiation of 10–30 keV energy, a process known as Bremsstrahlung.

The huge size of the Sun and stars means that the x-rays produced in this process will not escape and will deposit their energy back into the plasma. They are said to be opaque to x-rays. But any terrestrial fusion reactor will be optically thin for x-rays of this energy range. X-rays are difficult to reflect but they are effectively absorbed (and converted into heat) in less than mm thickness of stainless steel (which is part of a reactor's shield). This means the bremsstrahlung process is carrying energy out of the plasma, cooling it.

The ratio of fusion power produced to x-ray radiation lost to walls is an important figure of merit. This ratio is generally maximized at a much higher temperature than that which maximizes the power density. The following table shows estimates of the optimum temperature and the power ratio at that temperature for several reactions.

fuel	T_i (keV)	$P_{fusion}/P_{Bremsstrahlung}$
$^2_1$D-$^3_1$T	50	140
$^2_1$D-$^2_1$D	500	2.9
$^2_1$D-$^3_2$He	100	5.3
$^3_2$He-$^3_2$He	1000	0.72
p$^+$-$^6_3$Li	800	0.21
p$^+$-$^{11}_5$B	300	0.57

The actual ratios of fusion to Bremsstrahlung power will likely be significantly lower for several reasons. For one, the calculation assumes that the energy of the fusion products is transmitted completely to the fuel ions, which then lose energy to the electrons by collisions, which in turn lose energy by Bremsstrahlung. However, because the fusion products move much faster than the fuel ions, they will give up a significant fraction of their energy directly to the electrons. Secondly, the ions in the plasma are assumed to be purely fuel ions. In practice, there will be a significant proportion of impurity ions, which will then lower the ratio. In particular, the fusion products themselves *must* remain in the plasma until they have given up their energy, and *will* remain some time after that in any proposed confinement scheme. Finally, all channels of energy loss other than Bremsstrahlung have been neglected. The last two factors are related. On theoretical and experimental grounds, particle and energy confinement seem to be closely related. In a confinement scheme that does a good job of retaining energy, fusion products will build up. If the fusion products are efficiently ejected, then energy confinement will be poor, too.

The temperatures maximizing the fusion power compared to the Bremsstrahlung are in every case higher than the temperature that maximizes the power density and minimizes the required value of the fusion triple product. This will not change the optimum operating point for $^2_1$D-$^3_1$T very much because the Bremsstrahlung fraction is low, but it will push the other fuels into regimes where the power density relative to $^2_1$D-$^3_1$T is even lower and the required confinement even more difficult to achieve. For $^2_1$D-$^2_1$D and $^2_1$D-$^3_2$He, Bremsstrahlung losses will be a serious, possibly prohibitive problem. For $^3_2$He-$^3_2$He, p$^+$-$^6_3$Li and p$^+$-$^{11}_5$B the Bremsstrahlung losses appear to make a fusion reactor using these fuels with a quasineutral, isotropic plasma impossible. Some ways out of this dilemma are considered—and rejected—in *fundamental limitations on plasma fusion systems not in thermodynamic equilibrium*. This limitation does not apply to non-neutral and anisotropic plasmas; however, these have their own challenges to contend with.

Methods for Achieving Fusion

Thermonuclear Fusion

Thermonuclear fusion is a way to achieve nuclear fusion by using extremely high temperatures. There are two forms of thermonuclear fusion: *uncontrolled*, in which the resulting energy is released in an uncontrolled manner, as it is in thermonuclear weapons such as the "hydrogen bomb", and *controlled*, where the fusion reactions take place in an environment allowing some of the resulting energy to be harnessed for constructive purposes. This article focuses on the latter.

Temperature Requirements

Temperature is a measure of the average kinetic energy of particles, so by heating the material it will gain energy. After reaching sufficient temperature, given by the Lawson criterion, the energy of accidental collisions within the plasma is high enough to overcome the Coulomb barrier and the particles may fuse together.

In a deuterium–tritium fusion reaction, for example, the energy necessary to overcome the Coulomb barrier is 0.1 MeV. Converting between energy and temperature shows that the 0.1 MeV barrier would be overcome at a temperature in excess of 1.2 billion Kelvin.

There are two effects that lower the actual temperature needed. One is the fact that temperature is the *average* kinetic energy, implying that some nuclei at this temperature would actually have much higher energy than 0.1 MeV, while others would be much lower. It is the nuclei in the high-energy tail of the velocity distribution that account for most of the fusion reactions. The other effect is quantum tunnelling. The nuclei do not actually have to have enough energy to overcome the Coulomb barrier completely. If they have nearly enough energy, they can tunnel through the remaining barrier. For these reasons fuel at lower temperatures will still undergo fusion events, at a lower rate.

Thermonuclear fusion is one of the methods being researched in the attempts to produce fusion power. If Thermonuclear fusion becomes favorable to use, it would reduce the world's carbon footprint significantly.

Confinement

The key problem in achieving thermonuclear fusion is how to confine the hot plasma. Due to the high temperature, the plasma can not be in direct contact with any solid material, so in fact it has to be located in a vacuum. But as the high temperatures also imply high pressures, the plasma tends to expand immediately and some force is necessary to act against this thermal pressure. This force can be either gravitation in stars, magnetic forces in magnetic confinement fusion reactors, or the fusion reaction may occur before the plasma starts to expand, so in fact the plasma's inertia is keeping the material together.

Gravitational Confinement

One force capable of confining the fuel well enough to satisfy the Lawson criterion is gravity. The mass needed, however, is so great that gravitational confinement is only found in stars—the least massive stars capable of sustained fusion are red dwarfs, while brown dwarfs are able to fuse deuterium and lithium if they are of sufficient mass. In stars heavy enough, after the supply of hydrogen is exhausted in their cores, their cores (or a shell around the core) start fusing helium to carbon. In the most massive stars (at least 8–11 solar masses), the process is continued until some of their energy is produced by fusing lighter elements to iron. As iron has one of the highest binding energies, reactions producing heavier elements are generally endothermic. Therefore significant amounts of heavier elements are not formed during stable periods of massive star evolution, but are formed in supernova explosions. Some lighter stars also form these elements in the outer parts of the stars over long periods of time, by absorbing energy from fusion in the inside of the star, by absorbing neutrons that are emitted from the fusion process.

All of the elements heavier than iron have some potential energy to release, in theory. At the extremely heavy end of element production, these heavier elements can produce energy in the process of being split again back toward the size of iron, in the process of nuclear fission. Nuclear fission thus releases energy which has been stored, sometimes billions of years before, during stellar nucleosynthesis.

Magnetic Confinement

Electrically charged particles (such as fuel ions) will follow magnetic field lines. The fusion fuel can therefore be trapped using a strong magnetic field. A variety of magnetic configurations exist, including the toroidal geometries of tokamaks and stellarators and open-ended mirror confinement systems.

Inertial Confinement

A third confinement principle is to apply a rapid pulse of energy to a large part of the surface of a pellet of fusion fuel, causing it to simultaneously "implode" and heat to very high pressure and temperature. If the fuel is dense enough and hot enough, the fusion reaction rate will be high enough to burn a significant fraction of the fuel before it has dissipated. To achieve these extreme conditions, the initially cold fuel must be explosively compressed. Inertial confinement is used in the hydrogen bomb, where the driver is x-rays created by a fission bomb. Inertial confinement is also attempted in "controlled" nuclear fusion, where the driver is a laser, ion, or electron beam, or a Z-pinch. Another method is to use conventional high explosive material to compress a fuel to fusion conditions. The UTIAS explosive-driven-implosion facility was used to produce stable, centred and focused hemispherical implosions to generate neutrons from D-D reactions. The simplest and most direct method proved to be in a predetonated stoichiometric mixture of deuterium-oxygen. The other successful method was using a miniature Voitenko compressor, where a plane diaphragm was driven by the implosion wave into a secondary small spherical cavity that contained pure deuterium gas at one atmosphere.

Electrostatic Confinement

There are also electrostatic confinement fusion devices. These devices confine ions using electrostatic fields. The best known is the Fusor. This device has a cathode inside an anode wire cage. Positive ions fly towards the negative inner cage, and are heated by the electric field in the process. If they miss the inner cage they can collide and fuse. Ions typically hit the cathode, however, creating prohibitory high conduction losses. Also, fusion rates in fusors are very low due to competing physical effects, such as energy loss in the form of light radiation. Designs have been proposed to avoid the problems associated with the cage, by generating the field using a non-neutral cloud. These include a plasma oscillating device, a penning trap and the polywell. The technology is relatively immature, however, and many scientific and engineering questions remain.

Inertial Confinement Fusion

Inertial confinement fusion (ICF) is a type of fusion energy research that attempts to initiate nuclear fusion reactions by heating and compressing a fuel target, typically in the form of a pellet that most often contains a mixture of deuterium and tritium.

Inertial confinement fusion using lasers rapidly progressed in the late 1970s and early 1980s from being able to deliver only a few joules of laser energy (per pulse) to being able to deliver tens of kilojoules to a target. At this point, incredibly large scientific devices were needed for experimentation. Here, a view of the 10 beam LLNL Nova laser, shown shortly after the laser's completion in 1984. Around the time of the construction of its predecessor, the Shiva laser, laser fusion had entered the realm of "big science".

To compress and heat the fuel, energy is delivered to the outer layer of the target using high-energy beams of laser light, electrons or ions, although for a variety of reasons, almost all ICF devices as of 2015 have used lasers. The heated outer layer explodes outward, producing a reaction force against the remainder of the target, accelerating it inwards, compressing the target. This process is designed to create shock waves that travel inward through the target. A sufficiently powerful set of shock waves can compress and heat the fuel at the center so much that fusion reactions occur.

The energy released by these reactions will then heat the surrounding fuel, and if the heating is strong enough this could also begin to undergo fusion. The aim of ICF is to produce a condition known as *ignition*, where this heating process causes a chain reaction that burns a significant portion of the fuel. Typical fuel pellets are about the size of a pinhead and contain around 10 milligrams of fuel: in practice, only a small proportion of this fuel will undergo fusion, but if all this fuel were consumed it would release the energy equivalent to burning a barrel of oil.

ICF is one of two major branches of fusion energy research, the other being magnetic confinement fusion. When it was first proposed in the early 1970s, ICF appeared to be a practical approach to fusion power production and the field flourished. Experiments during the 1970s and '80s demonstrated that the efficiency of these devices was much lower than expected, and reaching ignition would not be easy. Throughout the 1980s and '90s, many experiments were conducted in order to understand the complex interaction of high-intensity laser light and plasma. These led to the design of newer machines, much larger, that would finally reach ignition energies.

The largest operational ICF experiment is the National Ignition Facility (NIF) in the US, designed using all of the decades-long experience of earlier experiments. Like those earlier experiments, however, NIF has failed to reach ignition and is, as of 2015, generating about $\frac{1}{3}$ of the required energy levels. As of October 7, 2013, this facility is understood to have achieved an important milestone towards commercialization of fusion, namely, for the first time a fuel capsule gave off more energy than was applied to it. This is a major step forward. A similar large-scale device in France, Laser Mégajoule, was officially inaugurated in October 2014. Experiments have started since then, albeit with low laser energies involved.

Description

Basic Fusion

Fusion reactions combine lighter atoms, such as hydrogen, together to form larger ones. Generally the reactions take place at such high temperatures that the atoms have been ionized, their electrons stripped off by the heat; thus, fusion is typically described in terms of "nuclei" instead of "atoms".

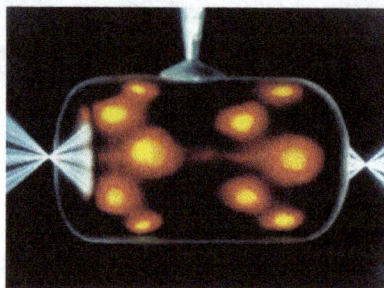

Indirect drive laser ICF uses a *hohlraum* which is irradiated with laser beam cones from either side on its inner surface to bathe a fusion microcapsule inside with smooth high intensity X-rays. The highest energy X-rays can be seen leaking through the hohlraum, represented here in orange/red.

Nuclei are positively charged, and thus repel each other due to the electrostatic force. Overcoming this repulsion costs a considerable amount of energy, which is known as the *Coulomb barrier* or *fusion barrier energy*. Generally, less energy will be needed to cause lighter nuclei to fuse, as they have less charge and thus a lower barrier energy, and when they do fuse, more energy will be released. As the mass of the nuclei increase, there is a point where the reaction no longer gives off net energy—the energy needed to overcome the energy barrier is greater than the energy released in the resulting fusion reaction.

The best fuel from an energy perspective is a one-to-one mix of deuterium and tritium; both are heavy isotopes of hydrogen. The D-T (deuterium & tritium) mix has a low barrier because of its high ratio of neutrons to protons. The presence of neutral neutrons in the nuclei helps pull them together via the nuclear force, while the presence of positively charged protons pushes the nuclei apart via electrostatic force. Tritium has one of the highest ratios of neutrons to protons of any stable or moderately unstable nuclide—two neutrons and one proton. Adding protons or removing neutrons increases the energy barrier.

A mix of D-T at standard conditions does not undergo fusion; the nuclei must be forced together before the nuclear force can pull them together into stable collections. Even in the hot, dense center of the sun, the average proton will exist for billions of years before it fuses. For practical fusion power systems, the rate must be dramatically increased; heated to tens of millions of degrees, and/or compressed to immense pressures. The temperature and pressure required for any particular fuel to fuse is known as the Lawson criterion. These conditions have been known since the 1950s when the first H-bombs were built. To meet the Lawson Criterion is extremely difficult on Earth, which explains why fusion research has taken many years to reach the current high state of technical prowess.

ICF Mechanism of Action

In a hydrogen bomb, the fusion fuel is compressed and heated with a separate fission bomb. A variety of mechanisms transfers the energy of the fission "trigger"'s explosion into the fusion fuel. The requirement of a fission bomb makes the method impractical for power generation. Not only would the triggers be prohibitively expensive to produce, but there is a minimum size that such a bomb can be built, defined roughly by the critical mass of the plutonium fuel used. Generally it seems difficult to build nuclear devices smaller than about 1 kiloton in yield, which would make it a difficult engineering problem to extract power from the resulting explosions.

As the explosion size is scaled down, so too is the amount of energy needed to start the reaction off. Studies from the late 1950s and early 1960s suggested that scaling down into the megajoule energy range would require energy levels that could be delivered by any number of means. This led to the idea of using a device that would "beam" the energy at the fusion fuel, ensuring mechanical separation. By the mid-1960s, it appeared that the laser would develop to the point where the required energy levels would be available.

Generally ICF systems use a single laser, the *driver*, whose beam is split up into a number of beams which are subsequently individually amplified by a trillion times or more. These are sent into the reaction chamber (called a target chamber) by a number of mirrors, positioned in order to illuminate the target evenly over its whole surface. The heat applied by the driver causes the outer layer of the target to explode, just as the outer layers of an H-bomb's fuel cylinder do when illuminated by the X-rays of the fission device.

The material exploding off the surface causes the remaining material on the inside to be driven inwards with great force, eventually collapsing into a tiny near-spherical ball. In modern ICF devices the density of the resulting fuel mixture is as much as one-hundred times the density of lead, around 1000 g/cm³. This density is not high enough to create any useful rate of fusion on its own. However, during the collapse of the fuel, shock waves also form and travel into the center of the fuel at high speed. When they meet their counterparts moving in from the other sides of the fuel in the center, the density of that spot is raised much further.

Given the correct conditions, the fusion rate in the region highly compressed by the shock wave can give off significant amounts of highly energetic alpha particles. Due to the high density of the surrounding fuel, they move only a short distance before being "thermalised", losing their energy to the fuel as heat. This additional energy will cause additional fusion reactions in the heated fuel, giving off more high-energy particles. This process spreads outward from the centre, leading to a kind of self-sustaining burn known as *ignition*.

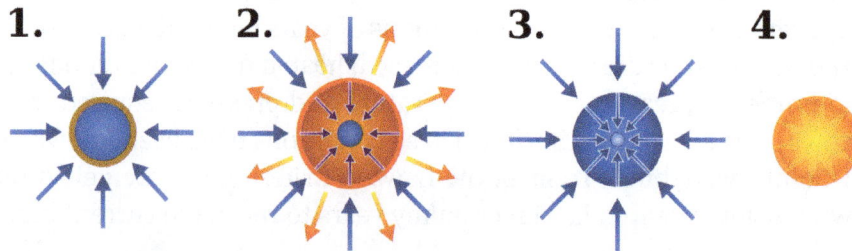

Schematic of the stages of inertial confinement fusion using lasers. The blue arrows represent radiation; orange is blowoff; purple is inwardly transported thermal energy.
1. Laser beams or laser-produced X-rays rapidly heat the surface of the fusion target, forming a surrounding plasma envelope.
2. Fuel is compressed by the rocket-like blowoff of the hot surface material.
3. During the final part of the capsule implosion, the fuel core reaches 20 times the density of lead and ignites at 100,000,000 °C.
4. Thermonuclear burn spreads rapidly through the compressed fuel, yielding many times the input energy.

Issues With Successful Achievement

The primary problems with increasing ICF performance since the early experiments in the 1970s have been of energy delivery to the target, controlling symmetry of the imploding fuel, preventing premature heating of the fuel (before maximum density is achieved), preventing premature mixing of hot and cool fuel by hydrodynamic instabilities and the formation of a 'tight' shockwave convergence at the compressed fuel center.

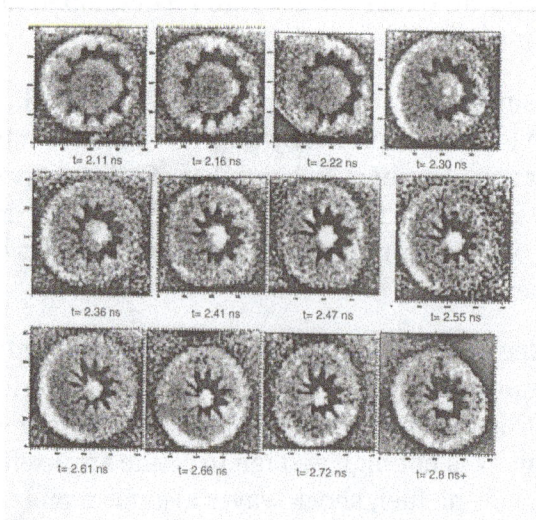

An Inertial confinement fusion target, which was a cylindrical hohlraum target of D-T, being compressed by the Nova Laser. This shot was done in 1995. The image shows the compression of the target, as well as the growth of the Rayleigh-Taylor instabilities.

In order to focus the shock wave on the center of the target, the target must be made with extremely high precision and sphericity with aberrations of no more than a few micrometres over its surface (inner and outer). Likewise the aiming of the laser beams must be extremely precise and the beams must arrive at the same time at all points on the target. Beam timing is a relatively simple issue though and is solved by using delay lines in the beams' optical path to achieve picosecond levels of timing accuracy. The other major problem plaguing the achievement of high symmetry and high temperatures/densities of the imploding target are so called "beam-beam" imbalance and beam anisotropy. These problems are, respectively, where the energy delivered by one beam may be higher or lower than other beams impinging on the target and of "hot spots" within a beam diameter hitting a target which induces uneven compression on the target surface, thereby forming Rayleigh–Taylor instabilities in the fuel, prematurely mixing it and reducing heating efficacy at the time of maximum compression. The Richtmyer-Meshkov instability is also formed during the process due to shock waves being formed.

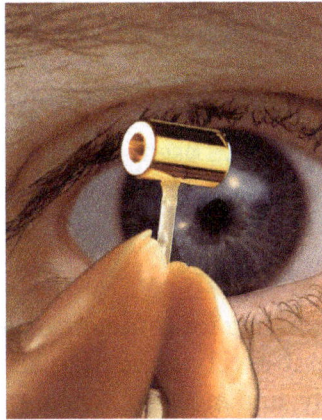

Mockup of a gold plated NIF hohlraum.

All of these problems have been substantially mitigated to varying degrees in the past two decades of research by using various beam smoothing techniques and beam energy diagnostics to balance beam to beam energy; however, RT instability remains a major issue. Target design has also improved tremendously over the years. Modern cryogenic hydrogen ice targets tend to freeze a thin layer of deuterium just on the inside of a plastic sphere while irradiating it with a low power IR laser to smooth its inner surface while monitoring it with a microscope equipped camera, thereby allowing the layer to be closely monitored ensuring its "smoothness". Cryogenic targets filled with a deuterium tritium (D-T) mixture are "self-smoothing" due to the small amount of heat created by the decay of the radioactive tritium isotope. This is often referred to as "beta-layering".

Certain targets are surrounded by a small metal cylinder which is irradiated by the laser beams instead of the target itself, an approach known as "*indirect drive*". In this approach the lasers are focused on the inner side of the cylinder, heating it to a superhot plasma which radiates mostly in X-rays. The X-rays from this plasma are then absorbed by the target surface, imploding it in the same way as if it had been hit with the lasers directly. The absorption of thermal x-rays by the target is more efficient than the direct absorption of laser light, however these *hohlraums* or "burning chambers" also take up considerable energy to heat on their own thus significantly reducing the overall efficiency of laser-to-target energy transfer. They are thus a debated feature even today; the equally numerous "*direct-drive*" design does not use them. Most often, indirect drive hohlraum targets are used to simulate thermonuclear weapons tests due to the fact that the fusion fuel in them is also imploded mainly by X-ray radiation.

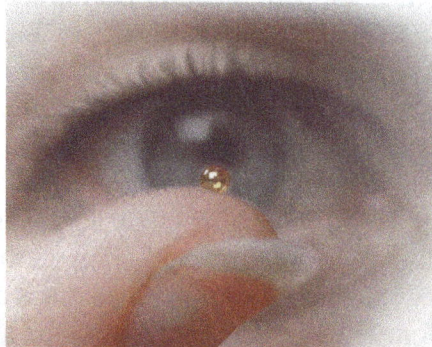

An inertial confinement fusion fuel microcapsule (sometimes called a "microballoon") of the size to be used on the NIF which can be filled with either deuterium and tritium gas or DT ice. The capsule can be either inserted in a hohlraum (as above) and imploded in the **indirect drive** mode or irradiated directly with laser energy in the direct drive configuration. Microcapsules used on previous laser systems were significantly smaller owing to the less powerful irradiation earlier lasers were capable of delivering to the target.

A variety of ICF drivers are being explored. Lasers have improved dramatically since the 1970s, scaling up in energy and power from a few joules and kilowatts to megajoules and hundreds of terawatts, using mostly frequency doubled or tripled light from neodymium glass amplifiers.

Heavy ion beams are particularly interesting for commercial generation, as they are easy to create, control, and focus. On the downside, it is very difficult to achieve the very high energy densities required to implode a target efficiently, and most ion-beam systems require the use of a hohlraum surrounding the target to smooth out the irradiation, reducing the overall efficiency of the coupling of the ion beam's energy to that of the imploding target further.

History of ICF

First Conception

In The US

Inertial confinement feasibility began in the mid-1950s by the inventor of modern television, Dr. Philo Farnsworth. During this time period, he began concerted efforts to recreate the generation of high energy plasma he found in his earlier "Multipactor" tube design. With scant funding, he was successful in developing three generations of his "Fusor" tube. During the last several tests of this device, notable levels of excess output were produced. Patents were approved, and the science of this type of "inertial confinement" fusion is well documented. While it has been relatively dormant, it remains the first true experimentation into the concept.

A second stream of inertial confinement fusion history can be traced back to a seminal meeting called by Edward Teller in 1957 on the topic of peaceful uses of atomic explosions. Among the many topics covered during the event, some consideration was given to using a hydrogen bomb to heat a water-filled underground cavern. The resulting steam would then be used to power conventional generators, and thereby provide electrical power.

This meeting led to the Operation Plowshare efforts, given this name in 1961. Three primary concepts were studied as part of Plowshare; energy generation under Project PACER, the use of large nuclear explosions for excavation, and as a sort of nuclear fracking for the natural gas industry. PACER was directly tested in December 1961 when the 3 kt Project Gnome device was emplaced in bedded salt in New Mexico. In spite of all theorizing and attempts to stop it, radioactive steam was released from the drill shaft, some distance from the test site. Further studies as part of Project PACER led to a number of engineered cavities replacing natural ones, but through this period the entire Plowshare efforts turned from bad to worse, especially after the failure of 1962's Sedan which released huge quantities of fallout. PACER nevertheless continued to receive some funding until 1975, when a 3rd party study demonstrated that the cost of electricity from PACER would be the equivalent to conventional nuclear plants with fuel costs over ten times as great as they were.

Another outcome of the Teller meeting was to prompt John Nuckolls to start considering what happens when the fusion side of the bomb, the "secondary," was scaled down to very small size. His earliest work concerned the study of how small a fusion bomb could be made while still having a large "gain" to provide net energy output. This work suggested that at very small sizes, on the order of milligrams, very little energy would be needed to ignite it, much less than a fission "primary". He proposed building, in effect, tiny all-fusion explosives using a tiny drop of D-T fuel suspended in the center of a metal shell, today known as a hohlraum. The shell provided the same effect as the bomb casing in an H-bomb, trapping x-rays inside so they irradiated the fuel. The main difference is that the x-rays would not be supplied by a primary within the shell, but some sort of external device that heated the shell from the outside until it was glowing in the x-ray region. The power would be delivered by a then-unidentified pulsed power source he referred to using bomb terminology, the "primary".

The main advantage to this scheme is the efficiency of the fusion process at high densities. According to the Lawson criterion, the amount of energy needed to heat the D-T fuel to break-even conditions at ambient pressure is perhaps 100 times greater than the energy needed to compress it to a pressure that would deliver the same rate of fusion. So, in theory, the ICF approach would be dramatically more efficient in terms of gain. This can be understood by considering the energy losses in a conventional scenario where the fuel is slowly heated, as in the case of magnetic fusion energy; the rate of energy loss to the environment is based on the temperature difference between the fuel and its surroundings, which continues to increase as the fuel is heated. In the ICF case, the entire hohlraum is filled with high-temperature radiation, limiting losses.

In Germany

Around the same time (in 1956) a meeting was organized at the Max Planck Institute in Germany by the fusion pioneer Carl Friedrich von Weizsäcker. At this meeting Friedwardt Winterberg proposed the non-fission ignition of a thermonuclear micro-explosion by a convergent shock wave driven with high explosives. Further reference to Winterberg's work in Germany on nuclear micro explosions (mininukes) is contained in a declassified report of the former East German Stasi (Staatsicherheitsdienst).

In 1964 Winterberg proposed that ignition could be achieved by an intense beam of microparticles accelerated to a velocity of 1000 km/s. And in 1968, he proposed to use intense electron and ion beams, generated by Marx generators, for the same purpose. The advantage of this proposal is that the generation of charged particle beams is not only less expensive than the generation of laser beams but also can entrap the charged fusion reaction products due to the strong self-magnetic beam field, drastically reducing the compression requirements for beam ignited cylindrical targets.

In USSR

In 1967 research fellow Gurgen Askaryan published article with proposition to use focused laser beam in fusion lithium deuteride or deuterium.

Early Research

Through the late 1950s, Nuckolls and collaborators at the Lawrence Livermore National Laboratory (LLNL) ran a number of computer simulations of the ICF concept. In early 1960 this produced a full simulation of the implosion of 1 mg of D-T fuel inside a dense shell. The simulation suggested that a 5 MJ power input to the hohlraum would produce 50 MJ of fusion output, a gain of 10. At the time the laser had not yet been invented, and a wide variety of possible drivers were considered, including pulsed power machines, charged particle accelerators, plasma guns, and hypervelocity pellet guns.

Through the year two key theoretical advances were made. New simulations considered the timing of the energy delivered in the pulse, known as "pulse shaping", leading to better implosion. Additionally, the shell was made much larger and thinner, forming a thin shell as opposed to an almost solid ball. These two changes dramatically increased the efficiency of the implosion, and thereby greatly lowered the energy required to compress it. Using these improvements, it was calculated that a driver of about 1 MJ would be needed, a five-fold improvement. Over the next two years several other theoretical advancements were proposed, notably Ray Kidder's development of an implosion system without a hohlraum, the so-called "direct drive" approach, and Stirling Colgate and Ron Zabawski's work on very small systems with as little as 1 µg of D-T fuel.

The introduction of the laser in 1960 at Hughes Research Laboratories in California appeared to present a perfect driver mechanism. Starting in 1962, Livermore's director John S. Foster, Jr. and Edward Teller began a small-scale laser study effort directed toward the ICF approach. Even at this early stage the suitability of the ICF system for weapons research was well understood, and the primary reason for its ability to gain funding. Over the next decade, LLNL made several small experimental devices for basic laser-plasma interaction studies.

Development Begins

In 1967 Kip Siegel started KMS Industries using the proceeds of the sale of his share of an earlier company, Conductron, a pioneer in holography. In the early 1970s he formed KMS Fusion to begin development of a laser-based ICF system. This development led to considerable opposition from the weapons labs, including LLNL, who put forth a variety of reasons that KMS should not be allowed to develop ICF in public. This opposition was funnelled through the Atomic Energy Commission, who demanded funding for their own efforts. Adding to the background noise were rumours of an aggressive Soviet ICF program, new higher-powered CO_2 and glass lasers, the electron beam driver concept, and the 1970s energy crisis which added impetus to many energy projects.

In 1972 Nuckolls wrote an influential public paper in *Nature* introducing ICF and suggesting that testbed systems could be made to generate fusion with drivers in the kJ range, and high-gain systems with MJ drivers.

In spite of limited resources and numerous business problems, KMS Fusion successfully demonstrated fusion from the ICF process on 1 May 1974. However, this success was followed not long after by Siegel's death, and the end of KMS fusion about a year later, having run the company on Siegel's life insurance policy. By this point several weapons labs and universities had started their own programs, notably the solid-state lasers (Nd:glass lasers) at LLNL and the University of Rochester, and krypton fluoride excimer lasers systems at Los Alamos and the Naval Research Laboratory.

Although KMS's success led to a major development effort, the advances that followed were, and still are, hampered by the seemingly intractable problems that characterize fusion research in general.

High-energy ICF

High energy ICF experiments (multi-hundred joules per shot and greater experiments) began in earnest in the early-1970s, when lasers of the required energy and power were first designed. This was some time after the successful design of magnetic confinement fusion systems, and around the time of the particularly successful tokamak design that was introduced in the early '70s. Nevertheless, high funding for fusion research stimulated by the multiple energy crises during the mid to late 1970s produced rapid gains in performance, and inertial designs were soon reaching the same sort of "below break-even" conditions of the best magnetic systems.

LLNL was, in particular, very well funded and started a major laser fusion development program. Their Janus laser started operation in 1974, and validated the approach of using Nd:glass lasers to generate very high power devices. Focusing problems were explored in the Long path laser and Cyclops laser, which led to the larger Argus laser. None of these were intended to be practical ICF devices, but each one advanced the state of the art to the point where there was some confidence the basic approach was valid. At the time it was believed that making a much larger device of the Cyclops type could both compress and heat the ICF targets, leading to ignition in the "short term". This was a misconception based on extrapolation of the fusion yields seen from experiments utilizing the so-called "exploding pusher" type of fuel capsules. During the period spanning the years of the late '70s and early '80s the estimates for laser energy on target needed to achieve ignition doubled almost yearly as the various plasma instabilities and laser-plasma energy coupling loss modes were gradually understood. The realization that the simple exploding pusher target designs and mere few kilojoule (kJ) laser irradiation intensities would never scale to high gain fusion yields led to the effort to increase laser energies to the 100 kJ level in the UV and to the production of advanced ablator and cryogenic DT ice target designs.

Shiva and Nova

One of the earliest serious and large scale attempts at an ICF driver design was the Shiva laser, a 20-beam neodymium doped glass laser system built at the Lawrence Livermore National Laboratory (LLNL) that started operation in 1978. Shiva was a "proof of concept" design intended to demonstrate compression of fusion fuel capsules to many times the liquid density of hydrogen. In this, Shiva succeeded and compressed its pellets to 100 times the liquid density of deuterium. However, due to the laser's strong coupling with hot electrons, premature heating of the dense plasma (ions) was problematic and fusion yields were low. This failure by Shiva to efficiently heat the compressed plasma pointed to the use of optical frequency multipliers as a solution which would frequency triple the infrared light from the laser into the ultraviolet at 351 nm. Newly discovered schemes to efficiently frequency triple high intensity laser light discovered at the Laboratory for Laser Energetics in 1980 enabled this method of target irradiation to be experimented with in the 24 beam OMEGA laser and the NOVETTE laser, which was followed by the Nova laser design with 10 times the energy of Shiva, the first design with the specific goal of reaching ignition conditions.

Nova also failed in its goal of achieving ignition, this time due to severe variation in laser intensity in its beams (and differences in intensity between beams) caused by filamentation which resulted in large non-uniformity in irradiation smoothness at the target and asymmetric implosion. The techniques pioneered earlier could not address these new issues. But again this failure led to a much greater understanding of the process of implosion, and the way forward again seemed clear, namely the increase in uniformity of irradiation, the reduction of hot-spots in the laser beams through beam smoothing techniques to reduce Rayleigh–Taylor instability imprinting on the target and increased laser energy on target by at least an order of magnitude. Funding for fusion research was severely constrained in the 80's, but Nova nevertheless successfully gathered enough information for a next generation machine.

National Ignition Facility

The resulting design, now known as the National Ignition Facility, started construction at LLNL in 1997. NIF's main objective will be to operate as the flagship experimental device of the so-called nuclear stewardship program, supporting LLNLs traditional bomb-making role. Completed in March 2009, NIF has now conducted experiments using all 192 beams, including experiments that set new records for power delivery by a laser. The first credible attempts at ignition were initially scheduled for 2010, but ignition was not achieved as of September 30, 2012. As of October 7, 2013, the facility is understood to have achieved an important milestone towards commercialization of fusion, namely, for the first time a fuel capsule gave off more energy than was applied to it. This is still a long way from satisfying the Lawson criterion, but is a major step forward.

Fast Ignition

A more recent development is the concept of "fast ignition," which may offer a way to directly heat the high density fuel after compression, thus decoupling the heating and compression phases of the implosion. In this approach the target is first compressed "normally" using a driver laser system, and then when the implosion reaches maximum density (at the stagnation point or "bang time"), a second ultra-short pulse ultra-high power petawatt (PW) laser delivers a single pulse focused on one side of the core, dramatically heating it and hopefully starting fusion ignition. The two types of fast ignition are the "plasma bore-through" method and the "cone-in-shell" method. In the first method the petawatt laser is simply expected to bore straight through the outer plasma of an imploding capsule and to impinge on and heat the dense core, whereas in the cone-in-shell method, the capsule is mounted on the end of a small high-z (high atomic number) cone such that the tip of the cone projects into the core of the capsule. In this second method, when the capsule is imploded, the petawatt has a clear view straight to the high density core and does not have to waste energy boring through a 'corona' plasma; however, the presence of the cone affects the implosion process in significant ways that are not fully understood. Several projects are currently underway to explore the fast ignition approach, including upgrades to the OMEGA laser at the University of

Rochester, the GEKKO XII device in Japan, and an entirely new £500 million facility, known as HiPER, proposed for construction in the European Union. If successful, the fast ignition approach could dramatically lower the total amount of energy needed to be delivered to the target; whereas NIF uses UV beams of 2 MJ, HiPER's driver is 200 kJ and heater 70 kJ, yet the predicted fusion gains are nevertheless even higher than on NIF.

Other Projects

Laser Mégajoule, the French project, has seen its first experimental line achieved in 2002, and was finally completed in 2014.

Using a different approach entirely is the z-pinch device. Z-pinch uses massive amounts of electric current which is switched into a cylinder comprising many of extremely fine wires. The wires vaporize to form an electrically conductive plasma that carries a very high current; the resulting circumferential magnetic field squeezes the plasma cylinder, imploding it and thereby generating a high-power x-ray pulse that can be used to drive the implosion of a fuel capsule. Challenges to this approach include relatively low drive temperatures, resulting in slow implosion velocities and potentially large instability growth, and preheat caused by high-energy x-rays.

Most recently, Winterberg has proposed the ignition of a deuterium microexplosion, with a gigavolt super-Marx generator, which is a Marx generator driven by up to 100 ordinary Marx generators.

As An Energy Source

Practical power plants built using ICF have been studied since the late 1970s when ICF experiments were beginning to ramp up to higher powers; they are known as inertial fusion energy, or IFE plants. These devices would deliver a successive stream of targets to the reaction chamber, several a second typically, and capture the resulting heat and neutron radiation from their implosion and fusion to drive a conventional steam turbine.

Technical Challenges

IFE faces continued technical challenges in reaching the conditions needed for ignition. But even if these were all to be solved, there are a significant number of practical problems that seem just as difficult to overcome. Laser-driven systems were initially believed to be able to generate commercially useful amounts of energy. However, as estimates of the energy required to reach ignition grew dramatically during the 1970s and '80s, these hopes were abandoned. Given the low efficiency of the laser amplification process (about 1 to 1.5%), and the losses in generation (steam-driven turbine systems are typically about 35% efficient), fusion gains would have to be on the order of 350 just to energetically break even. These sorts of gains appeared to be impossible to generate, and ICF work turned primarily to weapons research.

With the recent introduction of fast ignition and similar approaches, things have changed dramatically. In this approach gains of 100 are predicted in the first experimental device, HiPER. Given a gain of about 100 and a laser efficiency of about 1%, HiPER produces about the same amount of *fusion* energy as electrical energy was needed to create it. It also appears that an order of magnitude improvement in laser efficiency may be possible through the use of newer designs that replace the flash lamps with laser diodes that are tuned to produce most of their energy in a frequency range that is strongly absorbed. Initial experimental devices offer efficiencies of about 10%, and it is suggested that 20% is a real possibility with some additional development.

With "classical" devices like NIF about 330 MJ of electrical power are used to produce the driver beams, producing an expected yield of about 20 MJ, with the maximum credible yield of 45 MJ. Using the same sorts of numbers in a reactor combining fast ignition with newer lasers would offer dramatically improved performance. HiPER requires about 270 kJ of laser energy, so assuming a first-generation diode laser driver at 10% the reactor would require about 3 MJ of electrical power. This is expected to produce about 30 MJ of fusion power. Even a very poor conversion to electrical energy appears to offer real-world power output, and incremental improvements in yield and laser efficiency appear to be able to offer a commercially useful output.

Practical Problems

ICF systems face some of the same secondary power extraction problems as magnetic systems in generating useful power from their reactions. One of the primary concerns is how to successfully remove heat from the reaction chamber without interfering with the targets and driver beams. Another serious concern is that the huge number of neutrons released in the fusion reactions react with the plant, causing them to become intensely radioactive themselves, as well as mechanically weakening metals. Fusion plants built of conventional metals like steel would have a fairly short lifetime and the core containment vessels will have to be replaced frequently.

One current concept in dealing with both of these problems, as shown in the HYLIFE-II baseline design, is to use a "waterfall" of FLiBe, a molten mix of fluoride salts of lithium and beryllium, which both protect the chamber from neutrons and carry away heat. The FLiBe is then passed into a heat exchanger where it heats water for use in the turbines. Another, Sombrero, uses a reaction chamber built of Carbon-fiber-reinforced polymer which has a very low neutron cross section. Cooling is provided by a molten ceramic, chosen because of its ability to stop the neutrons from traveling any further, while at the same time being an efficient heat transfer agent.

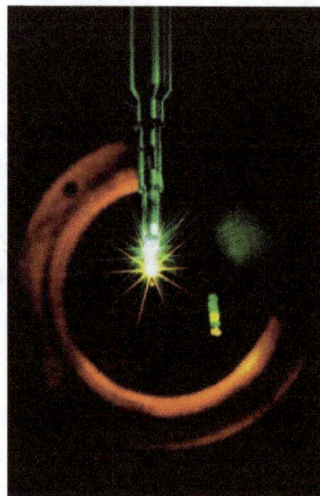

An inertial confinement fusion implosion in Nova, creating "micro sun" conditions of tremendously high density and temperature rivalling even those found at the core of our Sun.

Economic Viability

Even if these technical advances solve the considerable problems in IFE, another factor working against IFE is the cost of the fuel. Even as Nuckolls was developing his earliest detailed calculations on the idea, co-workers pointed this out: if an IFE machine produces 50 MJ of fusion energy, one might expect that a shot could produce perhaps 10 MJ of power for export. Converted to better known units, this is the equivalent of 2.8 kWh of electrical power. Wholesale rates for electrical power on the grid were about 0.3 cents/kWh at the time, which meant the monetary value of the shot was perhaps one cent. In the intervening 50 years the price of power has remained about even with the rate of inflation, and the rate in 2012 in Ontario, Canada was about 2.8 cents/kWh

Thus, in order for an IFE plant to be economically viable, fuel shots would have to cost considerably less than ten cents in year 2012 dollars. At the time this objection was first noted, Nuckolls suggested using liquid droplets sprayed into the hohlraum from an eye-dropper-like apparatus. Given the ever-increasing demands for higher uniformity of the targets, this approach does not appear practical, as even the inner ablator and fuel itself currently costs several orders of magnitude more than this. Moreover, Nuckolls' solution had the fuel dropped into a fixed hohlraum that would be re-used in a continual cycle, but at current energy levels the hohlraum is destroyed with every shot.

Direct-drive systems avoid the use of a hohlraum and thereby may be less expensive in fuel terms. However, these systems still require an ablator, and the accuracy and geometrical considerations are even more important. They are also far less developed than the indirect-drive systems, and face considerably more technical problems in terms of implosion physics. Currently there is no strong consensus whether a direct-drive system would actually be less expensive to operate.

Projected Development

The various phases of such a project are the following, the sequence of inertial confinement fusion development follows much the same outline:

- burning demonstration: reproducible achievement of some fusion energy release (not necessarily a Q factor of >1).

- high gain demonstration: experimental demonstration of the feasibility of a reactor with a sufficient energy gain.

- industrial demonstration: validation of the various technical options, and of the whole data needed to define a commercial reactor.

- commercial demonstration: demonstration of the reactor ability to work over a long period, while respecting all the requirements for safety, liability and cost.

At the moment, according to the available data, inertial confinement fusion experiments have not gone beyond the first phase, although Nova and others have repeatedly demonstrated operation within this realm.

In the short term a number of new systems are expected to reach the second stage.

For a true industrial demonstration, further work is required. In particular, the laser systems need to be able to run at high operating frequencies, perhaps one to ten times a second. Most of the laser systems mentioned in this article have trouble operating even as much as once a day. Parts of the HiPER budget are dedicated to research in this direction as well. Because they convert electricity into laser light with much higher efficiency, diode lasers also run cooler, which in turn allows them to be operated at much higher frequencies. HiPER is currently studying devices that operate at 1 MJ at 1 Hz, or alternately 100 kJ at 10 Hz.

Nuclear Weapons Program

The very hot and dense conditions encountered during an Inertial Confinement Fusion experiment are similar to those created in a thermonuclear weapon, and have applications to the nuclear weapons program. ICF experiments might be used, for example, to help determine how warhead performance will degrade as it ages, or as part of a program of designing new weapons. Retaining knowledge and corporate expertise in the nuclear weapons program is another motivation for pursuing ICF. Funding for the NIF in the United States is sourced from the 'Nuclear Weapons Stockpile Stewardship' program, and the goals of the program are oriented accordingly. It has been argued that some aspects of ICF research may violate the Comprehensive Test Ban Treaty or the Nuclear Non-Proliferation Treaty. In the long term, despite the formidable technical hurdles, ICF research might potentially lead to the creation of a "pure fusion weapon".

Neutron Source

Inertial confinement fusion has the potential to produce orders of magnitude more neutrons than spallation. Neutrons are capable of locating hydrogen atoms in molecules, resolving atomic thermal motion and studying collective excitations of photons more effectively than X-rays. Neutron scattering studies of molecular structures could resolve problems associated with protein folding, diffusion through membranes, proton transfer mechanisms, dynamics of molecular motors, etc. by modulating thermal neutrons into beams of slow neutrons. In combination with fissionable materials, neutrons produced by ICF can potentially be used in Hybrid Nuclear Fusion designs to produce electric power.

Muon-catalyzed Fusion

Muon-catalyzed fusion (μCF) is a process allowing nuclear fusion to take place at temperatures significantly lower than the temperatures required for thermonuclear fusion, even at room temperature or lower. It is one of the few known ways of catalyzing nuclear fusion reactions.

Muons are unstable subatomic particles. They are similar to electrons, but are about 207 times more massive. If a muon replaces one of the electrons in a hydrogen molecule, the nuclei are consequently drawn 207 times closer together than in a normal molecule. When the nuclei are this close together, the probability of nuclear fusion is greatly increased, to the point where a significant number of fusion events can happen at room temperature.

Current techniques for creating large numbers of muons require large amounts of energy, larger than the amounts produced by the catalyzed nuclear fusion reactions. This prevents it from becoming a practical power source. Moreover, each muon has about a 1% chance of "sticking" to the alpha particle produced by the nuclear fusion of a deuteron with a triton, removing the "stuck" muon from the catalytic cycle, meaning that each muon can only catalyze at most a few hundred deuterium tritium nuclear fusion reactions. So, these two factors, of muons being too expensive to make and then sticking too easily to alpha particles, limit muon-catalyzed fusion to a laboratory curiosity. To create useful room-temperature muon-catalyzed fusion, reactors would need a cheaper, more efficient muon source and/or a way for each individual muon to catalyze many more fusion reactions.

History

Andrei Sakharov and F.C. Frank predicted the phenomenon of muon-catalyzed fusion on theoretical grounds before 1950. Yakov Borisovich Zel'dovich also wrote about the phenomenon of muon-catalyzed fusion in 1954. Luis W. Alvarez *et al.*, when analyzing the outcome of some experiments with muons incident on a hydrogen bubble chamber at Berkeley in 1956, observed muon-catalysis of exothermic p-d, proton and deuteron, nuclear fusion, which results in a helion, a gamma ray, and a release of about 5.5 MeV of energy. The Alvarez experimental results, in particular, spurred John David Jackson to publish one of the first comprehensive theoretical studies of muon-catalyzed fusion in his ground-breaking 1957 paper. This paper contained the first serious speculations on useful energy release from muon-catalyzed fusion. Jackson concluded that it would be impractical as an energy source, unless the "alpha-sticking problem" could be solved, leading potentially to an energetically cheaper and more efficient way of utilizing the catalyzing muons.

Viability As A Power Source

Potential Benefits

If muon-catalyzed d-t nuclear fusion were able to be realized practically, it would be a much more attractive way of generating power than conventional nuclear fission reactors because muon-catalyzed d-t nuclear fusion (like most other types of nuclear fusion), produces far fewer harmful (and far less long-lived) radioactive wastes.

The large amount of neutrons produced in muon-catalyzed d-t nuclear fusions may be used to breed fissile fuels, from fertile material - for example, thorium-232 could breed uranium-233 in this way. The fissile fuels that have been bred can then be "burned," either in a conventional supercritical nuclear fission reactor or in an unconventional subcritical fission reactor, for example, a reactor using nuclear transmutation to process nuclear waste, or a reactor using the energy amplifier concept devised by Carlo Rubbia and others.

Problems Facing Practical Exploitation

Except for some refinements, little has changed since Jackson's assessment of the feasibility of muon-catalyzed fusion, other than Vesman's prediction of the hyperfine resonant formation of the muonic (d-μ-t)$^+$ molecular ion, which was subsequently experimentally observed. This helped spark renewed interest in the whole field of muon-catalyzed fusion, which remains an active area of research worldwide. As Jackson observed in his 1957 paper, muon-catalyzed fusion is "unlikely" to provide "useful power production... unless an energetically cheaper way of producing μ^--mesons can be found."

One practical problem with the muon-catalyzed fusion process is that muons are unstable, decaying in about 2.2 μs (in their rest frame). Hence, there needs to be some cheap means of producing muons, and the muons must be arranged to catalyze as many nuclear fusion reactions as possible before decaying.

Another, and in many ways more serious, problem is the "alpha-sticking" problem, which was recognized by Jackson in his 1957 paper. The α-sticking problem is the approximately 1% probability of the muon "sticking" to the alpha particle that results from deuteron-triton nuclear fusion, thereby effectively removing the muon from the muon-catalysis process altogether. Even if muons were absolutely stable, each muon could catalyze, on average, only about 100 d-t fusions before sticking to an alpha particle, which is only about one-fifth the number of muon catalyzed d-t fusions needed for break-even, where as much thermal energy is generated as electrical energy is consumed to produce the muons in the first place, according to Jackson's rough 1957 estimate.

More recent measurements seem to point to more encouraging values for the α-sticking probability, finding the α-sticking probability to be about 0.5% (or perhaps even about 0.4% or 0.3%), which could mean as many as about 200 (or perhaps even about 250 or about 333) muon-catalyzed d-t fusions per muon. Indeed, the team led by Steven E. Jones achieved 150 d-t fusions per muon (average) at the Los Alamos Meson Physics Facility. Unfortunately, 200 (or 250 or even 333) muon-catalyzed d-t fusions per muon is still not enough to reach break-even. Even with break-even, the conversion efficiency from *thermal* energy to *electrical* energy is only about 40% or so, further limiting viability. The best recent estimates of the *electrical* "energy cost" per muon is about 6 GeV with accelerators that are (coincidentally) about 40% efficient at transforming *electrical* energy from the power grid into acceleration of the deuterons.

As of 2012, no practical method of producing energy through this means has been published, although some discoveries using the Hall effect show promise.

Alternative Estimation of Breakeven

According to Gordon Pusch, a physicist at Argonne National Laboratory, various breakeven calculations on muon-catalyzed fusion omit the heat energy the muon beam itself deposits in the target. By taking this factor into account, muon-catalyzed fusion can already exceed breakeven; however, the recirculated power is usually very large compared to power out to the electrical grid (about 3-5 times as large, according to estimates). Despite this rather high recirculated power, the overall cycle efficiency is comparable to conventional fission reactors; however the need for 4-6 MW electrical generating capacity for each megawatt out to the grid probably represents an unacceptably large capital investment. Pusch suggested using Bogdan Maglich's "migma" self-colliding beam concept to significantly increase the muon production efficiency, by eliminating target losses, and using tritium nuclei as the driver beam, to optimize the number of negative muons.

Process

To create this effect, a stream of negative muons, most often created by decaying pions, is sent to a block that may be made up of all three hydrogen isotopes (protium, deuterium, and/or tritium), where the block is usually frozen, and the block may be at temperatures of about 3 kelvin (−270 degrees Celsius) or so. The muon may bump the electron from one of the hydrogen isotopes. The muon, 207 times more massive than the electron, effectively shields and reduces the electromagnetic repulsion between two nuclei and draws them much closer into a covalent bond than an electron can. Because the nuclei are so close, the strong nuclear force is able to kick in and bind both nuclei together. They fuse, release the catalytic muon (most of the time), and part of the original mass of both nuclei is released as energetic particles, as with any other type of nuclear fusion. The release of the catalytic muon is critical to continue the reactions. The majority of the muons continue to bond with other hydrogen isotopes and continue fusing nuclei together. However, not all of the muons are recycled: some bond with other debris emitted following the fusion of the nuclei (such as alpha particles and helions), removing the muons from the catalytic process. This gradually chokes off the reactions, as there are fewer and fewer muons with which the nuclei may bond. The number of reactions achieved in the lab can be as high as 150 d-t fusions per muon (average).

Deuterium-tritium (d-t or dt)

In the muon-catalyzed fusion of most interest, a positively charged deuteron (d), a positively charged triton (t), and a muon essentially form a positively charged muonic molecular heavy hydrogen ion $(d-\mu-t)^+$. The muon, with a rest mass about 207 times greater than the rest mass of an electron, is able to drag the more massive triton and deuteron about 207 times closer together to each other in the *muonic* $(d-\mu-t)^+$ molecular ion than can an electron in the corresponding *electronic* $(d-e-t)^+$ molecular ion. The average separation between the triton and the deuteron in the electronic molecular ion is about one angstrom (100 pm), so the average separation between the triton and the deuteron in the muonic molecular ion is about 207 times smaller than that. Due to the strong nuclear force, whenever the triton and the deuteron in the muonic molecular ion happen to get even closer to each other during their periodic vibrational motions, the probability is very greatly enhanced that the positively charged triton and the positively charged deuteron would undergo quantum tunnelling through the repulsive Coulomb barrier that acts to keep them apart. Indeed, the quantum mechanical tunnelling probability depends roughly exponentially on the average separation between the triton and the deuteron, allowing a single muon to catalyze the d-t nuclear fusion in less than about half a picosecond, once the muonic molecular ion is formed.

The formation time of the muonic molecular ion is one of the "rate-limiting steps" in muon-catalyzed fusion that can easily take up to ten thousand or more picoseconds in a liquid molecular deuterium and tritium mixture (D_2, DT, T_2), for example. Each catalyzing muon thus spends most of its ephemeral existence of about 2.2 microseconds, as measured in its rest frame wandering around looking for suitable deuterons and tritons with which to bind.

Another way of looking at muon-catalyzed fusion is to try to visualize the ground state orbit of a muon around either a deuteron or a triton. Suppose the muon happens to have fallen into an orbit around a deuteron initially, which it has about a 50% chance of doing if there are approximately equal numbers of deuterons and tritons present, forming an electrically neutral *muonic* deuterium atom $(d-\mu)^\circ$ that acts somewhat like a "fat, heavy neutron" due both to its relatively small size (again, about 207 times smaller than an electrically neutral *electronic* deuterium atom $(d-e)^\circ$) and to the very effective "shielding" by the muon of the positive charge of the proton in the deuteron. Even so, the muon still has a much greater chance of being *transferred* to any triton that comes near enough to the muonic deuterium than it does of forming a muonic molecular ion. The electrically neutral muonic tritium atom $(t-\mu)^\circ$ thus formed will act somewhat like an even "fatter, heavier neutron," but it will most likely hang on to its muon, eventually forming a muonic molecular ion, most likely due to the resonant formation of a hyperfine molecular state within an entire deuterium molecule D_2 ($d=e^2=d$), with the muonic molecular ion acting as a "fatter, heavier nucleus" of the "fatter, heavier" neutral "muonic/electronic" deuterium molecule ($[d-\mu-t]=e^2=d$), as predicted by Vesman, an Estonian graduate student, in 1967.

Once the muonic molecular ion state is formed, the shielding by the muon of the positive charges of the proton of the triton and the proton of the deuteron from each other allows the triton and the deuteron to tunnel through the coulomb barrier in time span of order of a nanosecond The muon survives the d-t muon-catalyzed nuclear fusion reaction and remains available (usually) to catalyze further d-t muon-catalyzed nuclear fusions. Each exothermic d-t nuclear fusion releases about 17.6 MeV of energy in the form of a "very fast" neutron having a kinetic energy of about 14.1 MeV and an alpha particle α (a helium-4 nucleus) with a kinetic energy of about 3.5 MeV. An additional 4.8 MeV can be gleaned by having the fast neutrons *moderated* in a suitable "blanket" surrounding the reaction chamber, with the blanket containing lithium-6, whose nuclei, known by some as "lithions," readily and exothermically absorb thermal neutrons, the lithium-6 being transmuted thereby into an alpha particle and a triton.

Deuterium-deuterium (d-d or dd) and Other Types

The first kind of muon-catalyzed fusion to be observed experimentally, by L.W. Alvarez *et al.*, was actually protium (H or $_1H^1$) and deuterium (D or $_1H^2$) muon-catalyzed fusion. The fusion rate for p-d (or pd) muon-catalyzed fusion has been estimated to be about a million times slower than the fusion rate for d-t muon-catalyzed fusion.

Of more practical interest, deuterium-deuterium muon-catalyzed fusion has been frequently observed and extensively studied experimentally, in large part because deuterium already exists in relative abundance and, like hydrogen, deuterium is not at all radioactive (Tritium rarely occurs naturally, and is radioactive with a half-life of about 12.5 years.)

The fusion rate for d-d muon-catalyzed fusion has been estimated to be only about 1% of the fusion rate for d-t muon-catalyzed fusion, but this still gives about one d-d nuclear fusion every 10 to 100 picoseconds or so. However, the energy released with every d-d muon-catalyzed fusion reaction is only about 20% or so of the energy released with every d-t muon-catalyzed fusion reaction. Moreover, the catalyzing muon has a probability of sticking to at least one of the d-d muon-catalyzed fusion reaction products that Jackson in this 1957 paper estimated to be at least 10 times greater than the corresponding probability of the catalyzing muon sticking to at least one of the d-t muon-catalyzed fusion reaction products, thereby preventing the muon from catalyzing any more nuclear fusions. Effectively, this means that each muon catalyzing d-d muon-catalyzed fusion reactions in pure deuterium is only able to catalyze about one-tenth of the number of d-t muon-catalyzed fusion reactions that each muon is able to catalyze in a mixture of equal amounts of deuterium and tritium, and each d-d fusion only yields about one-fifth of the yield of each d-t fusion, thereby making the prospects for useful energy release from d-d muon-catalyzed fusion at least 50 times worse than the already dim prospects for useful energy release from d-t muon-catalyzed fusion.

Potential "aneutronic" (or substantially aneutronic) nuclear fusion possibilities, which result in essentially no neutrons among the nuclear fusion products, are almost certainly not very amenable to muon-catalyzed fusion. This is somewhat disappointing because aneutronic nuclear fusion reactions typically produce substantially only energetic charged particles whose energy could potentially be converted to more useful *electrical* energy with a much higher efficiency than is the case with the conversion of *thermal* energy. One such essentially aneutronic nuclear fusion reaction involves a deuteron from deuterium fusing with a helion (h^{+2}) from helium-3, which yields an energetic alpha particle and a much more energetic proton, both positively charged (with a few neutrons coming from inevitable d-d nuclear fusion side reactions). However, one muon with only one negative electric charge is incapable of shielding both positive charges of a helion from the one positive charge of a deuteron. The chances of the requisite *two* muons being present simultaneously are exceptionally remote.

In Culture

The term "cold fusion" was coined to refer to muon-catalyzed fusion in a 1956 New York Times article about Luis W. Alvarez's paper.

In 1957 Theodore Sturgeon wrote a short story "The Pod in the Barrier" where humanity has ubiquitous cold fusion reactors that work with muons. The reaction is "When Hydrogen One and Hydrogen Two are in the presence of Mu Mesons, they fuse into Helium Three, with an energy yield in electron volts of 5.4 times ten to the fifth power". Unlike the thermonuclear bomb contained in the Pod (which is used to destroy the Barrier) they can become temporarily disabled by "concentrated disbelief" that muon fusion works.

Fusion Power

Fusion power is energy generated by nuclear fusion. Fusion reactions fuse two lighter atomic nuclei to form a heavier nucleus. It is a major area of plasma physics research that attempts to harness such reactions as a source of large scale sustainable energy. Fusion reactions are how stars transmute matter into energy.

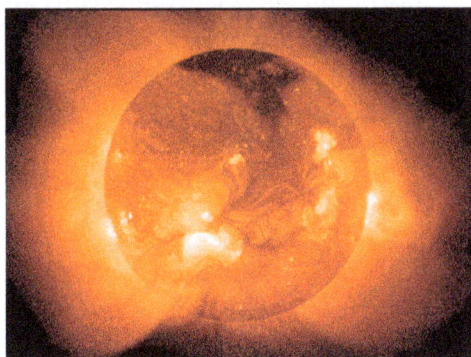

The Sun is a natural fusion reactor.

In most large scale commercial programs, heat from neutron scattering in a controlled reaction is used to operate a steam turbine that drives electric generators. Many fusion concepts are under investigation. The current leading designs are the tokamak and inertial confinement by laser. As of January 2016, these technologies were not viable, as they cannot produce more energy than is required to initiate and sustain a fusion reaction.

Alternative approaches rely on other means of energy transfer, mostly that capture energy without relying on neutron capture.

Background

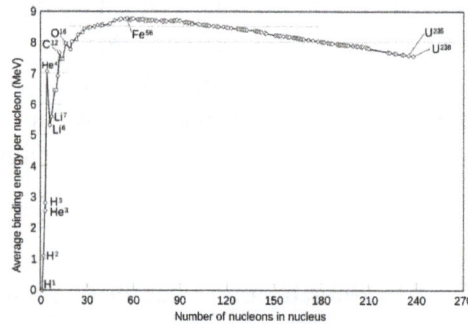

Binding energy for different atoms. Iron-56 has the highest, making it the most stable. Atoms to the left are likely to fuse; atoms to the right are likely to split.

Mechanism

Fusion reactions occur when two (or more) atomic nuclei come close enough for long enough that the strong nuclear force pulling them together exceeds the electrostatic force pushing them apart, fusing them into heavier nuclei. For nuclei lighter than iron-56, the reaction is exothermic, releasing energy. For nuclei heavier than iron-56, the reaction is endothermic, requiring an external source of energy. Hence, nuclei smaller than iron-56 are more likely to fuse while those heavier than iron-56 are more likely to break apart.

The strong force acts only over short distances. The repulsive electrostatic force acts over longer distances, so kinetic energy is needed to overcome this "Coulomb barrier" before the reaction can take place. Way of doing this include speeding up atoms in a particle accelerator, or heating them to high temperatures.

Once an atom is heated above its ionization energy, its electrons are stripped away (it is ionized), leaving just the bare nucleus (the ion). The result is a hot cloud of ions and the electrons formerly attached to them. This cloud is known as a plasma. Because the charges are separated, plasmas are electrically conductive and magnetically controllable. Many fusion devices take advantage of this to control the particles as they are heated.

Cross Section

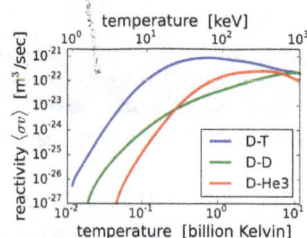

The fusion reaction rate increases rapidly with temperature until it maximizes and then gradually drops off. The deuterium-tritium fusion rate peaks at a lower temperature (about 70 keV, or 800 million kelvin) and at a higher value than other reactions commonly considered for fusion energy.

A reaction's cross section, denoted σ, is the measure of the probability that a fusion reaction will happen. This depends on the relative velocity of the two nuclei. Higher relative velocities increase the probability. Cross sections for many fusion reactions were measured (mainly in the 1970s) using particle beams.

In a plasma, particle velocity can be characterized using a probability distribution. If the plasma is thermalized, the distribution looks like a bell curve, or maxwellian distribution. In this case, it is useful to take the average cross section over the velocity distribution. This is entered into the volumetric fusion rate:

$$P_{fusion} = n_A n_B \langle \sigma v_{A,B} \rangle E_{fusion}$$

where:

- $_{fusion}$ is the energy made by fusion, per time and volume
- n is the number density of species A or B, the particles in the volume
- $\sigma v_{A,B}$ is the cross section of that reaction, average over all the velocities of the two species v
- E_{fusion} is the energy released by that fusion reaction.

Lawson Criterion

The Lawson Criterion shows how energy varies with temperature, density, speed of collision and fuel. This equation was central to John Lawson's analysis of fusion working with a hot plasma. Lawson assumed an energy balance, shown below.

*Net Power = Efficiency * (Fusion - Radiation Loss - Conduction Loss)*

- *Net Power* is the net power for any fusion power station.
- *Efficiency* how much energy is needed to drive the device and how well it collects power.
- *Fusion* is rate of energy generated by the fusion reactions.
- *Radiation* is the energy lost as light, leaving the plasma.
- *Conduction* is the energy lost, as momentum leaves the plasma.

Plasma clouds lose energy through conduction and radiation. Conduction occurs when ions, electrons or neutrals impact a surface and transfer a portion of their kinetic energy to the atoms of the surface. Radiation is energy that leaves the cloud as light in the visible, UV, IR, or X-ray spectra. Radiation increases with temperature. Fusion power technologies must overcome these losses.

Triple Product: Density, Temperature, Time

The Lawson criterion argues that a machine holding a thermalized (hot) and quasi-neutral plasma has to meet basic criteria to overcome radiation losses, conduction losses and reach efficiency of 30 percent. This became known as the "triple product": the plasma density, temperature and confinement time. Attempts to increase the triple product led to targeting larger plants. Larger plants move structural materials further away from the centre of the plasma, which reduces conduction and radiation losses since more of the radiation is internally reflected. This emphasis on $(nT\tau)$ as a metric of success has impacted other considerations such as cost, size, complexity and efficiency. This has led to larger, more complicated and more expensive machines such as ITER and NIF.

Plasma Behavior

Plasma is an ionized gas that conducts electricity. In bulk, it is modeled using hydrodynamics, which is a combination of the Navier-Stokes equations governing fluids and Maxwell's equations governing how magnetic and electric fields behave. Fusion exploits several plasma properties, including:

Self-organizing plasma conducts electric and magnetic fields. Its motions can generate fields that can in turn contain it.

Diamagnetic plasma can generate its own internal magnetic field. This can reject an externally applied magnetic field, making it diamagnetic.

Magnetic mirrors can reflect plasma when it moves from a low to high density field.

Energy Capture

Multiple approaches have been proposed for energy capture. The simplest is to heat a fluid. The neutrons generated by fusion can re-generate a spent fission fuel. Direct energy conversion was developed (at LLNL in the 1980s) as a method to maintain a voltage using the fusion reaction products. This has demonstrated energy capture efficiency of 48 percent.

Approaches

Magnetic Confinement

Tokamak: the most well-developed and well-funded approach to fusion energy. This method races hot plasma around in a magnetically confined, donut-shaped ring, with an internal current. When completed, ITER will be the world's largest tokamak. As of April 2012 an estimated 215 experimental tokamaks were either planned, decommissioned or currently operating (35) worldwide.

Spherical tokamak: A variation on the tokamak with a spherical shape.

Stellarator: Twisted rings of hot plasma. The stellarator attempts to create a natural twist plasma path, using external magnets, while tokamaks create those magnetic fields using an internal current. Stellarators were developed by Lyman Spitzer in 1950 and have four designs: Torsatron, Heliotron, Heliac and Helias. One example is Wendelstein 7-X, a German fusion device that produced its first plasma on December 10, 2015. It is the world's largest stellarator, designed to investigate the suitability of this type of device for a power station.

Levitated Dipole Experiment (LDX): These use a solid superconducting torus. This is magnetically levitated inside the reactor chamber. The superconductor forms an axisymmetric magnetic field that contains the plasma. The LDX was developed by MIT and Columbia University after 2000 by Jay Kesner and Michael E. Mauel.

Magnetic mirror: Developed by Richard F. Post and teams at LLNL in the 1960s. Magnetic mirrors reflected hot plasma back and forth in a line. Variations included the magnetic bottle and the biconic cusp. A series of well-funded, large, mirror machines were built by the US government in the 1970s and 1980s.

Field-reversed configuration: This device traps plasma in a self-organized quasi-stable structure; where the particle motion makes an internal magnetic field which then traps itself.

Reversed field pinch: Here the plasma moves inside a ring. It has an internal magnetic field. Moving out from the center of this ring, the magnetic field reverses direction.

Inertial Confinement

Direct drive: In this technique, lasers directly blast a pellet of fuel. The goal is to ignite a fusion chain reaction. Ignition was first suggested by John Nuckolls, in 1972. Notable direct drive experiments have been conducted at the Laboratory for Laser Energetics, Laser Mégajoule and the GEKKO XII facilities. Good implosions require fuel pellets with close to a perfect shape in order to generate a symmetrical inward shock wave that produces the high-density plasma.

Fast ignition: This method uses two laser blasts. The first blast compresses the fusion fuel, while the second high energy pulse ignites it. Experiments have been conducted at the Laboratory for Laser Energetics using the Omega and Omega EP systems and at the GEKKO XII laser at the Institute for Laser Engineering in Osaka Japan.

Indirect drive: In this technique, lasers blasts a structure around the pellet of fuel. This structure is known as a Hohlraum. As it disintegrates the pellet is bathed in a more uniform x-ray light, creating better compression. The largest system using this method is the National Ignition Facility.

Magneto-inertial fusion or *Magnetized Liner Inertial Fusion:* This combines a laser pulse with a magnetic pinch. The pinch community refers to it as magnetized liner Inertial fusion while the ICF community refers to it as magneto-inertial fusion.

Heavy Ion Beams There are also proposals to do inertial confinement fusion with ion beams instead of laser beams. The main difference is the mass of the beam has momentum, whereas lasers do not.

Magnetic or Electric Pinches

Z-Pinch: This method sends a strong current (in the z-direction) through the plasma. The current generates a magnetic field that squeezes the plasma to fusion conditions. Pinches were the first method for man-made controlled fusion. Some examples include the Dense plasma focus and the Z machine at Sandia National Laboratories.

Theta-Pinch: This method sends a current inside a plasma, in the theta direction.

Screw Pinch: This method combines a theta and z-pinch for improved stabilization.

Inertial Electrostatic Confinement

Fusor: This method uses an electric field to heat ions to fusion conditions. The machine typically uses two spherical cages, a cathode inside the anode, inside a vacuum. These machines are not considered a viable approach to net power because of their high conduction and radiation losses. They are simple enough to build that amateurs have fused atoms using them.

Polywell: This designs attempts to combine magnetic confinement with electrostatic fields, to avoid the conduction losses generated by the cage.

Other

Magnetized target fusion: This method confines hot plasma using a magnetic field and squeezes it using inertia. Examples include LANL FRX-L machine, General Fusion and the plasma liner experiment.

Uncontrolled: Fusion has been initiated by man, using uncontrolled fission explosions to ignite so-called Hydrogen Bombs. Early proposals for fusion power included using bombs to initiate reactions.

Beam fusion: A beam of high energy particles can be fired at another beam or target and fusion will occur. This was used in the 1970s and 1980s to study the cross sections of high energy fusion reactions.

Bubble fusion: This was a fusion reaction that was supposed to occur inside extraordinarily large collapsing gas bubbles, created during acoustic liquid cavitation. This approach was discredited.

Cold fusion: This is a hypothetical type of nuclear reaction that would occur at, or near, room temperature. Cold fusion is discredited and gained a reputation as pathological science.

Muon-catalyzed fusion: Muons allow atoms to get much closer and thus reduce the kinetic energy required to initiate fusion. Muons require more energy to produce than can be obtained from muon-catalysed fusion, making this approach impractical for power generation.

Gravitational-confinement fusion (GCF) Direct Photo-Electric Conversion: Also known as Space-Based Solar Power argues that a majority of available fusion fuels exists within the sphere of the Sun where it is gravitationally confined, and that a tractable way to accomplish large-scale fusion power is to build very large space-borne platforms that capture energy via photons rather than via a carnot cycle. The theoretical limit of power via this means is a type-2 civilization via a Dyson Sphere.

Common Tools

Heating

Gas must be first heated to form a plasma. This then needs to be hot enough to start fusion reactions. A number of heating schemes have been explored:

Radiofrequency Heating A radio wave is applied to the plasma, causing it to oscillate. This is basically the same concept as a microwave oven. This is also known as electron cyclotron resonance heating or Dielectric heating.

Electrostatic Heating An electric field can do work on charged ions or electrons, heating them.

Neutral Beam Injection An external source of hydrogen is ionized and accelerated by an electric field to form a charged beam which is shone through a source of neutral hydrogen gas towards the plasma which itself is ionized and contained in the reactor by a magnetic field. Some of the intermediate hydrogen gas is accelerated towards the plasma by collisions with the charged beam while remaining neutral: this neutral beam is thus unaffected by the magnetic field and so shines through it into the plasma. Once inside the plasma the neutral beam transmits energy to the plasma by collisions as a result of which it becomes ionized and thus contained by the magnetic field thereby both heating and refuelling the reactor in one operation. The remainder of the charged beam is diverted by magnetic fields onto cooled beam dumps.

Magnetic Oscillations

Measurement

Thomson Scattering Light scatters from plasma. This light can be detected and used to reconstruct the plasmas' behavior. This technique can be used to find its density and temperature. It is common in Inertial confinement fusion, Tokamaks and fusors. In ICF systems, this can be done by firing a second beam into a gold foil adjacent to the target. This makes x-rays that scatter or traverse the plasma. In Tokamaks, this can be done using mirrors and detectors to reflect light across a plane (two dimensions) or in a line (one dimension).

Langmuir probe This is a metal object placed in a plasma. A potential is applied to it, giving it a positive or negative voltage against the surrounding plasma. The metal collects charged particles, drawing a current. As the voltage changes, the current changes. This makes a IV Curve. The IV-curve can be used to determine the local plasma density, potential and temperature.

Geiger counter Deuterium or tritium fusion produces neutrons. Geiger counters record the rate of neutron production, so they are an essential tool for demonstrating success.

Flux loop A loop of wire is inserted into the magnetic field. As the field passes through the loop, a current is made. The current is measured and used to find the total magnetic flux through that loop. This has been used on the National Compact Stellarator Experiment, the polywell and the LDX machines.

X-ray detector All plasma loses energy by emitting light. This covers the whole spectrum: visible, IR, UV, and X-rays. This occurs anytime a particle changes speed, for any reason. If the reason is deflection by a magnetic field, the radiation is Cyclotron radiation at low speeds and Synchrotron radiation at high speeds. If the reason is deflection by another particle, plasma radiates X-rays, known as Bremsstrahlung radiation. X-rays are termed in both hard and soft, based on their energy.

Power Production

Steam turbines It has been proposed that steam turbines be used to convert the heat from the fusion chamber into electricity. The heat is transferred into a working fluid that turns into steam, driving electric generators.

Neutron blankets Deuterium and tritium fusion generates neutrons. This varies by technique (NIF has a record of 3E14 neutrons per second while a typical fusor produces 1E5–1E9 neutrons per second). It has been proposed to use these neutrons as a way to regenerate spent fission fuel or as a way to breed tritium from a liquid lithium blanket.

Direct conversion This is a method where the kinetic energy of a particle is converted into voltage. It was first suggested by Richard F. Post in conjunction with magnetic mirrors, in the late sixties. It has also been suggested for Field-Reversed Configurations. The process takes the plasma, expands it, and converts a large fraction of the random energy of the fusion products into directed motion. The particles are then collected on electrodes at various large electrical potentials. This method has demonstrated an experimental efficiency of 48 percent.

Confinement

Confinement refers to all the conditions necessary to keep a plasma dense and hot long enough to undergo fusion. Here are some general principles.

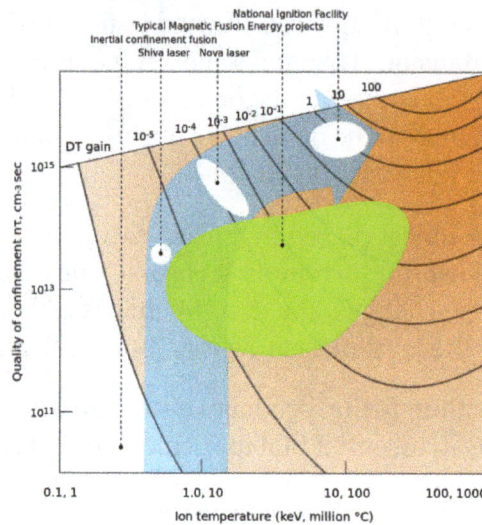

Parameter space occupied by inertial fusion energy and magnetic fusion energy devices as of the mid 1990s. The regime allowing thermonuclear ignition with high gain lies near the upper right corner of the plot.

- Equilibrium: The forces acting on the plasma must be balanced for containment. One exception is inertial confinement, where the relevant physics must occur faster than the disassembly time.

- Stability: The plasma must be so constructed so that disturbances will not lead to the plasma disassembling.

- Transport or conduction: The loss of material must be sufficiently slow. The plasma carries off energy with it, so rapid loss of material will disrupt any machines power balance. Material can be lost by transport into different regions or conduction through a solid or liquid.

To produce self-sustaining fusion, the energy released by the reaction (or at least a fraction of it) must be used to heat new reactant nuclei and keep them hot long enough that they also undergo fusion reactions.

Unconfined

The first human-made, large-scale fusion reaction was the test of the hydrogen bomb, Ivy Mike, in 1952. As part of the PACER project, it was once proposed to use hydrogen bombs as a source of power by detonating them in underground caverns and then generating electricity from the heat produced, but such a power station is unlikely ever to be constructed.

Magnetic Confinement

At the temperatures required for fusion, the fuel is heated to a plasma state. In this state it has a very good electrical conductivity. This opens the possibility of confining the plasma with magnetic fields. This is the case of magnetized plasma, where the magnetic fields and plasma intermix. This is generally known as magnetic confinement. The field lines put a Lorentz force on the plasma. The force works perpendicular to the magnetic fields, so one problem in magnetic confinement is preventing the plasma from leaking out the ends of the field lines. A general measure of magnetic trapping in fusion is the beta ratio:

$$\beta = \frac{p}{p_{mag}} = \frac{nk_B T}{(B^2 / 2\mu_0)}$$

This is the ratio of the externally applied field to the internal pressure of the plasma. A value of 1 is ideal trapping. Some examples of beta vales include:

1. The START machine: 0.32

2. The Levitated dipole experiment: 0.26

3. Spheromaks: $\approx$ 0.1, Maximum 0.2 based on Mercier limit.

4. The DIII-D machine: 0.126

5. The Gas Dynamic Trap a magnetic mirror: 0.6 for 5E-3 seconds.

Magnetic Mirror One example of magnetic confinement is with the magnetic mirror effect. If a particle follows the field line and enters a region of higher field strength, the particles can be reflected. There are several devices that try to use this effect. The most famous was the magnetic mirror machines, which was a series of large, expensive devices built at the Lawrence Livermore National Laboratory from the 1960s to mid 1980s. Some other examples include the magnetic bottles and Biconic cusp. Because the mirror machines were straight, they had some advantages over a ring shape. First, mirrors were easier to construct and maintain and second direct conversion energy capture, was easier to implement. As the confinement achieved in experiments was poor, this approach was abandoned.

Magnetic Loops Another example of magnetic confinement is to bend the field lines back on themselves, either in circles or more commonly in nested toroidal surfaces. The most highly developed system of this type is the *tokamak*, with the *stellarator* being next most advanced, followed by the Reversed field pinch. Compact toroids, especially the *Field-Reversed Configuration* and the spheromak, attempt to combine the advantages of toroidal magnetic surfaces with those of a simply connected (non-toroidal) machine, resulting in a mechanically simpler and smaller confinement area.

Inertial Confinement

Inertial confinement is the use of rapidly imploding shell to heat and confine plasma. The shell is imploded using a direct laser blast (direct drive) or a secondary x-ray blast (indirect drive) or heavy ion beams. Theoretically, fusion using lasers would be done using tiny pellets of fuel that explode several times a second. To induce the explosion, the pellet must be compressed to about 30 times solid density with energetic beams. If direct drive is used—the beams are focused directly on the pellet—it can in principle be very efficient, but in practice is difficult to obtain the needed uniformity. The alternative approach, indirect drive, uses beams to heat a shell, and then the shell radiates x-rays, which then implode the pellet. The beams are commonly laser beams, but heavy and light ion beams and electron beams have all been investigated.

Electrostatic Confinement

There are also electrostatic confinement fusion devices. These devices confine ions using electrostatic fields. The best known is the Fusor. This device has a cathode inside an anode wire cage. Positive ions fly towards the negative inner cage, and are heated by the electric field in the process. If they miss the inner cage they can collide and fuse. Ions typically hit the cathode, however, creating prohibitory high conduction losses. Also, fusion rates in fusors are very low because of competing physical effects, such as energy loss in the form of light radiation. Designs have been proposed to avoid the problems associated with the cage, by generating the field using a non-neutral cloud. These include a plasma oscillating device, a magnetically-shielded-grid a penning trap and the polywell. The technology is relatively immature, however, and many scientific and engineering questions remain.

History of Research

1920s

Research into nuclear fusion started in the early part of the 20th century. In 1920 the British physicist Francis William Aston discovered that the total mass equivalent of four hydrogen atoms (two protons and two neutrons) are heavier than the total mass of one helium atom (He-4), which implied that net energy can be released by combining hydrogen atoms together to form helium, and provided the first hints of a mechanism by which stars could produce energy in the quantities being measured. Through the 1920s, Arthur Stanley Eddington became a major proponent of the proton–proton chain reaction (PP reaction) as the primary system running the Sun.

1930s

A theory was verified by Hans Bethe in 1939 showing that beta decay and quantum tunneling in the Sun's core might convert one of the protons into a neutron and thereby producing deuterium rather than a diproton. The deuterium would then fuse through other reactions to further increase the energy output. For this work, Bethe won the Nobel Prize in Physics.

1940s

In 1942, nuclear fusion research was subsumed into the Manhattan Project when the secrecy surrounding the field obscured by the science. The first patent related to a fusion reactor was registered in 1946 by the United Kingdom Atomic Energy Authority. The inventors were Sir George Paget Thomson and Moses Blackman. This was the first detailed examination of the Z-pinch concept.

Z-pinch is based on the fact that plasmas are electrically conducting. Running a current through the plasma, will generate a magnetic field around the plasma. This field will, according to Lenz's law, create an inward directed force that causes the plasma to collapse inward, raising its density. Denser plasmas generate denser magnetic fields, increasing the inward force, leading to a chain reaction. If the conditions are correct, this can lead to the densities and temperatures needed for fusion. The difficulty is getting the current into the plasma, which would normally melt any sort of mechanical electrode. A solution emerges again because of the conducting nature of the plasma; by placing the plasma in the middle of an electromagnet, induction can be used to generate the current.

Starting in 1947, two UK teams carried out small experiments and began building a series of ever-larger experiments. When the Huemul results hit the news, James L. Tuck, a UK physicist working at Los Alamos, introduced the pinch concept in the US and produced a series of machines known as the Perhapsatron. The Soviet Union, unbeknownst to the West, was also building a series of similar machines. All of these devices quickly demonstrated a series of instabilities when the pinch was applied. This broke up the plasma column long before it reached the densities and temperatures required for fusion.

1950s

The first successful man-made fusion device was the boosted fission weapon tested in 1951 in the Greenhouse Item test. This was followed by true fusion weapons in 1952's Ivy Mike, and the first practical examples in 1954's Castle Bravo. This was uncontrolled fusion. In these devices, the energy released by the fission explosion is used to compress and heat fusion fuel, starting a fusion reaction. Fusion releases neutrons. These neutrons hit the surrounding fission fuel, causing the atoms to split apart much faster than normal fission processes—almost instantly by comparison. This increases the effectiveness of bombs: normal fission weapons blow themselves apart before all their fuel is used; fusion/fission weapons do not have this practical upper limit.

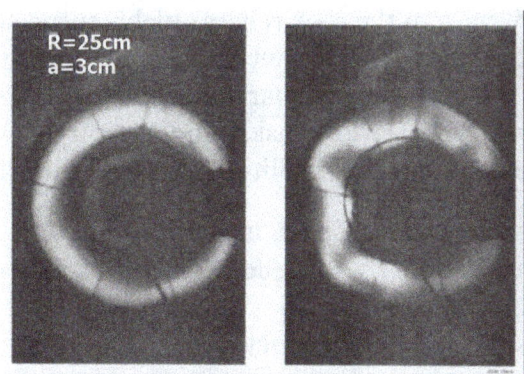

Early photo of plasma inside a pinch machine (imperial college 1950/1951)

In 1949 an expatriate German, Ronald Richter, proposed the Huemul Project in Argentina, announcing positive results in 1951. These turned out to be fake, but it prompted considerable interest in the concept as a whole. In particular, it prompted Lyman Spitzer to begin considering ways to solve some of the more obvious problems involved in confining a hot plasma, and, unaware of the z-pinch efforts, he developed a new solution to the problem known as the stellarator. Spitzer applied to the US Atomic Energy Commission for funding to build a test device. During this period, Jim Tuck who had worked with the UK teams had been introducing the z-pinch concept to his coworkers at his new job at Los Alamos National Laboratory (LANL). When he heard of Spitzer's pitch for funding, he applied to build a machine of his own, the Perhapsatron.

Spitzer's idea won funding and he began work on the stellarator under the code name Project Matterhorn. His work led to the creation of the Princeton Plasma Physics Laboratory. Tuck returned to LANL and arranged local funding to build his machine. By this time, however, it was clear that all of the pinch machines were suffering from the same issues involving stability, and progress stalled. In 1953, Tuck and others suggested a number of solutions to the stability problems. This led to the design of a second series of pinch machines, led by the UK ZETA and Sceptre devices.

Spitzer had planned an aggressive development project of four machines, A, B, C, and D. A and B were small research devices, C would be the prototype of a power-producing machine, and D would be the prototype of a commercial device. A worked without issue, but even by the time B was being used it was clear the stellarator was also suffering from instabilities and plasma leakage. Progress on C slowed as attempts were made to correct for these problems.

By the mid-1950s it was clear that the simple theoretical tools being used to calculate the performance of all fusion machines were simply not predicting their actual behavior. Machines invariably leaked their plasma from their confinement area at rates far higher than predicted. In 1954, Edward Teller held a gathering of fusion researchers at the Princeton Gun Club, near the Project Matterhorn (now known as Project Sherwood) grounds. Teller started by pointing out the problems that everyone was having, and suggested that any system where the plasma was confined within concave fields was doomed to fail. Attendees remember him saying something to the effect that the fields were like rubber bands, and they would attempt to snap back to a straight configuration whenever the power was increased, ejecting the plasma. He went on to say that it appeared the only way to confine the plasma in a stable configuration would be to use convex fields, a "cusp" configuration.

When the meeting concluded, most of the researchers quickly turned out papers saying why Teller's concerns did not apply to their particular device. The pinch machines did not use magnetic fields in this way at all, while the mirror and stellarator seemed to have various ways out. This was soon followed by a paper by Martin David Kruskal and Martin Schwarzschild discussing pinch machines, however, which demonstrated instabilities in those devices were inherent to the design.

The largest "classic" pinch device was the ZETA, including all of these suggested upgrades, starting operations in the UK in 1957. In early 1958, John Cockcroft announced that fusion had been achieved in the ZETA, an announcement that made headlines around the world. When physicists in the US expressed concerns about the claims they were initially dismissed. US experiments soon demonstrated the same neutrons, although temperature measurements suggested these could not be from fusion reactions. The neutrons seen in the UK were later demonstrated to be from different versions of the same instability processes that plagued earlier machines. Cockcroft was forced to retract the fusion claims, and the entire field was tainted for years. ZETA ended its experiments in 1968.

The first controlled fusion experiment was accomplished using Scylla I at the Los Alamos National Laboratory in 1958. This was a pinch machine, with a cylinder full of deuterium. Electric current shot down the sides of the cylinder. The current made magnetic fields that compressed the plasma to 15 million degrees Celsius, squeezed the gas, fused it and produced neutrons.

In 1950–1951 I.E. Tamm and A.D. Sakharov in the Soviet Union, first discussed a tokamak-like approach. Experimental research on those designs began in 1956 at the Kurchatov Institute in Moscow by a group of Soviet scientists led by Lev Artsimovich. The tokamak essentially combined a low-power pinch device with a low-power simple stellarator. The key was to combine the fields in such a way that the particles orbited within the reactor a particular number of times, today known as the "safety factor". The combination of these fields dramatically improved confinement times and densities, resulting in huge improvements over existing devices.

1960s

A key plasma physics text was published by Lyman Spitzer at Princeton in 1963. Spitzer took the ideal gas laws and adopted them to an ionized plasma, developing many of the fundamental equations used to model a plasma.

Laser fusion was suggested in 1962 by scientists at Lawrence Livermore National Laboratory, shortly after the invention of the laser itself in 1960. At the time, Lasers were low power machines, but low-level research began as early as 1965. Laser fusion, formally known as inertial confinement fusion, involves imploding a target by using laser beams. There are two ways to do this: indirect drive and direct drive. In direct drive, the laser blasts a pellet of fuel. In indirect drive, the lasers blast a structure around the fuel. This makes x-rays that squeeze the fuel. Both methods compress the fuel so that fusion can take place.

At the 1964 World's Fair, the public was given its first demonstration of nuclear fusion. The device was a θ-pinch from General Electric. This was similar to the Scylla machine developed earlier at Los Alamos.

The magnetic mirror was first published in 1967 by Richard F. Post and many others at the Lawrence Livermore National Laboratory. The mirror consisted of two large magnets arranged so they had strong fields within them, and a weaker, but connected, field between them. Plasma introduced in the area between the two magnets would "bounce back" from the stronger fields in the middle.

The A.D. Sakharov group constructed the first tokamaks, the most successful being the T-3 and its larger version T-4. T-4 was tested in 1968 in Novosibirsk, producing the world's first quasistationary fusion reaction. When this were first announced, the international community was highly skeptical. A British team was invited to see T-3, however, and after measuring it in depth they released their results that confirmed the Soviet claims. A burst of activity followed as many planned devices were abandoned and new tokamaks were introduced in their place — the C model stellarator, then under construction after many redesigns, was quickly converted to the Symmetrical Tokamak.

In his work with vacuum tubes, Philo Farnsworth observed that electric charge would accumulate in regions of the tube. Today, this effect is known as the Multipactor effect. Farnsworth reasoned that if ions were concentrated high enough they could collide and fuse. In 1962, he filed a patent on a design using a positive inner cage to concentrate plasma, in order to achieve nuclear fusion. During this time, Robert L. Hirsch joined the Farnsworth Television labs and began work on what became the fusor. Hirsch patented the design in 1966 and published the design in 1967.

1970s

In 1972, John Nuckolls outlined the idea of ignition. This is a fusion chain reaction. Hot helium made during fusion reheats the fuel and starts more reactions. John argued that ignition would require lasers of about 1 kJ. This turned out to be wrong. Nuckolls's paper started a major development effort. Several laser systems were built at LLNL. These included the argus, the Cyclops, the Janus, the long path, the Shiva laser and the Nova in 1984. This prompted the UK to build the Central Laser Facility in 1976.

Shiva laser, 1977, the largest ICF laser system built in the seventies

During this time, great strides in understanding the tokamak system were made. A number of improvements to the design are now part of the "advanced tokamak" concept, which includes non-circular plasma, internal diverters and limiters, often superconducting magnets, and operate in the so-called "H-mode" island of increased stability. Two other designs have also become fairly well studied; the compact tokamak is wired with the magnets on the inside of the vacuum chamber, while the spherical tokamak reduces its cross section as much as possible.

The Tandem Mirror Experiment (TMX) in 1979

In 1974 a study of the ZETA results demonstrated an interesting side-effect; after an experimental run ended, the plasma would enter a short period of stability. This led to the reversed field pinch concept, which has seen some level of development since. On May 1, 1974, the KMS fusion company (founded by Kip Siegel) achieves the world's first laser induced fusion in a deuterium-tritium pellet.

In the mid-1970s, Project PACER, carried out at Los Alamos National Laboratory (LANL) explored the possibility of a fusion power system that would involve exploding small hydrogen bombs (fusion bombs) inside an underground cavity. As an energy source, the system is the only fusion power system that could be demonstrated to work using existing technology. It would also require a large, continuous supply of nuclear bombs, however, making the economics of such a system rather questionable.

In 1976, the two beam Argus laser becomes operational at livermore. In 1977, The 20 beam Shiva laser at Livermore is completed, capable of delivering 10.2 kilojoules of infrared energy on target. At a price of $25 million and a size approaching that of a football field, Shiva is the first of the megalasers. That same year, the JET project is approved by the European Commission and a site is selected.

1980s

As a result of advocacy, the cold war, and the 1970s energy crisis a massive magnetic mirror program was funded by the US federal government in the late 1970s and early 1980s. This program resulted in a series of large magnetic mirror devices including: 2X, Baseball I, Baseball II, the Tandem Mirror Experiment, the Tandem mirror experiment upgrade, the Mirror Fusion Test Facility and the MFTF-B. These machines were built and tested at Livermore from the late 1960s to the mid 1980s. A number of institutions collaborated on these machines, conducting experiments. These included the Institute for Advanced Study and the University of Wisconsin–Madison. The last machine, the Mirror Fusion Test Facility cost 372 million dollars and was, at that time, the most expensive project in Livermore history. It opened on February 21, 1986 and was promptly shut down. The reason given was to balance the United States federal budget. This program was supported from within the Carter and early Reagan administrations by Edwin E. Kintner, a US Navy captain, under Alvin Trivelpiece.

Magnetic mirrors suffered from end losses, requiring high power, complex magnetic designs, such as the baseball coil pictured here.

In Laser fusion progressed: in 1983, the NOVETTE laser was completed. The following December 1984, the ten beam NOVA laser was finished. Five years later, NOVA would produce a maximum of 120 kilojoules of infrared light, during a nanosecond pulse. Meanwhile, efforts focused on either fast delivery or beam smoothness. Both tried to deliver the energy uniformly to implode the target. One early problem was that the light in the infrared wavelength, lost lots of energy before hitting the fuel. Breakthroughs were made at the Laboratory for Laser Energetics at the University of Rochester. Rochester scientists used frequency-tripling crystals to transform the infrared laser beams into ultraviolet beams. In 1985, Donna Strickland and Gérard Mourou invented a method to amplify lasers pulses by "chirping". This method changes a single wavelength into a full spectrum. The system then amplifies the laser at each wavelength and then reconstitutes the beam into one color. Chirp pulsed amplification became instrumental in building the National Ignition Facility and the Omega EP system. Most research into ICF was towards weapons research, because the implosion is relevant to nuclear weapons.

The magnetic mirror test facility during construction

During this time Los Alamos National Laboratory constructed a series of laser facilities. This included Gemini (a two beam system), Helios (eight beams), Antares (24 beams) and Aurora (96 beams). The program ended in the early nineties with a cost on the order of one billion dollars.

The Novette target chamber (metal sphere with diagnostic devices protruding radially), which was reused from the Shiva project and two newly built laser chains visible in background.

In 1987, Akira Hasegawa noticed that in a dipolar magnetic field, fluctuations tended compress the plasma without energy loss. This effect was noticed in data taken by Voyager 2, when it encountered Uranus. This observation would become the basis for a fusion approach known as the Levitated dipole.

In Tokamaks, the Tore Supra was under construction over the middle of the eighties (1983 to 1988). This was a Tokamak built in Cadarache, France. In 1983, the JET was completed and first plasmas achieved. In 1985, the Japanese tokamak, JT-60 was completed. In 1988, the T-15 a Soviet tokamak was completed. It was the first industrial fusion reactor to use superconducting magnets to control the plasma. These were Helium cooled.

In 1989, Pons and Fleischmann submitted papers to the *Journal of Electroanalytical Chemistry* claiming that they had observed fusion in a room temperature device and disclosing their work in a press release. Some scientists reported excess heat, neutrons, tritium, helium and other nuclear effects in so-called cold fusion systems, which for a time gained interest as showing promise. Hopes fell when replication failures were weighed in view of several reasons cold fusion is not likely to occur, the discovery of possible sources of experimental error, and finally the discovery that Fleischmann and Pons had not actually detected nuclear reaction byproducts. By late 1989, most scientists considered cold fusion claims dead, and cold fusion subsequently gained a reputation as pathological science. However, a small community of researchers continues to investigate cold fusion claiming to replicate Fleishmann and Pons' results including nuclear reaction byproducts. Claims related to cold fusion are largely disbelieved in the mainstream scientific community. In 1989, the majority of a review panel organized by the US Department of Energy (DOE) found that the evidence for the discovery of a new nuclear process was not persuasive. A second DOE review, convened in 2004 to look at new research, reached conclusions similar to the first.

In 1984, Martin Peng of ORNL proposed an alternate arrangement of the magnet coils that would greatly reduce the aspect ratio while avoiding the erosion issues of the compact tokamak: a Spherical tokamak. Instead of wiring each magnet coil separately, he proposed using a single large conductor in the center, and wiring the magnets as half-rings off of this conductor. What was once a series of individual rings passing through the hole in the center of the reactor was reduced to a single post, allowing for aspect ratios as low as 1.2. The ST concept appeared to represent an enormous advance in tokamak design. However, it was being proposed during a period when US fusion research budgets were being dramatically scaled back. ORNL was provided with funds to develop a suitable central column built out of a high-strength copper alloy called "Glidcop". However, they were unable to secure funding to build a demonstration machine, "STX". Failing to build an ST at ORNL, Peng began a worldwide effort to interest other teams in the ST concept and get a test machine built. One way to do this quickly would be to convert a spheromak machine to the Spherical tokamak layout. Peng's advocacy also caught the interest of Derek Robinson, of the United Kingdom Atomic Energy Authority fusion center at Culham. Robinson was able to gather together a team and secure funding on the order of 100,000 pounds to build an experimental machine, the Small Tight Aspect Ratio Tokamak, or START. Several parts of the machine were recycled from earlier projects, while others were loaned from other labs, including a 40 keV neutral beam injector from ORNL. Construction of START began in 1990, it was assembled rapidly and started operation in January 1991.

1990s

Z Machine (a pinch at SNL) went through a number of upgrades during the mid to late nineties

In 1991 the Preliminary Tritium Experiment at the Joint European Torus in England achieved the world's first controlled release of fusion power.

In 1992, a major article was published in Physics Today by Robert McCory at the Laboratory for laser energetics outlying the current state of ICF and advocating for a national ignition facility. This was followed up by a major review article, from John Lindl in 1995, advocating for NIF. During this time a number of ICF subsystems were developing, including target manufacturing, cryogenic handling systems, new laser designs (notably the NIKE laser at NRL) and improved diagnostics like time of flight analyzers and Thomson scattering. This work was done at the NOVA laser system, General Atomics, Laser Mégajoule and the GEKKO XII system in Japan. Through this work and lobbying by groups like the fusion power associates and John Sethian at NRL, a vote was made in congress, authorizing funding for the NIF project in the late nineties.

In the early nineties, theory and experimental work regarding fusors and polywells was published. In response, Todd Rider at MIT developed general models of these devices. Rider argued that all plasma systems at thermodynamic equilibrium were fundamentally limited. In 1995, William Nevins published a criticism arguing that the particles inside fusors and polywells would build up angular momentum, causing the dense core to degrade.

In 1995, the University of Wisconsin–Madison built a large fusor, known as HOMER, which is still in operation. Meanwhile, Dr George H. Miley at Illinois, built a small fusor that has produced neutrons using deuterium gas and discovered the "star mode" of fusor operation. The following year, the first "US-Japan Workshop on IEC Fusion", was conducted. At this time in Europe, an IEC device was developed as a commercial neutron source by Daimler-Chrysler and NSD Fusion.

In 1996, the Z-machine was upgraded and opened to the public by the US Army in August 1998 in Scientific American. The key attributes of Sandia's Z machine are its 18 million amperes and a discharge time of less than 100 nanoseconds. This generates a magnetic pulse, inside a large oil tank, this strikes an array of tungsten wires called a *liner*. Firing the Z-machine has become a way to test very high energy, high temperature (2 billion degrees) conditions. In 1996, the Tore Supra creates a plasma for two minutes with a current of almost 1 million amperes driven non-inductively by 2.3 MW of lower hybrid frequency waves. This is 280 MJ of injected and extracted energy. This result was possible because of the actively cooled plasma-facing components

In 1997, JET produced a peak of 16.1MW of fusion power (65% of heat to plasma), with fusion power of over 10MW sustained for over 0.5 sec. Its successor, the International Thermonuclear Experimental Reactor (ITER), was officially announced as part of a seven-party consortium (six countries and the EU). ITER is designed to produce ten times more fusion power than the power put into the plasma. ITER is currently under construction in Cadarache, France.

In the late nineties, a team at Columbia University and MIT developed the Levitated dipole a fusion device which consisted of a superconducting electromagnet, floating in a saucer shaped vacuum chamber. Plasma swirled around this donut and fused along the center axis.

2000s

In the March 8, 2002 issue of the peer-reviewed journal *Science*, Rusi P. Taleyarkhan and colleagues at the Oak Ridge National Laboratory (ORNL) reported that acoustic cavitation experiments conducted with deuterated acetone (C_3D_6O) showed measurements of tritium and neutron output consistent with the occurrence of fusion. Taleyarkhan was later found guilty of misconduct, the Office of Naval Research debarred him for 28 months from receiving Federal Funding, and his name was listed in the 'Excluded Parties List'.

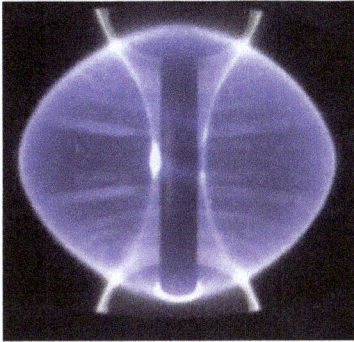

The Mega Ampere Spherical Tokamak became operational in the UK in 1999

Starting in 1999, a growing number of amateurs have been able to fuse atoms using homemade fusors, shown here.

"Fast ignition" was developed in the late nineties, and was part of a push by the Laboratory for Laser Energetics for building the Omega EP system. This system was finished in 2008. Fast ignition showed such dramatic power savings that ICF appears to be a useful technique for energy production. There are even proposals to build an experimental facility dedicated to the fast ignition approach, known as HiPER.

In April 2005, a team from UCLA announced it had devised a way of producing fusion using a machine that "fits on a lab bench", using lithium tantalate to generate enough voltage to smash deuterium atoms together. The process, however, does not generate net power. Such a device would be useful in the same sort of roles as the fusor. In 2006, China's EAST test reactor is completed. This was the first tokamak to use superconducting magnets to generate both the toroidal and poloidal fields.

In the early 2000s, Researchers at LANL reasoned that a plasma oscillating could be at local thermodynamic equilibrium. This prompted the POPS and Penning trap designs. At this time, researchers at MIT became interested in fusors for space propulsion and powering space vehicles. Specifically, researchers developed fusors with multiple inner cages. Greg Piefer graduated from Madison and founded Phoenix Nuclear Labs, a company that developed the fusor into a neutron source for the mass production of medical isotopes. Robert Bussard began speaking openly about the Polywell in 2006. He attempted to generate interest in the research, before his death. In 2008, Taylor Wilson achieved notoriety for achieving nuclear fusion at 14, with a homemade fusor.

In 2009, a high-energy laser system, the National Ignition Facility (NIF), was finished in the US, which can heat hydrogen atoms to temperatures only existing in nature in the cores of stars. The new laser is expected to have the ability to produce, for the first time, more energy from controlled, inertially confined nuclear fusion than was required to initiate the reaction.

2010s

In 2010, NIF researchers were conducting a series of "tuning" shots to determine the optimal target design and laser parameters for high-energy ignition experiments with fusion fuel in the following months. Two firing tests were performed on October 31, 2010 and November 2, 2010. In early 2012, NIF director Mike Dunne expected the laser system to generate fusion with net energy gain by the end of 2012. However, it was delayed and not achieved by that date.

The Wendelstein7X under construction

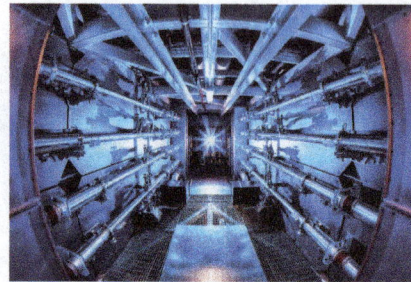

The preamplifiers of the National Ignition Facility. In 2012, the NIF achieved a 500-terawatt shot.

Inertial (laser) confinement is being developed at the United States National Ignition Facility (NIF) based at Lawrence Livermore National Laboratory in California, the French Laser Mégajoule, and the planned European Union High Power laser Energy Research (HiPER) facility. NIF reached initial operational status in 2010 and has been in the process of increasing the power and energy of its "shots", with fusion ignition tests to follow. A three-year goal announced in 2009 to produce net energy from fusion by 2012 was missed; in September 2013, however, the facility announced a significant milestone from an August 2013 test that produced more energy from the fusion reaction than had been provided to the fuel pellet. This was reported as the first time this had been accomplished in fusion power research. The facility reported that their next step involved improving the system to prevent the hohlraum from either breaking up asymmetrically or too soon.

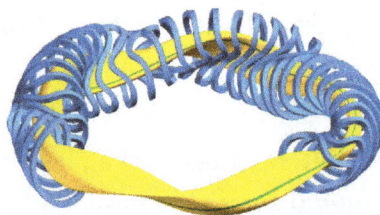

Example of a stellarator design: A coil system (blue) surrounds plasma (yellow). A magnetic field line is highlighted in green on the yellow plasma surface.

A 2012 paper demonstrated that a dense plasma focus had achieved temperatures of 1.8 billion degrees Celsius, sufficient for boron fusion, and that fusion reactions were occurring primarily within the contained plasmoid, a necessary condition for net power. The focus consists of two coaxial cylindrical electrodes made from copper or beryllium and housed in a vacuum chamber containing a low-pressure fusible gas. An electrical pulse is applied across the electrodes, heating the gas into a plasma. The current forms into a minuscule vortex along the axis of the machine, which then kinks into a cage of current with an associated magnetic field. The cage of current and magnetic-field-entrapped plasma is called a plasmoid. The acceleration of the electrons about the magnetic field lines heats the nuclei within the plasmoid to fusion temperatures.

In April 2014, Lawrence Livermore National Laboratory ended the Laser Inertial Fusion Energy (LIFE) program and redirected their efforts towards NIF. In August 2014, Phoenix Nuclear Labs announced the sale of a high-yield neutron generator that could sustain 5×10^{11} deuterium fusion reactions per second over a 24-hour period. In October 2014, Lockheed Martin's Skunk Works announced the development of a high-beta fusion reactor that they hope to yield a functioning 100-megawatt prototype by 2017 and to be ready for regular operation by 2022.

Deep-space exploration, as well as higher-velocity lower-cost space transport services in general would be enabled by this compact fusion reactor technology.

In January 2015, the polywell was presented at Microsoft Research.

In August, 2015, MIT announced a tokamak it named ARC fusion reactor design using rare-earth barium-copper oxide (REBCO) superconducting tapes to produce high-magnetic field coils that it claimed produce comparable magnetic field strength in a smaller configuration than other designs.

In October 2015, researchers at the Max Planck Institute of Plasma Physics completed building the largest stellarator to date, named Wendelstein 7-X. On December 10, they successfully produced the first helium plasma, and on February 3, 2016 produced the device's first Hydrogen plasma. With plasma discharges lasting up to 30 minutes, Wendelstein 7-X will try to demonstrate the essential stellarator attribute: continuous operation of a high-temperature hydrogen plasma.

Fuels

By firing particle beams at targets, many fusion reactions have been tested, while the fuels considered for power have all been light elements like the isotopes of hydrogen—deuterium and tritium. Other reactions like the deuterium and Helium³ reaction or the Helium³ and Helium³ reactions, would require a supply of Helium³. This can either come from other nuclear reactions or from extraterrestrial sources. Finally, researchers hope to do the p-11 B
reaction, because it does not directly produce neutrons, though side reactions can.

Deuterium, Tritium

The easiest nuclear reaction, at the lowest energy, is:

$$_1^2 D + {}_1^3 T \rightarrow {}_2^4 He + {}_0^1 n$$

This reaction is common in research, industrial and military applications, usually as a convenient source of neutrons. Deuterium is a naturally occurring isotope of hydrogen and is commonly available. The large mass ratio of the hydrogen isotopes makes their separation easy compared to the difficult uranium enrichment process. Tritium is a natural isotope of hydrogen, but because it has a short half-life of 12.32 years, it is hard to find, store, produce, and is expensive. Consequently, the deuterium-tritium fuel cycle requires the breeding of tritium from lithium using one of the following reactions:

$$\mathrm{^1_0n} + \mathrm{^6_3Li} \rightarrow \mathrm{^3_1T} + \mathrm{^4_2He}$$

$$\mathrm{^1_0n} + \mathrm{^7_3Li} \rightarrow \mathrm{^3_1T} + \mathrm{^4_2He} + \mathrm{^1_0n}$$

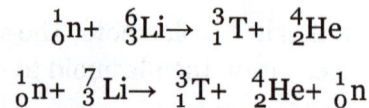

The reactant neutron is supplied by the D-T fusion reaction shown above, and the one that has the greatest yield of energy. The reaction with ^{6}Li is exothermic, providing a small energy gain for the reactor. The reaction with ^{7}Li is endothermic but does not consume the neutron. At least some ^{7}Li reactions are required to replace the neutrons lost to absorption by other elements. Most reactor designs use the naturally occurring mix of lithium isotopes.

Several drawbacks are commonly attributed to D-T fusion power:

1. It produces substantial amounts of neutrons that result in the neutron activation of the reactor materials.

2. Only about 20% of the fusion energy yield appears in the form of charged particles with the remainder carried off by neutrons, which limits the extent to which direct energy conversion techniques might be applied.

3. It requires the handling of the radioisotope tritium. Similar to hydrogen, tritium is difficult to contain and may leak from reactors in some quantity. Some estimates suggest that this would represent a fairly large environmental release of radioactivity.

The neutron flux expected in a commercial D-T fusion reactor is about 100 times that of current fission power reactors, posing problems for material design. After a series of D-T tests at JET, the vacuum vessel was sufficiently radioactive that remote handling was required for the year following the tests.

In a production setting, the neutrons would be used to react with lithium in order to create more tritium. This also deposits the energy of the neutrons in the lithium, which would then be transferred to drive electrical production. The lithium neutron absorption reaction protects the outer portions of the reactor from the neutron flux. Newer designs, the advanced tokamak in particular, also use lithium inside the reactor core as a key element of the design. The plasma interacts directly with the lithium, preventing a problem known as "recycling". The advantage of this design was demonstrated in the Lithium Tokamak Experiment.

Deuterium

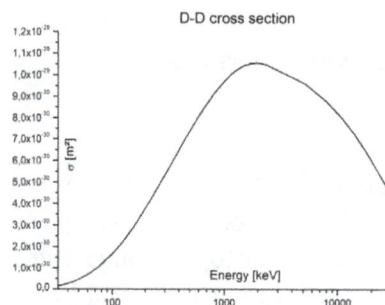

Deuterium fusion cross section (in square meters) at different ion collision energies.

This is the second easiest fusion reaction, fusing deuterium with itself. The reaction has two branches that occur with nearly equal probability:

$$D + D \rightarrow T + {}^1H$$
$$D + D \rightarrow {}^3He + n$$

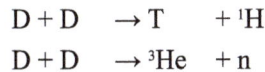

This reaction is also common in research. The optimum energy to initiate this reaction is 15 keV, only slightly higher than the optimum for the D-T reaction. The first branch does not produce neutrons, but it does produce tritium, so that a D-D reactor will not be completely tritium-free, even though it does not require an input of tritium or lithium. Unless the tritons can be quickly removed, most of the tritium produced would be burned before leaving the reactor, which would reduce the handling of tritium, but would produce more neutrons, some of which are very energetic. The neutron from the second branch has an energy of only 2.45 MeV (0.393 pJ), whereas the neutron from the D-T reaction has an energy of 14.1 MeV (2.26 pJ), resulting in a wider range of isotope production and material damage. When the tritons are removed quickly while allowing the ${}^3He$ to react, the fuel cycle is called "tritium suppressed fusion" The removed tritium decays to ${}^3He$ with a 12.5 year half life. By recycling the ${}^3He$ produced from the decay of tritium back into the fusion reactor, the fusion reactor does not require materials resistant to fast 14.1 MeV (2.26 pJ) neutrons.

Assuming complete tritium burn-up, the reduction in the fraction of fusion energy carried by neutrons would be only about 18%, so that the primary advantage of the D-D fuel cycle is that tritium breeding would not be required. Other advantages are independence from scarce lithium resources and a somewhat softer neutron spectrum. The disadvantage of D-D compared to D-T is that the energy confinement time (at a given pressure) must be 30 times longer and the power produced (at a given pressure and volume) would be 68 times less .

Assuming complete removal of tritium and recycling of ${}^3He$, only 6% of the fusion energy is carried by neutrons. The tritium-suppressed D-D fusion requires an energy confinement that is 10 times longer compared to D-T and a plasma temperature that is twice as high.

Deuterium, Helium 3

A second generation approach to controlled fusion power involves combining helium-3 (${}^3He$) and deuterium (${}^2H$):

$$D + {}^3He \rightarrow {}^4He + {}^1H$$

This reaction produces a helium-4 nucleus (${}^4He$) and a high-energy proton. As with the p-${}^{11}B$ aneutronic fusion fuel cycle, most of the reaction energy is released as charged particles, reducing activation of the reactor housing and potentially allowing more efficient energy harvesting (via any of several speculative technologies). In practice, D-D side reactions produce a significant number of neutrons, resulting in p-${}^{11}B$ being the preferred cycle for aneutronic fusion.

Proton, Boron 11

If aneutronic fusion is the goal, then the most promising candidate may be the Hydrogen-1 (proton)/boron reaction, which releases alpha (helium) particles, but does not rely on neutron scattering for energy transfer.

$$ {}^1H + {}^{11}B \rightarrow 3\ {}^4He$$

Under reasonable assumptions, side reactions will result in about 0.1% of the fusion power being carried by neutrons. At 123 keV, the optimum temperature for this reaction is nearly ten times higher than that for the pure hydrogen reactions, the energy confinement must be 500 times better than that required for the D-T reaction, and the power density will be 2500 times lower than for D-T.

Because the confinement properties of conventional approaches to fusion such as the tokamak and laser pellet fusion are marginal, most proposals for aneutronic fusion are based on radically different confinement concepts, such as the Polywell and the Dense Plasma Focus. Results have been extremely promising:

> "In the October 2013 edition of Nature Communications, a research team led by Christine Labaune at École Polytechnique in Palaiseau, France, reported a new record fusion rate: an estimated 80 million fusion reactions during the 1.5 nanoseconds that the laser fired, which is at least 100 times more than any previous proton-boron experiment. "

Material Selection

Considerations

Any power station using hot plasma, is going to have plasma facing walls. In even the simplest plasma approaches, the material will get blasted with matter and energy. This leads to a minimum list of considerations, including dealing with:

- A heating and cooling cycle, up to a 10 MW/m² thermal load.

- Neutron radiation, which over time leads to neutron activation and embrittlement.

- High energy ions leaving at tens to hundreds of electronvolts.

- Alpha particles leaving at millions of electronvolts.

- Electrons leaving at high energy.

- Light radiation (IR, visible, UV, X-ray).

Depending on the approach, these effects may be higher or lower than typical fission reactors like the pressurized water reactor (PWR). One estimate put the radiation at 100 times the (PWR). Materials need to be selected or developed that can withstand these basic conditions. Depending on the approach, however, there may be other considerations such as electrical conductivity, magnetic permeability and mechanical strength. There is also a need for materials whose primary components and impurities do not result in long-lived radioactive wastes.

Durability

For long term use, each atom in the wall is expected to be hit by a neutron and displaced about a hundred times before the material is replaced. High-energy neutrons will produce hydrogen and helium by way of various nuclear reactions that tends to form bubbles at grain boundaries and result in swelling, blistering or embrittlement.

Selection

One can choose either a low-Z material, such as graphite or beryllium, or a high-Z material, usually tungsten with molybdenum as a second choice. Use of liquid metals (lithium, gallium, tin) has also been proposed, e.g., by injection of 1–5 mm thick streams flowing at 10 m/s on solid substrates.

If graphite is used, the gross erosion rates due to physical and chemical sputtering would be many meters per year, so one must rely on redeposition of the sputtered material. The location of the redeposition will not exactly coincide with the location of the sputtering, so one is still left with erosion rates that may be prohibitive. An even larger problem is the tritium co-deposited with the redeposited graphite. The tritium inventory in graphite layers and dust in a reactor could quickly build up to many kilograms, representing a waste of resources and a serious radiological hazard in case of an accident. The consensus of the fusion community seems to be that graphite, although a very attractive material for fusion experiments, cannot be the primary PFC material in a commercial reactor.

The sputtering rate of tungsten by the plasma fuel ions is orders of magnitude smaller than that of carbon, and tritium is much less incorporated into redeposited tungsten, making this a more attractive choice. On the other hand, tungsten impurities in a plasma are much more damaging than carbon impurities, and self-sputtering of tungsten can be high, so it will be necessary to ensure that the plasma in contact with the tungsten is not too hot (a few tens of eV rather than hundreds of eV). Tungsten also has disadvantages in terms of eddy currents and melting in off-normal events, as well as some radiological issues.

Safety and The Environment

Accident Potential

Nuclear fusion is unlike nuclear fission: fusion requires extremely precise and controlled temperature, pressure and magnetic field parameters for any net energy to be produced. If a reactor suffers damage or loses even a small degree of required control, fusion reactions and heat generation would rapidly cease. Additionally, fusion reactors contain relatively small amounts of fuel, enough to "burn" for minutes, or in some cases, microseconds. Unless they are actively refueled, the reactions will quickly end. Therefore, fusion reactors are considered extremely safe.

Runaway reactions cannot occur in a fusion reactor. The plasma is burnt at optimal conditions, and any significant change will quench the reactions. The reaction process is so delicate that this level of safety is inherent. Although the plasma in a fusion power station is expected to have a volume of 1,000 cubic metres (35,000 cu ft) or more, the plasma density is low and the total amount of fusion fuel in the vessel typically only a few grams. If the fuel supply is closed, the reaction stops within seconds. In comparison, a fission reactor is typically loaded with enough fuel for several months or years, and no additional fuel is necessary to continue the reaction. It is this large amount of fuel that gives rise to the possibility of a meltdown; nothing analogous exists in a fusion reactor.

In the magnetic approach, strong fields are developed in coils that are held in place mechanically by the reactor structure. Failure of this structure could release this tension and allow the magnet to "explode" outward. The severity of this event would be similar to any other industrial accident or an MRI machine quench/explosion, and could be effectively stopped with a containment building similar to those used in existing (fission) nuclear generators. The laser-driven inertial approach is generally lower-stress because of the increased size of the reaction chamber. Although failure of the reaction chamber is possible, simply stopping fuel delivery would prevent any sort of catastrophic failure.

Most reactor designs rely on liquid hydrogen as both a coolant and a method for converting stray neutrons from the reaction into tritium, which is fed back into the reactor as fuel. Hydrogen is highly flammable, and in the case of a fire it is possible that the hydrogen stored on-site could be burned up and escape. In this case, the tritium contents of the hydrogen would be released into the atmosphere, posing a radiation risk. Calculations suggest that at about 1 kg the total amount of tritium and other radioactive gases in a typical power station would be so small that they would have diluted to legally acceptable limits by the time they blew as far as the station's perimeter fence.

The likelihood of *small industrial* accidents including the local release of radioactivity and injury to staff cannot be estimated yet. These would include accidental releases of lithium or tritium or mis-handling of decommissioned radioactive components of the reactor itself.

Magnet Quench

A quench is an abnormal termination of magnet operation that occurs when part of the superconducting coil enters the normal (resistive) state. This can occur because the field inside the magnet is too large, the rate of change of field is too large (causing eddy currents and resultant heating in the copper support matrix), or a combination of the two.

More rarely a defect in the magnet can cause a quench. When this happens, that particular spot is subject to rapid Joule heating from the enormous current, which raises the temperature of the surrounding regions. This pushes those regions into the normal state as well, which leads to more heating in a chain reaction. The entire magnet rapidly becomes normal (this can take several seconds, depending on the size of the superconducting coil). This is accompanied by a loud bang as the energy in the magnetic field is converted to heat, and rapid boil-off of the cryogenic fluid. The abrupt decrease of current can result in kilovolt inductive voltage spikes and arcing. Permanent damage to the magnet is rare, but components can be damaged by localized heating, high voltages, or large mechanical forces.

In practice, magnets usually have safety devices to stop or limit the current when the beginning of a quench is detected. If a large magnet undergoes a quench, the inert vapor formed by the evaporating cryogenic fluid can present a significant asphyxiation hazard to operators by displacing breathable air.

A large section of the superconducting magnets in CERN's Large Hadron Collider unexpectedly quenched during start-up operations in 2008, necessitating the replacement of a number of magnets. In order to mitigate against potentially destructive quenches, the superconducting magnets that form the LHC are equipped with fast-ramping heaters which are activated once a quench event is detected by the complex quench protection system. As the dipole bending magnets are connected in series, each power circuit includes 154 individual magnets, and should a quench event occur, the entire combined stored energy of these magnets must be dumped at once. This energy is transferred into dumps that are massive blocks of metal which heat up to several hundreds of degrees Celsius—because of resistive heating—in a matter of seconds. Although undesirable, a magnet quench is a "fairly routine event" during the operation of a particle accelerator.

Effluents

The natural product of the fusion reaction is a small amount of helium, which is completely harmless to life. Of more concern is tritium, which, like other isotopes of hydrogen, is difficult to retain completely. During normal operation, some amount of tritium will be continually released.

Although tritium is volatile and biologically active, the health risk posed by a release is much lower than that of most radioactive contaminants, because of tritium's short half-life (12.32 years) and very low decay energy (~14.95 keV), and because it does not bioaccumulate (instead being cycled out of the body as water, with a biological half-life of 7 to 14 days). Current ITER designs are investigating total containment facilities for any tritium.

Waste Management

The large flux of high-energy neutrons in a reactor will make the structural materials radioactive. The radioactive inventory at shut-down may be comparable to that of a fission reactor, but there are important differences.

The half-life of the radioisotopes produced by fusion tends to be less than those from fission, so that the inventory decreases more rapidly. Unlike fission reactors, whose waste remains radioactive for thousands of years, most of the radioactive material in a fusion reactor would be the reactor core itself, which would be dangerous for about 50 years, and low-level waste for another 100. Although this waste will be considerably more radioactive during those 50 years than fission waste, the very short half-life makes the process very attractive, as the waste management is fairly straightforward. By 500 years the material would have the same radiotoxicity as coal ash.

Additionally, the choice of materials used in a fusion reactor is less constrained than in a fission design, where many materials are required for their specific neutron cross-sections. This allows a fusion reactor to be designed using materials that are selected specifically to be "low activation", materials that do not easily become radioactive. Vanadium, for example, would become much less radioactive than stainless steel. Carbon fiber materials are also low-activation, as well as being strong and light, and are a promising area of study for laser-inertial reactors where a magnetic field is not required.

In general terms, fusion reactors would create far less radioactive material than a fission reactor, the material it would create is less damaging biologically, and the radioactivity "burns off" within a time period that is well within existing engineering capabilities for safe long-term waste storage.

Nuclear Proliferation

Although fusion power uses nuclear technology, the overlap with nuclear weapons would be limited. A huge amount of tritium could be produced by a fusion power station; tritium is used in the trigger of hydrogen bombs and in a modern boosted fission weapon, but it can also be produced by nuclear fission. The energetic neutrons from a fusion reactor could be used to breed weapons-grade plutonium or uranium for an atomic bomb (for example by transmutation of U^{238} to Pu^{239}, or Th^{232} to U^{233}).

A study conducted 2011 assessed the risk of three scenarios:

- *Use in small-scale fusion station*: As a result of much higher power consumption, heat dissipation and a more recognizable design compared to enrichment gas centrifuges this choice would be much easier to detect and therefore implausible.

- *Modifications to produce weapon-usable material in a commercial facility:* The production potential is significant. But no fertile or fissile substances necessary for the production of weapon-usable materials needs to be present at a civil fusion system at all. If not shielded, a detection of these materials can be done by their characteristic gamma radiation. The underlying redesign could be detected by regular design information verifications. In the (technically more feasible) case of solid breeder blanket modules, it would be necessary for incoming components to be inspected for the presence of fertile material, otherwise plutonium for several weapons could be produced each year.

- *Prioritizing a fast production of weapon-grade material regardless of secrecy:* The fastest way to produce weapon usable material was seen in modifying a prior civil fusion power station. Unlike in some nuclear power stations, there is no weapon compatible material during civil use. Even without the need for covert action this modification would still take about 2 months to start the production and at least an additional week to generate a significant amount for weapon production. This was seen as enough time to detect a military use and to react with diplomatic or military means. To stop the production, a military destruction of inevitable parts of the facility leaving out the reactor itself would be sufficient. This, together with the intrinsic safety of fusion power would only bear a low risk of radioactive contamination.

Another study concludes that "[..]large fusion reactors – even if not designed for fissile material breeding – could easily produce several hundred kg Pu per year with high weapon quality and very low source material requirements." It was emphasized that the implementation of features for intrinsic proliferation resistance might only be possible at this phase of research and development. The theoretical and computational tools needed for hydrogen bomb design are closely related to those needed for inertial confinement fusion, but have very little in common with the more scientifically developed magnetic confinement fusion.

Energy Source

Large-scale reactors using neutronic fuels (e.g. ITER) and thermal power production (turbine based) are most comparable to fission power from an engineering and economics viewpoint. Both fission and fusion power stations involve a relatively compact heat source powering a conventional steam turbine-based power station, while producing enough neutron radiation to make activation of the station materials problematic. The main distinction is that fusion power produces no high-level radioactive waste (though activated station materials still need to be disposed of). There are some power station ideas that may significantly lower the cost or size of such stations; however, research in these areas is nowhere near as advanced as in tokamaks.

Fusion power commonly proposes the use of deuterium, an isotope of hydrogen, as fuel and in many current designs also use lithium. Assuming a fusion energy output equal to the 1995 global power output of about 100 EJ/yr (= 1×10^{20} J/yr) and that this does not increase in the future, which is unlikely, then the known current lithium reserves would last 3000 years. Lithium from sea water would last 60 million years, however, and a more complicated fusion process using only deuterium from sea water would have fuel for 150 billion years. To put this in context, 150 billion years is close to 30 times the remaining lifespan of the sun, and more than 10 times the estimated age of the universe.

Economics

While fusion power is still in early stages of development, substantial sums have been and continue to be invested in research. In the EU almost €10 billion was spent on fusion research up to the end of the 1990s, and the new ITER reactor alone is budgeted at €6.6 billion total for the timeframe between 2008 and 2020.

It is estimated that up to the point of possible implementation of electricity generation by nuclear fusion, R&D will need further promotion totalling around €60–80 billion over a period of 50 years or so (of which €20–30 billion within the EU) based on a report from 2002. Nuclear fusion research receives €750 million (excluding ITER funding) from the European Union, compared with €810 million for sustainable energy research, putting research into fusion power well ahead of that of any single rivaling technology. Indeed, the size of the investments and time frame of the expected results mean that fusion research is almost exclusively publicly funded, while research in other forms of energy can be done by the private sector. In spite of that, a number of start-up companies active in the field of fusion power have managed to attract private money.

Advantages

Fusion power would provide more energy for a given weight of fuel than any fuel-consuming energy source currently in use, and the fuel itself (primarily deuterium) exists abundantly in the Earth's ocean: about 1 in 6500 hydrogen atoms in seawater is deuterium. Although this may seem a low proportion (about 0.015%), because nuclear fusion reactions are so much more energetic than chemical combustion and seawater is easier to access and more plentiful than fossil fuels, fusion could potentially supply the world's energy needs for millions of years.

Despite being technically non-renewable, fusion power has many of the benefits of renewable energy sources (such as being a long-term energy supply and emitting no greenhouse gases) as well as some of the benefits of the resource-limited energy sources as hydrocarbons and nuclear fission (without reprocessing). Like these currently dominant energy sources, fusion could provide very high power-generation density and uninterrupted power delivery (because it is not dependent on the weather, unlike wind and solar power).

Another aspect of fusion energy is that the cost of production does not suffer from diseconomies of scale. The cost of water and wind energy, for example, goes up as the optimal locations are developed first, while further generators must be sited in less ideal conditions. With fusion energy the production cost will not increase much even if large numbers of stations are built, because the raw resource (seawater) is abundant and widespread.

Some problems that are expected to be an issue in this century, such as fresh water shortages, can alternatively be regarded as problems of energy supply. For example, in desalination stations, seawater can be purified through distillation or reverse osmosis. Nonetheless, these processes are energy intensive. Even if the first fusion stations are not competitive with alternative sources, fusion could still become competitive if large-scale desalination requires more power than the alternatives are able to provide.

A scenario has been presented of the effect of the commercialization of fusion power on the future of human civilization. ITER and later Demo are envisioned to bring online the first commercial nuclear fusion energy reactor by 2050. Using this as the starting point and the history of the uptake of nuclear fission reactors as a guide, the scenario depicts a rapid take up of nuclear fusion energy starting after the middle of this century.

Fusion power could be used in interstellar space, where solar energy is not available.

Criticism

Because commercial fusion projects are very large and complex, and ongoing funding is a political issue, such projects usually involve cost overruns and missed deadlines. For example, the construction of the National Ignition Facility cost $5 billion and took seven years longer than expected. ITER's expected cost has gone from $5 billion to $20 billion, and the date for full power operation has been put back to 2027, from the original estimate of 2016.

Nuclear Fission

In nuclear physics and nuclear chemistry, nuclear fission is either a nuclear reaction or a radioactive decay process in which the nucleus of an atom splits into smaller parts (lighter nuclei). The fission process often produces free neutrons and gamma photons, and releases a very large amount of energy even by the energetic standards of radioactive decay.

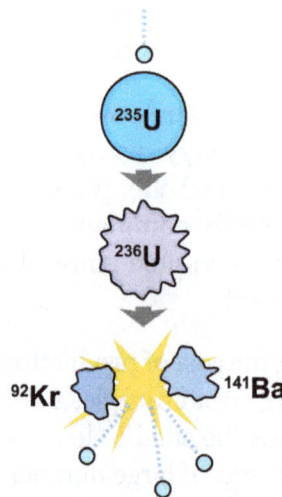

An induced fission reaction. A neutron is absorbed by a uranium-235 nucleus, turning it briefly into an excited uranium-236 nucleus, with the excitation energy provided by the kinetic energy of the neutron plus the forces that bind the neutron. The uranium-236, in turn, splits into fast-moving lighter elements (fission products) and releases three free neutrons. At the same time, one or more "prompt gamma rays" (not shown) are produced, as well.

Nuclear fission of heavy elements was discovered on December 17, 1938 by German Otto Hahn and his assistant Fritz Strassmann, and explained theoretically in January 1939 by Lise Meitner and her nephew Otto Robert Frisch. Frisch named the process by analogy with biological fission of living cells. It is an exothermic reaction which can release large amounts of energy both as electromagnetic radiation and as kinetic energy of the fragments (heating the bulk material where fission takes place). In order for fission to produce energy, the total binding energy of the resulting elements must be less negative (higher energy) than that of the starting element.

Fission is a form of nuclear transmutation because the resulting fragments are not the same element as the original atom. The two nuclei produced are most often of comparable but slightly different sizes, typically with a mass ratio of products of about 3 to 2, for common fissile isotopes. Most fissions are binary fissions (producing two charged fragments), but occasionally (2 to 4 times per 1000 events), *three* positively charged fragments are produced, in a ternary fission. The smallest of these fragments in ternary processes ranges in size from a proton to an argon nucleus.

Apart from fission induced by a neutron, harnessed and exploited by humans, a natural form of spontaneous radioactive decay (not requiring a neutron) is also referred to as fission, and occurs especially in very high-mass-number isotopes. Spontaneous fission was discovered in 1940 by Flyorov, Petrzhak and Kurchatov in Moscow, when they decided to confirm that, without bombardment by neutrons, the fission rate of uranium was indeed negligible, as predicted by Niels Bohr; it wasn't.

The unpredictable composition of the products (which vary in a broad probabilistic and somewhat chaotic manner) distinguishes fission from purely quantum-tunnelling processes such as proton emission, alpha decay, and cluster decay, which give the same products each time. Nuclear fission produces energy for nuclear power and drives the explosion of nuclear weapons. Both uses are possible because certain substances called nuclear fuels undergo fission when struck by fission neutrons, and in turn emit neutrons when they break apart. This makes possible a self-sustaining nuclear chain reaction that releases energy at a controlled rate in a nuclear reactor or at a very rapid uncontrolled rate in a nuclear weapon.

The amount of free energy contained in nuclear fuel is millions of times the amount of free energy contained in a similar mass of chemical fuel such as gasoline, making nuclear fission a very dense source of energy. The products of nuclear fission, however, are on average far more radioactive than the heavy elements which are normally fissioned as fuel, and remain so for significant amounts of time, giving rise to a nuclear waste problem. Concerns over nuclear waste accumulation and over the destructive potential of nuclear weapons may counterbalance the desirable qualities of fission as an energy source, and give rise to ongoing political debate over nuclear power.

Physical Overview

Mechanism

Nuclear fission can occur without neutron bombardment as a type of radioactive decay. This type of fission (called spontaneous fission) is rare except in a few heavy isotopes. In engineered nuclear devices, essentially all nuclear fission occurs as a "nuclear reaction" — a bombardment-driven process that results from the collision of two subatomic particles. In nuclear reactions, a subatomic particle collides with an atomic nucleus and causes changes to it. Nuclear reactions are thus driven by the mechanics of bombardment, not by the relatively constant exponential decay and half-life characteristic of spontaneous radioactive processes.

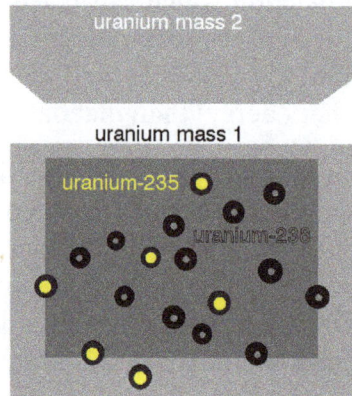

A visual representation of an induced nuclear fission event where a slow-moving neutron is absorbed by the nucleus of a uranium-235 atom, which fissions into two fast-moving lighter elements (fission products) and additional neutrons. Most of the energy released is in the form of the kinetic velocities of the fission products and the neutrons.

Many types of nuclear reactions are currently known. Nuclear fission differs importantly from other types of nuclear reactions, in that it can be amplified and sometimes controlled via a nuclear chain reaction (one type of general chain reaction). In such a reaction, free neutrons released by each fission event can trigger yet more events, which in turn release more neutrons and cause more fissions.

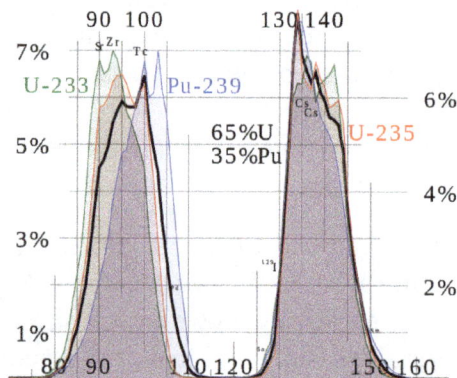

Fission product yields by mass for thermal neutron fission of U-235, Pu-239, a combination of the two typical of current nuclear power reactors, and U-233 used in the thorium cycle.

The chemical element isotopes that can sustain a fission chain reaction are called nuclear fuels, and are said to be *fissile*. The most common nuclear fuels are ^{235}U (the isotope of uranium with an atomic mass of 235 and of use in nuclear reactors) and ^{239}Pu (the isotope of plutonium with an atomic mass of 239). These fuels break apart into a bimodal range of chemical elements with atomic masses centering near 95 and 135 u (fission products). Most nuclear fuels undergo spontaneous fission only very slowly, decaying instead mainly via an alpha-beta decay chain over periods of millennia to eons. In a nuclear reactor or nuclear weapon, the overwhelming majority of fission events are induced by bombardment with another particle, a neutron, which is itself produced by prior fission events.

Nuclear fissions in fissile fuels are the result of the nuclear excitation energy produced when a fissile nucleus captures a neutron. This energy, resulting from the neutron capture, is a result of the attractive nuclear force acting between the neutron and nucleus. It is enough to deform the nucleus into a double-lobed "drop," to the point that nuclear fragments exceed the distances at which the nuclear force can hold two groups of charged nucleons together and, when this happens, the two fragments complete their separation and then are driven further apart by their mutually repulsive charges, in a process which becomes irreversible with greater and greater distance. A similar process occurs in fissionable isotopes (such as uranium-238), but in order to fission, these isotopes require additional energy provided by fast neutrons (such as those produced by nuclear fusion in thermonuclear weapons).

The liquid drop model of the atomic nucleus predicts equal-sized fission products as an outcome of nuclear deformation. The more sophisticated nuclear shell model is needed to mechanistically explain the route to the more energetically favorable outcome, in which one fission product is slightly smaller than the other. A theory of the fission based on shell model has been formulated by Maria Goeppert Mayer.

The most common fission process is binary fission, and it produces the fission products noted above, at 95 ± 15 and 135 ± 15 u. However, the binary process happens merely because it is the most probable. In anywhere from 2 to 4 fissions per 1000 in a nuclear reactor, a process called ternary fission produces three positively charged fragments (plus neutrons) and the smallest of these may range from so small a charge and mass as a proton (Z=1), to as large a fragment as argon (Z=18). The most common small fragments, however, are composed of 90% helium-4 nuclei with more energy than alpha particles from alpha decay (so-called "long range alphas" at ~ 16 MeV), plus helium-6 nuclei, and tritons (the nuclei of tritium). The ternary process is less common, but still ends up producing significant helium-4 and tritium gas buildup in the fuel rods of modern nuclear reactors.

Energetics

Input

The fission of a heavy nucleus requires a total input energy of about 7 to 8 million electron volts (MeV) to initially overcome the nuclear force which holds the nucleus into a spherical or nearly spherical shape, and from there, deform it into a two-lobed ("peanut") shape in which the lobes are able to continue to separate from each other, pushed by their mutual positive charge, in the most common process of binary fission (two positively charged fission products + neutrons). Once the nuclear lobes have been pushed to a critical distance, beyond which the short range strong force can no longer hold them together, the process of their separation proceeds from the energy of the (longer range) electromagnetic repulsion between the fragments. The result is two fission fragments moving away from each other, at high energy.

About 6 MeV of the fission-input energy is supplied by the simple binding of an extra neutron to the heavy nucleus via the strong force; however, in many fissionable isotopes, this amount of energy is not enough for fission. Uranium-238, for example, has a near-zero fission cross section for neutrons of less than one MeV energy. If no additional energy is supplied by any other mechanism, the nucleus will not fission, but will merely absorb the neutron, as happens when U-238 absorbs slow and even some fraction of fast neutrons, to become U-239. The remaining energy to initiate fission can be supplied by two other mechanisms: one of these is more kinetic energy of the incoming neutron, which is increasingly able to fission a fissionable heavy nucleus as it exceeds a kinetic energy of one MeV or more (so-called fast neutrons). Such high energy neutrons are able to fission U-238 directly. However, this process cannot happen to a great extent in a nuclear reactor, as too small a fraction of the fission neutrons produced by any type of fission have enough energy to efficiently fission U-238 (fission neutrons have a mode energy of 2 MeV, but a median of only 0.75 MeV, meaning half of them have less than this insufficient energy).

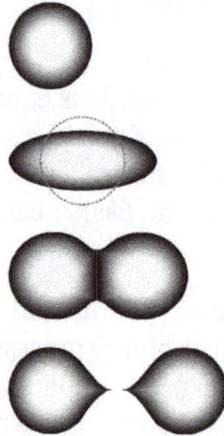

The stages of binary fission in a liquid drop model. Energy input deforms the nucleus into a fat "cigar" shape, then a "peanut" shape, followed by binary fission as the two lobes exceed the short-range nuclear force attraction distance, then are pushed apart and away by their electrical charge. In the liquid drop model, the two fission fragments are predicted to be the same size. The nuclear shell model allows for them to differ in size, as usually experimentally observed.

Among the heavy actinide elements, however, those isotopes that have an odd number of neutrons (such as U-235 with 143 neutrons) bind an extra neutron with an additional 1 to 2 MeV of energy over an isotope of the same element with an even number of neutrons (such as U-238 with 146 neutrons). This extra binding energy is made available as a result of the mechanism of neutron pairing effects. This extra energy results from the Pauli exclusion principle allowing an extra neutron to occupy the same nuclear orbital as the last neutron in the nucleus, so that the two form a pair. In such isotopes, therefore, no neutron kinetic energy is needed, for all the necessary energy is supplied by absorption of any neutron, either of the slow or fast variety (the former are used in moderated nuclear reactors, and the latter are used in fast neutron reactors, and in weapons). As noted above, the subgroup of fissionable elements that may be fissioned efficiently with their own fission neutrons (thus potentially causing a nuclear chain reaction in relatively small amounts of the pure material) are termed "fissile." Examples of fissile isotopes are U-235 and plutonium-239.

Output

Typical fission events release about two hundred million eV (200 MeV) of energy for each fission event. The exact isotope which is fissioned, and whether or not it is fissionable or fissile, has only a small impact on the amount of energy released. This can be easily seen by examining the curve of binding energy (image below), and noting that the average binding energy of the actinide nuclides beginning with uranium is around 7.6 MeV per nucleon. Looking further left on the curve of binding energy, where the fission products cluster, it is easily observed that the binding energy of the fission products tends to center around 8.5 MeV per nucleon. Thus, in any fission event of an isotope in the actinide's range of mass, roughly 0.9 MeV is released per nucleon of the starting element. The fission of U235 by a slow neutron yields nearly identical energy to the fission of U238 by a fast neutron. This energy release profile holds true for thorium and the various minor actinides as well.

By contrast, most chemical oxidation reactions (such as burning coal or TNT) release at most a few eV per event. So, nuclear fuel contains at least ten million times more usable energy per unit mass than does chemical fuel. The energy of nuclear fission is released as kinetic energy of the fission products and fragments, and as electromagnetic radiation in the form of gamma rays; in a nuclear reactor, the energy is converted to heat as the particles and gamma rays collide with the atoms that make up the reactor and its working fluid, usually water or occasionally heavy water or molten salts.

When a uranium nucleus fissions into two daughter nuclei fragments, about 0.1 percent of the mass of the uranium nucleus appears as the fission energy of ~200 MeV. For uranium-235 (total mean fission energy 202.5 MeV), typically ~169 MeV appears as the kinetic energy of the daughter nuclei, which fly apart at about 3% of the speed of light, due to Coulomb repulsion. Also, an average of 2.5 neutrons are emitted, with a mean kinetic energy per neutron of ~2 MeV (total of 4.8 MeV). The fission reaction also releases ~7 MeV in prompt gamma ray photons. The latter figure means that a nuclear fission explosion or criticality accident emits about 3.5% of its energy as gamma rays, less than 2.5% of its energy as fast neutrons (total of both types of radiation ~ 6%), and the rest as kinetic energy of fission fragments (this appears almost immediately when the fragments impact surrounding matter, as simple heat). In an atomic bomb, this heat may serve to raise the temperature of the bomb core to 100 million kelvin and cause secondary emission of soft X-rays, which convert some of this energy to ionizing radiation. However, in nuclear reactors, the fission fragment kinetic energy remains as low-temperature heat, which itself causes little or no ionization.

So-called neutron bombs (enhanced radiation weapons) have been constructed which release a larger fraction of their energy as ionizing radiation (specifically, neutrons), but these are all thermonuclear devices which rely on the nuclear fusion stage to produce the extra radiation. The energy dynamics of pure fission bombs always remain at about 6% yield of the total in radiation, as a prompt result of fission.

The total *prompt fission* energy amounts to about 181 MeV, or ~ 89% of the total energy which is eventually released by fission over time. The remaining ~ 11% is released in beta decays which have various half-lives, but begin as a process in the fission products immediately; and in delayed gamma emissions associated with these beta decays. For example, in uranium-235 this delayed energy is divided into about 6.5 MeV in betas, 8.8 MeV in antineutrinos (released at the same time as the betas), and finally, an additional 6.3 MeV in delayed gamma emission from the excited beta-decay products (for a mean total of ~10 gamma ray emissions per fission, in all). Thus, about 6.5% of the total energy of fission is released some time after the event, as non-prompt or delayed ionizing radiation, and the delayed ionizing energy is about evenly divided between gamma and beta ray energy.

In a reactor that has been operating for some time, the radioactive fission products will have built up to steady state concentrations such that their rate of decay is equal to their rate of formation, so that their fractional total contribution to reactor heat (via beta decay) is the same as these radioisotopic fractional contributions to the energy of fission. Under these conditions, the 6.5% of fission which appears as delayed ionizing radiation (delayed gammas and betas from radioactive fission products) contributes to the steady-state reactor heat production under power. It is this output fraction which remains when the reactor is suddenly shut down (undergoes scram). For this reason, the reactor decay heat output begins at 6.5% of the full reactor steady state fission power, once the reactor is shut down. However, within hours, due to decay of these isotopes, the decay power output is far less.

The remainder of the delayed energy (8.8 MeV/202.5 MeV = 4.3% of total fission energy) is emitted as antineutrinos, which as a practical matter, are not considered "ionizing radiation." The reason is that energy released as antineutrinos is not captured by the reactor material as heat, and escapes directly through all materials (including the Earth) at nearly the speed of light, and into interplanetary space (the amount absorbed is minuscule). Neutrino radiation is ordinarily not classed as ionizing radiation, because it is almost entirely not absorbed and therefore does not produce effects (although the very rare neutrino event is ionizing). Almost all of the rest of the radiation (6.5% delayed beta and gamma radiation) is eventually converted to heat in a reactor core or its shielding.

Some processes involving neutrons are notable for absorbing or finally yielding energy — for example neutron kinetic energy does not yield heat immediately if the neutron is captured by a uranium-238 atom to breed plutonium-239, but this energy is emitted if the plutonium-239 is later fissioned. On the other hand, so-called delayed neutrons emitted as radioactive decay products with half-lives up to several minutes, from fission-daughters, are very important to reactor control, because they give a characteristic "reaction" time for the total nuclear reaction to double in size, if the reaction is run in a "delayed-critical" zone which deliberately relies on these neutrons for a supercritical chain-reaction (one in which each fission cycle yields more neutrons than it absorbs). Without their existence, the nuclear chain-reaction would be prompt critical and increase in size faster than it could be controlled by human intervention. In this case, the first experimental atomic reactors would have run away to a dangerous and messy "prompt critical reaction" before their operators could have manually shut them down (for this reason, designer Enrico Fermi included radiation-counter-triggered control rods, suspended by electromagnets, which could automatically drop into the center of Chicago Pile-1). If these delayed neutrons are captured without producing fissions, they produce heat as well.

Product Nuclei and Binding Energy

In fission there is a preference to yield fragments with even proton numbers, which is called the odd-even effect on the fragments charge distribution. However, no odd-even effect is observed on fragment mass number distribution. This result is attributed to nucleon pair breaking.

In nuclear fission events the nuclei may break into any combination of lighter nuclei, but the most common event is not fission to equal mass nuclei of about mass 120; the most common event (depending on isotope and process) is a slightly unequal fission in which one daughter nucleus has a mass of about 90 to 100 u and the other the remaining 130 to 140 u. Unequal fissions are energetically more favorable because this allows one product to be closer to the energetic minimum near mass 60 u (only a quarter of the average fissionable mass), while the other nucleus with mass 135 u is still not far out of the range of the most tightly bound nuclei (another statement of this, is that the atomic binding energy curve is slightly steeper to the left of mass 120 u than to the right of it).

Origin of The Active Energy and The Curve of Binding Energy

Nuclear fission of heavy elements produces energy because the specific binding energy (binding energy per mass) of intermediate-mass nuclei with atomic numbers and atomic masses close to ^{62}Ni and ^{56}Fe is greater than the nucleon-specific binding energy of very heavy nuclei, so that energy is released when heavy nuclei are broken apart. The total rest masses of the fission products (**Mp**) from a single reaction is less than the mass of the original fuel nucleus (M). The excess mass $\Delta m = M - Mp$ is the invariant mass of the energy that is released as photons (gamma rays) and kinetic energy of the fission fragments, according to the mass-energy equivalence formula $E = mc^2$.

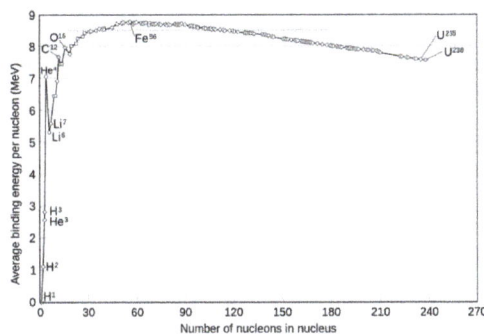

The "curve of binding energy": A graph of binding energy per nucleon of common isotopes.

The variation in specific binding energy with atomic number is due to the interplay of the two fundamental forces acting on the component nucleons (protons and neutrons) that make up the nucleus. Nuclei are bound by an attractive nuclear force between nucleons, which overcomes the electrostatic repulsion between protons. However, the nuclear force acts only over relatively short ranges (a few nucleon diameters), since it follows an exponentially decaying Yukawa potential which makes it insignificant at longer distances. The electrostatic repulsion is of longer range, since it decays by an inverse-square rule, so that nuclei larger than about 12 nucleons in diameter reach a point that the total electrostatic repulsion overcomes the nuclear force and causes them to be spontaneously unstable. For the same reason, larger nuclei (more than about eight nucleons in diameter) are less tightly bound per unit mass than are smaller nuclei; breaking a large nucleus into two or more intermediate-sized nuclei releases energy. The origin of this energy is the nuclear force, which intermediate-sized nuclei allows to act more efficiently, because each nucleon has more neighbors which are within the short range attraction of this force. Thus less energy is needed in the smaller nuclei and the difference to the state before is set free.

Also because of the short range of the strong binding force, large stable nuclei must contain proportionally more neutrons than do the lightest elements, which are most stable with a 1 to 1 ratio of protons and neutrons. Nuclei which have more than 20 protons cannot be stable unless they have more than an equal number of neutrons. Extra neutrons stabilize heavy elements because they add to strong-force binding (which acts between all nucleons) without adding to proton–proton repulsion. Fission products have, on average, about the same ratio of neutrons and protons as their parent nucleus, and are therefore usually unstable to beta decay (which changes neutrons to protons) because they have proportionally too many neutrons compared to stable isotopes of similar mass.

This tendency for fission product nuclei to beta-decay is the fundamental cause of the problem of radioactive high level waste from nuclear reactors. Fission products tend to be beta emitters, emitting fast-moving electrons to conserve electric charge, as excess neutrons convert to protons in the fission-product atoms.

Chain Reactions

Several heavy elements, such as uranium, thorium, and plutonium, undergo both spontaneous fission, a form of radioactive decay and *induced fission*, a form of nuclear reaction. Elemental isotopes that undergo induced fission when struck by a free neutron are called fissionable; isotopes that undergo fission when struck by a slow-moving thermal neutron are also called fissile. A few particularly fissile and readily obtainable isotopes (notably ^{233}U, ^{235}U and ^{239}Pu) are called nuclear fuels because they can sustain a chain reaction and can be obtained in large enough quantities to be useful.

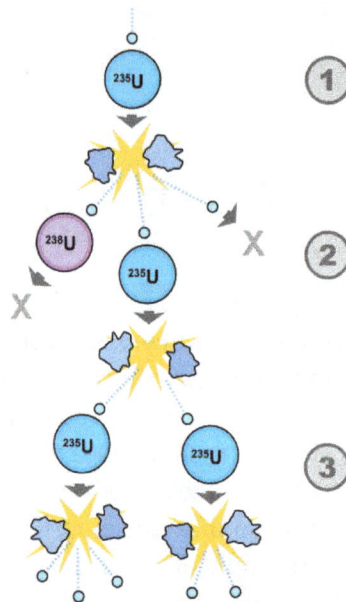

A schematic nuclear fission chain reaction. 1. A uranium-235 atom absorbs a neutron and fissions into two new atoms (fission fragments), releasing three new neutrons and some binding energy. 2. One of those neutrons is absorbed by an atom of uranium-238 and does not continue the reaction. Another neutron is simply lost and does not collide with anything, also not continuing the reaction. However, the one neutron does collide with an atom of uranium-235, which then fissions and releases two neutrons and some binding energy. 3. Both of those neutrons collide with uranium-235 atoms, each of which fissions and releases between one and three neutrons, which can then continue the reaction.

All fissionable and fissile isotopes undergo a small amount of spontaneous fission which releases a few free neutrons into any sample of nuclear fuel. Such neutrons would escape rapidly from the fuel and become a free neutron, with a mean lifetime of about 15 minutes before decaying to protons and beta particles. However, neutrons almost invariably impact and are absorbed by other nuclei in the vicinity long before this happens (newly created fission neutrons move at about 7% of the speed of light, and even moderated neutrons move at about 8 times the speed of sound). Some neutrons will impact fuel nuclei and induce further fissions, releasing yet more neutrons. If enough nuclear fuel is assembled in one place, or if the escaping neutrons are sufficiently contained, then these freshly emitted neutrons outnumber the neutrons that escape from the assembly, and a *sustained nuclear chain reaction* will take place.

An assembly that supports a sustained nuclear chain reaction is called a critical assembly or, if the assembly is almost entirely made of a nuclear fuel, a critical mass. The word "critical" refers to a cusp in the behavior of the differential equation that governs the number of free neutrons present in the fuel: if less than a critical mass is present, then the amount of neutrons is determined by radioactive decay, but if a critical mass or more is present, then the amount of neutrons is controlled instead by the physics of the chain reaction. The actual mass of a *critical mass* of nuclear fuel depends strongly on the geometry and surrounding materials.

Not all fissionable isotopes can sustain a chain reaction. For example, ^{238}U, the most abundant form of uranium, is fissionable but not fissile: it undergoes induced fission when impacted by an energetic neutron with over 1 MeV of kinetic energy. However, too few of the neutrons produced by ^{238}U fission are energetic enough to induce further fissions in ^{238}U, so no chain reaction is possible with this isotope. Instead, bombarding ^{238}U with slow neutrons causes it to absorb them (becoming ^{239}U) and decay by beta emission to ^{239}Np which then decays again by the same process to ^{239}Pu; that process is used to manufacture ^{239}Pu in breeder reactors. In-situ plutonium production also contributes to the neutron chain reaction in other types of reactors after sufficient plutonium-239 has been produced, since plutonium-239 is also a fissile element which serves as fuel. It is estimated that up to half of the power produced by a standard "non-breeder" reactor is produced by the fission of plutonium-239 produced in place, over the total life-cycle of a fuel load.

Fissionable, non-fissile isotopes can be used as fission energy source even without a chain reaction. Bombarding ^{238}U with fast neutrons induces fissions, releasing energy as long as the external neutron source is present. This is an important effect in all reactors where fast neutrons from the fissile isotope can cause the fission of nearby ^{238}U nuclei, which means that some small part of the ^{238}U is "burned-up" in all nuclear fuels, especially in fast breeder reactors that operate with higher-energy neutrons. That same fast-fission effect is used to augment the energy released by modern thermonuclear weapons, by jacketing the weapon with ^{238}U to react with neutrons released by nuclear fusion at the center of the device. But the explosive effects of nuclear fission chain reactions can be reduced by using substances like moderators which slow down the speed of secondary neutrons.

Fission Reactors

Critical fission reactors are the most common type of nuclear reactor. In a critical fission reactor, neutrons produced by fission of fuel atoms are used to induce yet more fissions, to sustain a controllable amount of energy release. Devices that produce engineered but non-self-sustaining fission reactions are subcritical fission reactors. Such devices use radioactive decay or particle accelerators to trigger fissions.

The cooling towers of the Philippsburg Nuclear Power Plant, in Germany.

Critical fission reactors are built for three primary purposes, which typically involve different engineering trade-offs to take advantage of either the heat or the neutrons produced by the fission chain reaction:

- *power reactors* are intended to produce heat for nuclear power, either as part of a generating station or a local power system such as a nuclear submarine.

- *research reactors* are intended to produce neutrons and/or activate radioactive sources for scientific, medical, engineering, or other research purposes.

- *breeder reactors* are intended to produce nuclear fuels in bulk from more abundant isotopes. The better known fast breeder reactor makes ^{239}Pu (a nuclear fuel) from the naturally very abundant ^{238}U (not a nuclear fuel). Thermal breeder reactors previously tested using ^{232}Th to breed the fissile isotope ^{233}U (thorium fuel cycle) continue to be studied and developed.

While, in principle, all fission reactors can act in all three capacities, in practice the tasks lead to conflicting engineering goals and most reactors have been built with only one of the above tasks in mind. (There are several early counter-examples, such as the Hanford N reactor, now decommissioned). Power reactors generally convert the kinetic energy of fission products into heat, which is used to heat a working fluid and drive a heat engine that generates mechanical or electrical power. The working fluid is usually water with a steam turbine, but some designs use other materials such as gaseous helium. Research reactors produce neutrons that are used in various ways, with the heat of fission being treated as an unavoidable waste product. Breeder reactors are a specialized form of research reactor, with the caveat that the sample being irradiated is usually the fuel itself, a mixture of ^{238}U and ^{235}U. For a more detailed description of the physics and operating principles of critical fission reactors.

Fission Bombs

One class of nuclear weapon, a *fission bomb*, otherwise known as an *atomic bomb* or *atom bomb*, is a fission reactor designed to liberate as much energy as possible as rapidly as possible, before the released energy causes the reactor to explode (and the chain reaction to stop). Development of nuclear weapons was the motivation behind early research into nuclear fission which the Manhattan Project during World War II (September 1, 1939–September 2, 1945) carried out most of the early scientific work on fission chain reactions, culminating in the three events involving fission bombs that occurred during the war. The first fission bomb, codenamed "The Gadget", was detonated during the Trinity Test in the desert of New Mexico on July 16, 1945. Two other fission bombs, codenamed "Little Boy" and "Fat Man", were used in combat against the Japanese cities of Hiroshima and Nagasaki in on August 6 and 9, 1945 respectively.

The mushroom cloud of the atomic bomb dropped on the Japanese city of Nagasaki on August 9, 1945, rose some 18 kilometres (11 mi) above the bomb's hypocenter. An estimated 39,000 people were killed by the atomic bomb, of whom 23,145-28,113 were Japanese factory workers, 2,000 were Korean slave laborers, and 150 were Japanese combatants.

Even the first fission bombs were thousands of times more explosive than a comparable mass of chemical explosive. For example, Little Boy weighed a total of about four tons (of which 60 kg was nuclear fuel) and was 11 feet (3.4 m) long; it also yielded an explosion equivalent to about 15 kilotons of TNT, destroying a large part of the city of Hiroshima. Modern nuclear weapons (which include a thermonuclear *fusion* as well as one or more fission stages) are hundreds of times more energetic for their weight than the first pure fission atomic bombs, so that a modern single missile warhead bomb weighing less than 1/8 as much as Little Boy has a yield of 475,000 tons of TNT, and could bring destruction to about 10 times the city area.

While the fundamental physics of the fission chain reaction in a nuclear weapon is similar to the physics of a controlled nuclear reactor, the two types of device must be engineered quite differently. A nuclear bomb is designed to release all its energy at once, while a reactor is designed to generate a steady supply of useful power. While overheating of a reactor can lead to, and has led to, meltdown and steam explosions, the much lower uranium enrichment makes it impossible for a nuclear reactor to explode with the same destructive power as a nuclear weapon. It is also difficult to extract useful power from a nuclear bomb, although at least one rocket propulsion system, Project Orion, was intended to work by exploding fission bombs behind a massively padded and shielded spacecraft.

The strategic importance of nuclear weapons is a major reason why the technology of nuclear fission is politically sensitive. Viable fission bomb designs are, arguably, within the capabilities of many, being relatively simple from an engineering viewpoint. However, the difficulty of obtaining fissile nuclear material to realize the designs is the key to the relative unavailability of nuclear weapons to all but modern industrialized governments with special programs to produce fissile materials.

History

Discovery of Nuclear Fission

The discovery of nuclear fission occurred in 1938 in the buildings of Kaiser Wilhelm Society for Chemistry, today part of the Free University of Berlin, following nearly five decades of work on the science of radioactivity and the elaboration of new nuclear physics that described the components of atoms. In 1911, Ernest Rutherford proposed a model of the atom in which a very small, dense and positively charged nucleus of protons (the neutron had not yet been discovered) was surrounded by orbiting, negatively charged electrons (the Rutherford model). Niels Bohr improved upon this in 1913 by reconciling the quantum behavior of electrons (the Bohr model). Work by Henri Becquerel, Marie Curie, Pierre Curie, and Rutherford further elaborated that the nucleus, though tightly bound, could undergo different forms of radioactive decay, and thereby transmute into other elements. (For example, by alpha decay: the emission of an alpha particle—two protons and two neutrons bound together into a particle identical to a helium nucleus.)

Some work in nuclear transmutation had been done. In 1917, Rutherford was able to accomplish transmutation of nitrogen into oxygen, using alpha particles directed at nitrogen $^{14}N + \alpha \rightarrow {}^{17}O + p$. This was the first observation of a nuclear reaction, that is, a reaction in which particles from one decay are used to transform another atomic nucleus. Eventually, in 1932, a fully artificial nuclear reaction and nuclear transmutation was achieved by Rutherford's colleagues Ernest Walton and John Cockcroft, who used artificially accelerated protons against lithium-7, to split this nucleus into two alpha particles. The feat was popularly known as "splitting the atom", although it was not the modern nuclear fission reaction later discovered in heavy elements, which is discussed below. Meanwhile, the possibility of *combining* nuclei—nuclear fusion—had been studied in connection with understanding the processes which power stars. The first artificial fusion reaction had been achieved by Mark Oliphant in 1932, using two accelerated deuterium nuclei (each consisting of a single proton bound to a single neutron) to create a helium-3 nucleus.

After English physicist James Chadwick discovered the neutron in 1932, Enrico Fermi and his colleagues in Rome studied the results of bombarding uranium with neutrons in 1934. Fermi concluded that his experiments had created new elements with 93 and 94 protons, which the group dubbed ausonium and hesperium. However, not all were convinced by Fermi's analysis of his results. The German chemist Ida Noddack notably suggested in print in 1934 that instead of creating a new, heavier element 93, that "it is conceivable that the nucleus breaks up into several large fragments." However, Noddack's conclusion was not pursued at the time.

After the Fermi publication, Otto Hahn, Lise Meitner, and Fritz Strassmann began performing similar experiments in Berlin. Meitner, an Austrian Jew, lost her citizenship with the "Anschluss", the occupation and annexation of Austria into Nazi Germany in March 1938, but she fled in July 1938 to Sweden and started a correspondence by mail with Hahn in Berlin. By coincidence, her nephew Otto Robert Frisch, also a refugee, was also in Sweden when Meitner received a letter from

Hahn dated 19 December describing his chemical proof that some of the product of the bombardment of uranium with neutrons was barium. Hahn suggested a *bursting* of the nucleus, but he was unsure of what the physical basis for the results were. Barium had an atomic mass 40% less than uranium, and no previously known methods of radioactive decay could account for such a large difference in the mass of the nucleus. Frisch was skeptical, but Meitner trusted Hahn's ability as a chemist. Marie Curie had been separating barium from radium for many years, and the techniques were well-known. According to Frisch:

The experimental apparatus with which Otto Hahn and Fritz Strassmann discovered nuclear fission in 1938

Was it a mistake? No, said Lise Meitner; Hahn was too good a chemist for that. But how could barium be formed from uranium? No larger fragments than protons or helium nuclei (alpha particles) had ever been chipped away from nuclei, and to chip off a large number not nearly enough energy was available. Nor was it possible that the uranium nucleus could have been cleaved right across. A nucleus was not like a brittle solid that can be cleaved or broken; George Gamow had suggested early on, and Bohr had given good arguments that a nucleus was much more like a liquid drop. Perhaps a drop could divide itself into two smaller drops in a more gradual manner, by first becoming elongated, then constricted, and finally being torn rather than broken in two? We knew that there were strong forces that would resist such a process, just as the surface tension of an ordinary liquid drop tends to resist its division into two smaller ones. But nuclei differed from ordinary drops in one important way: they were electrically charged, and that was known to counteract the surface tension.

The charge of a uranium nucleus, we found, was indeed large enough to overcome the effect of the surface tension almost completely; so the uranium nucleus might indeed resemble a very wobbly unstable drop, ready to divide itself at the slightest provocation, such as the impact of a single neutron. But there was another problem. After separation, the two drops would be driven apart by their mutual electric repulsion and would acquire high speed and hence a very large energy, about 200 MeV in all; where could that energy come from? ...Lise Meitner... worked out that the two nuclei formed by the division of a uranium nucleus together would be lighter than the original uranium nucleus by about one-fifth the mass of a proton. Now whenever mass disappears energy is created, according to Einstein's formula $E = mc^2$, and one-fifth of a proton mass was just equivalent to 200 MeV. So here was the source for that energy; it all fitted!

In short, Meitner and Frisch had correctly interpreted Hahn's results to mean that the nucleus of uranium had split roughly in half. Frisch suggested the process be named "nuclear fission," by analogy to the process of living cell division into two cells, which was then called binary fission.

Just as the term nuclear "chain reaction" would later be borrowed from chemistry, so the term "fission" was borrowed from biology.

On 22 December 1938, Hahn and Strassmann sent a manuscript to *Naturwissenschaften* reporting that they had discovered the element barium after bombarding uranium with neutrons. Simultaneously, they communicated these results to Meitner in Sweden. She and Frisch correctly interpreted the results as evidence of nuclear fission. Frisch confirmed this experimentally on 13 January 1939. For proving that the barium resulting from his bombardment of uranium with neutrons was the product of nuclear fission, Hahn was awarded the Nobel Prize for Chemistry in 1944 (the sole recipient) "for his discovery of the fission of heavy nuclei". (The award was actually given to Hahn in 1945, as "the Nobel Committee for Chemistry decided that none of the year's nominations met the criteria as outlined in the will of Alfred Nobel." In such cases, the Nobel Foundation's statutes permit that year's prize be reserved until the following year.)

German stamp honoring Otto Hahn and his discovery of nuclear fission (1979)

News spread quickly of the new discovery, which was correctly seen as an entirely novel physical effect with great scientific—and potentially practical—possibilities. Meitner's and Frisch's interpretation of the discovery of Hahn and Strassmann crossed the Atlantic Ocean with Niels Bohr, who was to lecture at Princeton University. I.I. Rabi and Willis Lamb, two Columbia University physicists working at Princeton, heard the news and carried it back to Columbia. Rabi said he told Enrico Fermi; Fermi gave credit to Lamb. Bohr soon thereafter went from Princeton to Columbia to see Fermi. Not finding Fermi in his office, Bohr went down to the cyclotron area and found Herbert L. Anderson. Bohr grabbed him by the shoulder and said: "Young man, let me explain to you about something new and exciting in physics." It was clear to a number of scientists at Columbia that they should try to detect the energy released in the nuclear fission of uranium from neutron bombardment. On 25 January 1939, a Columbia University team conducted the first nuclear fission experiment in the United States, which was done in the basement of Pupin Hall; the members of the team were Herbert L. Anderson, Eugene T. Booth, John R. Dunning, Enrico Fermi, G. Norris Glasoe, and Francis G. Slack. The experiment involved placing uranium oxide inside of an ionization chamber and irradiating it with neutrons, and measuring the energy thus released. The results confirmed that fission was occurring and hinted strongly that it was the isotope uranium 235 in particular that was fissioning. The next day, the Fifth Washington Conference on Theoretical Physics began in Washington, D.C. under the joint auspices of the George Washington University and the Carnegie Institution of Washington. There, the news on nuclear fission was spread even further, which fostered many more experimental demonstrations.

During this period the Hungarian physicist Leó Szilárd, who was residing in the United States at the time, realized that the neutron-driven fission of heavy atoms could be used to create a nuclear chain reaction. Such a reaction using neutrons was an idea he had first formulated in 1933, upon reading Rutherford's disparaging remarks about generating power from his team's 1932 experiment using protons to split lithium. However, Szilárd had not been able to achieve a neutron-driven chain reaction with neutron-rich light atoms. In theory, if in a neutron-driven chain reaction the number of secondary neutrons produced was greater than one, then each such reaction could trigger multiple additional reactions, producing an exponentially increasing number of reactions. It was thus a possibility that the fission of uranium could yield vast amounts of energy for civilian or military purposes (i.e., electric power generation or atomic bombs).

Szilard now urged Fermi (in New York) and Frédéric Joliot-Curie (in Paris) to refrain from publishing on the possibility of a chain reaction, lest the Nazi government become aware of the possibilities on the eve of what would later be known as World War II. With some hesitation Fermi agreed to self-censor. But Joliot-Curie did not, and in April 1939 his team in Paris, including Hans von Halban and Lew Kowarski, reported in the journal *Nature* that the number of neutrons emitted with nuclear fission of ^{235}U was then reported at 3.5 per fission. (They later corrected this to 2.6 per fission.) Simultaneous work by Szilard and Walter Zinn confirmed these results. The results suggested the possibility of building nuclear reactors (first called "neutronic reactors" by Szilard and Fermi) and even nuclear bombs. However, much was still unknown about fission and chain reaction systems.

Fission Chain Reaction Realized

"Chain reactions" at that time were a known phenomenon in *chemistry*, but the analogous process in nuclear physics, using neutrons, had been foreseen as early as 1933 by Szilárd, although Szilárd at that time had no idea with what materials the process might be initiated. Szilárd considered that neutrons would be ideal for such a situation, since they lacked an electrostatic charge.

Drawing of the first artificial reactor, Chicago Pile-1.

With the news of fission neutrons from uranium fission, Szilárd immediately understood the possibility of a nuclear chain reaction using uranium. In the summer, Fermi and Szilard proposed the idea of a nuclear reactor (pile) to mediate this process. The pile would use natural uranium as fuel. Fermi had shown much earlier that neutrons were far more effectively captured by atoms if they were of low energy (so-called "slow" or "thermal" neutrons), because for quantum reasons it made the atoms look like much larger targets to the neutrons. Thus to slow down the secondary neutrons released by the fissioning uranium nuclei, Fermi and Szilard proposed a graphite "moderator," against which the fast, high-energy secondary neutrons would collide, effectively slowing them down. With enough uranium, and with pure-enough graphite, their "pile" could theoretically sustain a slow-neutron chain reaction. This would result in the production of heat, as well as the creation of radioactive fission products.

In August 1939, Szilard and fellow Hungarian refugees physicists Teller and Wigner thought that the Germans might make use of the fission chain reaction and were spurred to attempt to attract the attention of the United States government to the issue. Towards this, they persuaded German-Jewish refugee Albert Einstein to lend his name to a letter directed to President Franklin Roosevelt. The Einstein–Szilárd letter suggested the possibility of a uranium bomb deliverable by ship, which would destroy "an entire harbor and much of the surrounding countryside." The President received the letter on 11 October 1939 — shortly after World War II began in Europe, but two years before U.S. entry into it. Roosevelt ordered that a scientific committee be authorized for overseeing uranium work and allocated a small sum of money for pile research.

In England, James Chadwick proposed an atomic bomb utilizing natural uranium, based on a paper by Rudolf Peierls with the mass needed for critical state being 30–40 tons. In America, J. Robert Oppenheimer thought that a cube of uranium deuteride 10 cm on a side (about 11 kg of uranium) might "blow itself to hell." In this design it was still thought that a moderator would need to be used for nuclear bomb fission (this turned out not to be the case if the fissile isotope was separated). In December, Werner Heisenberg delivered a report to the German Ministry of War on the possibility of a uranium bomb. Most of these models were still under the assumption that the bombs would be powered by slow neutron reactions—and thus be similar to a reactor undergoing a meltdown.

In Birmingham, England, Frisch teamed up with Peierls, a fellow German-Jewish refugee. They had the idea of using a purified mass of the uranium isotope ^{235}U, which had a cross section just determined, and which was much larger than that of ^{238}U or natural uranium (which is 99.3% the latter isotope). Assuming that the cross section for fast-neutron fission of ^{235}U was the same as for slow neutron fission, they determined that a pure ^{235}U bomb could have a critical mass of only 6 kg instead of tons, and that the resulting explosion would be tremendous. (The amount actually turned out to be 15 kg, although several times this amount was used in the actual uranium (Little Boy) bomb). In February 1940 they delivered the Frisch–Peierls memorandum. Ironically, they were still officially considered "enemy aliens" at the time. Glenn Seaborg, Joseph W. Kennedy, Arthur Wahl, and Italian-Jewish refugee Emilio Segrè shortly thereafter discovered ^{239}Pu in the decay products of ^{239}U produced by bombarding ^{238}U with neutrons, and determined it to be a fissile material, like ^{235}U.

The possibility of isolating uranium-235 was technically daunting, because uranium-235 and uranium-238 are chemically identical, and vary in their mass by only the weight of three neutrons. However, if a sufficient quantity of uranium-235 could be isolated, it would allow for a fast neutron fission chain reaction. This would be extremely explosive, a true "atomic bomb." The discovery that plutonium-239 could be produced in a nuclear reactor pointed towards another approach to a fast neutron fission bomb. Both approaches were extremely novel and not yet well understood, and there was considerable scientific skepticism at the idea that they could be developed in a short amount of time.

On June 28, 1941, the Office of Scientific Research and Development was formed in the U.S. to mobilize scientific resources and apply the results of research to national defense. In September,

Fermi assembled his first nuclear "pile" or reactor, in an attempt to create a slow neutron-induced chain reaction in uranium, but the experiment failed to achieve criticality, due to lack of proper materials, or not enough of the proper materials which were available.

Producing a fission chain reaction in natural uranium fuel was found to be far from trivial. Early nuclear reactors did not use isotopically enriched uranium, and in consequence they were required to use large quantities of highly purified graphite as neutron moderation materials. Use of ordinary water (as opposed to heavy water) in nuclear reactors requires enriched fuel — the partial separation and relative enrichment of the rare ^{235}U isotope from the far more common ^{238}U isotope. Typically, reactors also require inclusion of extremely chemically pure neutron moderator materials such as deuterium (in heavy water), helium, beryllium, or carbon, the latter usually as graphite. (The high purity for carbon is required because many chemical impurities such as the boron-10 component of natural boron, are very strong neutron absorbers and thus poison the chain reaction and end it prematurely.)

Production of such materials at industrial scale had to be solved for nuclear power generation and weapons production to be accomplished. Up to 1940, the total amount of uranium metal produced in the USA was not more than a few grams, and even this was of doubtful purity; of metallic beryllium not more than a few kilograms; and concentrated deuterium oxide (heavy water) not more than a few kilograms. Finally, carbon had never been produced in quantity with anything like the purity required of a moderator.

The problem of producing large amounts of high purity uranium was solved by Frank Spedding using the thermite or "Ames" process. Ames Laboratory was established in 1942 to produce the large amounts of natural (unenriched) uranium metal that would be necessary for the research to come. The critical nuclear chain-reaction success of the Chicago Pile-1 (December 2, 1942) which used unenriched (natural) uranium, like all of the atomic "piles" which produced the plutonium for the atomic bomb, was also due specifically to Szilard's realization that very pure graphite could be used for the moderator of even natural uranium "piles". In wartime Germany, failure to appreciate the qualities of very pure graphite led to reactor designs dependent on heavy water, which in turn was denied the Germans by Allied attacks in Norway, where heavy water was produced. These difficulties—among many others— prevented the Nazis from building a nuclear reactor capable of criticality during the war, although they never put as much effort as the United States into nuclear research, focusing on other technologies.

Manhattan Project and Beyond

In the United States, an all-out effort for making atomic weapons was begun in late 1942. This work was taken over by the U.S. Army Corps of Engineers in 1943, and known as the Manhattan Engineer District. The top-secret Manhattan Project, as it was colloquially known, was led by General Leslie R. Groves. Among the project's dozens of sites were: Hanford Site in Washington state, which had the first industrial-scale nuclear reactors; Oak Ridge, Tennessee, which was primarily concerned with uranium enrichment; and Los Alamos, in New Mexico, which was the scientific hub for research on bomb development and design. Other sites, notably the Berkeley Radiation Laboratory and the Metallurgical Laboratory at the University of Chicago, played important contributing roles. Overall scientific direction of the project was managed by the physicist J. Robert Oppenheimer.

In July 1945, the first atomic explosive device, dubbed "Trinity", was detonated in the New Mexico desert. It was fueled by plutonium created at Hanford. In August 1945, two more atomic devices – "Little Boy", a uranium-235 bomb, and "Fat Man", a plutonium bomb – were used against the Japanese cities of Hiroshima and Nagasaki.

In the years after World War II, many countries were involved in the further development of nuclear fission for the purposes of nuclear reactors and nuclear weapons. The UK opened the first commercial nuclear power plant in 1956. In 2013, there are 437 reactors in 31 countries.

Natural Fission Chain-reactors on Earth

Criticality in nature is uncommon. At three ore deposits at Oklo in Gabon, sixteen sites (the so-called Oklo Fossil Reactors) have been discovered at which self-sustaining nuclear fission took place approximately 2 billion years ago. Unknown until 1972 (but postulated by Paul Kuroda in 1956), when French physicist Francis Perrin discovered the Oklo Fossil Reactors, it was realized that nature had beaten humans to the punch. Large-scale natural uranium fission chain reactions, moderated by normal water, had occurred far in the past and would not be possible now. This ancient process was able to use normal water as a moderator only because 2 billion years before the present, natural uranium was richer in the shorter-lived fissile isotope ^{235}U (about 3%), than natural uranium available today (which is only 0.7%, and must be enriched to 3% to be usable in light-water reactors).

Radioactive Decay

Radioactive decay (also known as nuclear decay or radioactivity) is the process by which the nucleus of an unstable atom loses energy by emitting radiation, including alpha particles, beta particles, gamma rays, and conversion electrons. A material that spontaneously emits such radiation is considered radioactive.

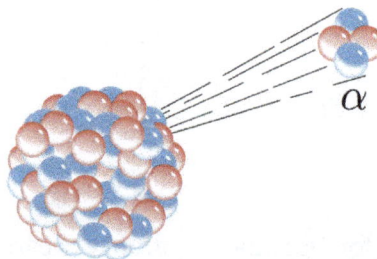

Alpha decay is one type of radioactive decay, in which an atomic nucleus emits an alpha particle, and thereby transforms (or "decays") into an atom with a mass number decreased by 4 and atomic number decreased by 2.

Radioactive decay is a stochastic (i.e. random) process at the level of single atoms, in that, according to quantum theory, it is impossible to predict when a particular atom will decay, regardless of how long the atom has existed. For a collection of atoms however, the collection's decay rate can be calculated from their measured decay constants or half-lives. This is the basis of radiometric dating. The half-lives of radioactive atoms have no known lower or upper limit, spanning a time range of over 55 orders of magnitude, from nearly instantaneous to far longer than the age of the universe. A radioactive source emits its decay products isotropically (all directions and without bias) in the absence of external influence.

There are many different types of radioactive decay. A decay, or loss of energy from the nucleus, results when an atom with an initial type of nucleus, called the *parent radionuclide* (or *parent radioisotope*), transforms into a *daughter nuclide*. The transformation produces an atom in a different state (a nucleus containing a different number of protons and neutrons). In some decays, the parent and the daughter nuclides are different chemical elements, and thus the decay process results in the creation of an atom of a different element. This is known as a nuclear transmutation.

The first decay processes to be discovered were alpha decay, beta decay, and gamma decay. Alpha decay occurs when the nucleus ejects an alpha particle (helium nucleus). This is the most common process of emitting nucleons, but in rarer types of decays, nuclei can eject protons, or in the case of cluster decay specific nuclei of other elements. Beta decay occurs when the nucleus emits an electron or positron and a neutrino, in a process that changes a proton to a neutron or the other way about. Highly excited neutron-rich nuclei, formed as the product of other types of decay, occasionally lose energy by way of neutron emission, resulting in a change from one isotope to another of the same element.. The nucleus may capture an orbiting electron, causing a proton to convert into a neutron in a process called electron capture. All of these processes result in a well-defined nuclear transmutation.

By contrast, there are radioactive decay processes that do not result in a nuclear transmutation. The energy of an excited nucleus may be emitted as a gamma ray in a process called gamma decay, or that energy may be lost when the nucleus interacts with an orbital electron causing its ejection from the atom, in a process called internal conversion.

Another type of radioactive decay results in products that are not defined, but appear in a range of "pieces" of the original nucleus. This decay, called spontaneous fission, happens when a large unstable nucleus spontaneously splits into two (and occasionally three) smaller daughter nuclei, and generally leads to the emission of gamma rays, neutrons, or other particles from those products.

There exist twenty-nine chemical elements on Earth that are radioactive. They are those that contain thirty-four radionuclides that date before the time of formation of the solar system, and are known as primordial nuclides. Well-known examples are uranium and thorium, but also included are naturally occurring long-lived radioisotopes such as potassium-40. Another fifty or so shorter-lived radionuclides, such as radium and radon, found on Earth, are the products of decay chains that began with the primordial nuclides, or are the product of ongoing cosmogenic processes, such as the production of carbon-14 from nitrogen-14 by cosmic rays. Radionuclides may also be produced artificially in particle accelerators or nuclear reactors, resulting in 650 of these with half-lives of over an hour, and several thousand more with even shorter half-lives.

History of Discovery

Radioactivity was discovered in 1896 by the French scientist Henri Becquerel, while working with phosphorescent materials. These materials glow in the dark after exposure to light, and he suspected that the glow produced in cathode ray tubes by X-rays might be associated with phosphorescence. He wrapped a photographic plate in black paper and placed various phosphorescent salts on it. All results were negative until he used uranium salts. The uranium salts caused a blackening of the plate in spite of the plate being wrapped in black paper. These radiations were given the name "Becquerel Rays".

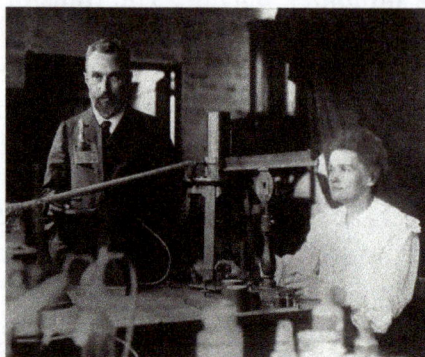

Pierre and Marie Curie in their Paris laboratory, before 1907

It soon became clear that the blackening of the plate had nothing to do with phosphorescence, as the blackening was also produced by non-phosphorescent salts of uranium and metallic uranium. It became clear from these experiments that there was a form of invisible radiation that could pass through paper and was causing the plate to react as if exposed to light.

At first, it seemed as though the new radiation was similar to the then recently discovered X-rays. Further research by Becquerel, Ernest Rutherford, Paul Villard, Pierre Curie, Marie Curie, and others showed that this form of radioactivity was significantly more complicated. Rutherford was the first to realize that all such elements decay in accordance with the same mathematical exponential formula. Rutherford and his student Frederick Soddy were the first to realize that many decay processes resulted in the transmutation of one element to another. Subsequently, the radioactive displacement law of Fajans and Soddy was formulated to describe the products of alpha and beta decay.

The early researchers also discovered that many other chemical elements, besides uranium, have radioactive isotopes. A systematic search for the total radioactivity in uranium ores also guided Pierre and Marie Curie to isolate two new elements: polonium and radium. Except for the radioactivity of radium, the chemical similarity of radium to barium made these two elements difficult to distinguish.

Marie and Pierre Curie's study of radioactivity is an important factor in science and medicine. After their research on Becquerel's rays led them to the discovery on both radium and polonium, they coined the term "radioactivity." Their research on the penetrating rays in uranium and discovery of radium launched an era of using radium for treatment of cancer. Their exploration of radium could be seen as the first peaceful use of nuclear energy and the start of modern nuclear medicine.

Early Health Dangers

The dangers of ionizing radiation due to radioactivity and X-rays were not immediately recognized.

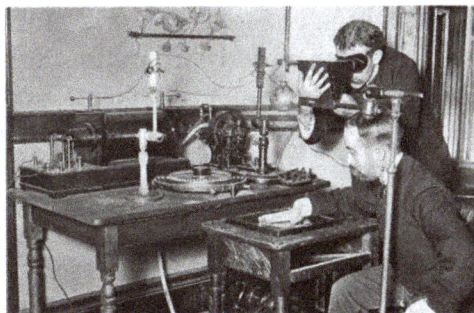

Taking an X-ray image with early Crookes tube apparatus in 1896. The Crookes tube is visible in the centre. The standing man is viewing his hand with a fluoroscope screen; this was a common way of setting up the tube. No precautions against radiation exposure are being taken; its hazards were not known at the time.

X-rays

The discovery of xrays by Wilhelm Röntgen in 1895 led to widespread experimentation by scientists, physicians, and inventors. Many people began recounting stories of burns, hair loss and worse in technical journals as early as 1896. In February of that year, Professor Daniel and Dr. Dudley of Vanderbilt University performed an experiment involving X-raying Dudley's head that resulted in his hair loss. A report by Dr. H.D. Hawks, of his suffering severe hand and chest burns in an X-ray demonstration, was the first of many other reports in *Electrical Review*.

Other experimenters including Elihu Thomson, and Nikola Tesla also reported burns. Thomson deliberately exposed a finger to an X-ray tube over a period of time and suffered pain, swelling, and blistering. Other effects, including ultraviolet rays and ozone were sometimes blamed for the damage, and many physicians still claimed that there were no effects from X-ray exposure at all.

Despite this, there were some early systematic hazard investigations, and as early as 1902 William Herbert Rollins wrote almost despairingly that his warnings about the dangers involved in careless use of X-rays was not being heeded, either by industry or by his colleagues. By this time Rollins had proved that X-rays could kill experimental animals, could cause a pregnant guinea pig to abort, and that they could kill a fetus. He also stressed that "animals vary in susceptibility to the external action of X-light" and warned that these differences be considered when patients were treated by means of X-rays.

Radioactive Substances

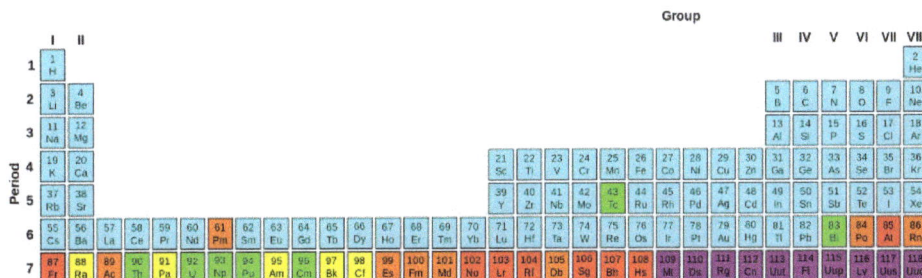

Radioactivity is characteristic of elements with large atomic number. Elements with at least one stable isotope are shown in light blue. Green shows elements whose most stable isotope has a half-life measured in millions of years. Yellow and orange are progressively less stable, with half-lives in thousands or hundreds of years, down toward one day. Red and purple show highly and extremely radioactive elements where the most stable isotopes exhibit half-lives measured on the order of one day and much less.

However, the biological effects of radiation due to radioactive substances were less easy to gauge. This gave the opportunity for many physicians and corporations to market radioactive substances as patent medicines. Examples were radium enema treatments, and radium-containing waters to be drunk as tonics. Marie Curie protested against this sort of treatment, warning that the effects of radiation on the human body were not well understood. Curie later died from aplastic anaemia, likely caused by exposure to ionizing radiation. By the 1930s, after a number of cases of bone necrosis and death of radium treatment enthusiasts, radium-containing medicinal products had been largely removed from the market (radioactive quackery).

Radiation Protection

Only a year after Röntgen's discovery of X rays, the American engineer Wolfram Fuchs (1896) gave what is probably the first protection advice, but it was not until 1925 that the first International Congress of Radiology (ICR) was held and considered establishing international protection standards. The effects of radiation on genes, including the effect of cancer risk, were recognized much later. In 1927, Hermann Joseph Muller published research showing genetic effects and, in 1946, was awarded the Nobel Prize in Physiology or Medicine for his findings.

The second ICR was held in Stockholm in 1928 and proposed the adoption of the rontgen unit, and the 'International X-ray and Radium Protection Committee' (IXRPC) was formed. Rolf Sievert was named Chairman, but a driving force was George Kaye of the British National Physical Laboratory. The committee met in 1931, 1934 and 1937.

After World War II the increased range and quantity of radioactive substances being handled as a result of military and civil nuclear programmes led to large groups of occupational workers and the public being potentially exposed to harmful levels of ionising radiation. This was considered at the first post-war ICR convened in London in 1950, when the present International Commission on Radiological Protection (ICRP) was born. Since then the ICRP has developed the present international system of radiation protection, covering all aspects of radiation hazard.

Units of Radioactivity

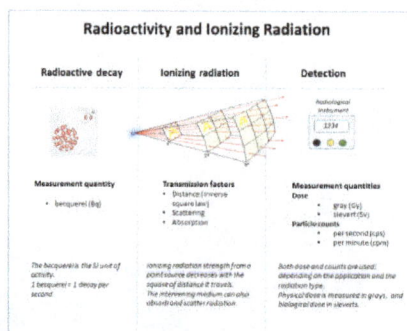

Graphic showing relationships between radioactivity and detected ionizing radiation

The International System of Units (SI) unit of radioactive activity is the becquerel (Bq), named in honour of the scientist Henri Becquerel. One Bq is defined as one transformation (or decay or disintegration) per second.

An older unit of radioactivity is the curie, Ci, which was originally defined as "the quantity or mass of radium emanation in equilibrium with one gram of radium (element)". Today, the curie is defined as 3.7×10^{10} disintegrations per second, so that 1 curie (Ci) = 3.7×10^{10} Bq. For radiological protection purposes, although the United States Nuclear Regulatory Commission permits the use of the unit curie alongside SI units, the European Union European units of measurement directives required that its use for "public health ... purposes" be phased out by 31 December 1985.

Types of Decay

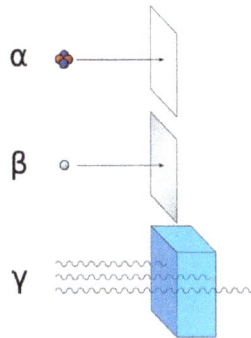

Alpha particles may be completely stopped by a sheet of paper, beta particles by aluminium shielding. Gamma rays can only be reduced by much more substantial mass, such as a very thick layer of lead.

Early researchers found that an electric or magnetic field could split radioactive emissions into three types of beams. The rays were given the names alpha, beta, and gamma, in order of their ability to penetrate matter. While alpha decay was observed only in heavier elements of atomic number 52 (tellurium) and greater, the other two types of decay were produced by all of the elements. Lead, atomic number 82, is the heaviest element to have any isotopes stable (to the limit of measurement) to radioactive decay. Radioactive decay is seen in all isotopes of all elements of atomic number 83 (bismuth) or greater. Bismuth, however, is only very slightly radioactive, with a half-life greater than the age of the universe; radioisotopes with extremely long half-lives are considered effectively stable for practical purposes.

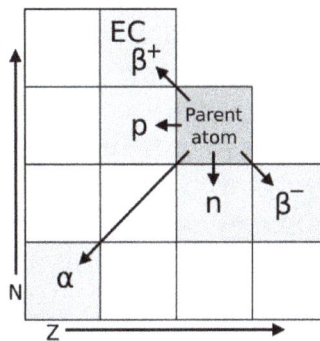

Transition diagram for decay modes of a radionuclide, with neutron number N and atomic number Z (shown are α, $\beta^{\pm}$, p^{+}, and n^{0} emissions, EC denotes electron capture).

In analysing the nature of the decay products, it was obvious from the direction of the electromagnetic forces applied to the radiations by external magnetic and electric fields that alpha particles carried a positive charge, beta particles carried a negative charge, and gamma rays were neutral. From the magnitude of deflection, it was clear that alpha particles were much more massive than beta particles. Passing alpha particles through a very thin glass window and trapping them in a discharge tube allowed researchers to study the emission spectrum of the captured particles, and ultimately proved that alpha particles are helium nuclei. Other experiments showed beta radiation, resulting from decay and cathode rays, were high-speed electrons. Likewise, gamma radiation and X-rays were found to be high-energy electromagnetic radiation.

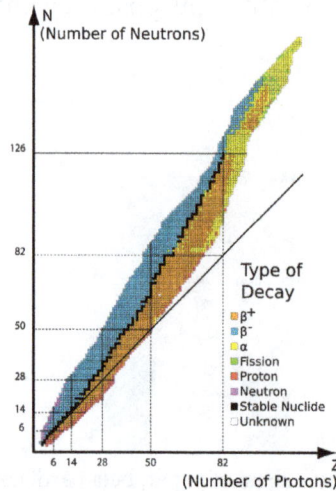

Types of radioactive decay related to N and Z numbers

The relationship between the types of decays also began to be examined: For example, gamma decay was almost always found to be associated with other types of decay, and occurred at about the same time, or afterwards. Gamma decay as a separate phenomenon, with its own half-life (now termed isomeric transition), was found in natural radioactivity to be a result of the gamma decay of excited metastable nuclear isomers, which were in turn created from other types of decay.

Although alpha, beta, and gamma radiations were most commonly found, other types of emission were eventually discovered. Shortly after the discovery of the positron in cosmic ray products, it was realized that the same process that operates in classical beta decay can also produce positrons (positron emission), along with neutrinos (classical beta decay produces antineutrinos). In a more common analogous process, called electron capture, some proton-rich nuclides were found to capture their own atomic electrons instead of emitting positrons, and subsequently these nuclides emit only a neutrino and a gamma ray from the excited nucleus (and often also Auger electrons and characteristic X-rays, as a result of the re-ordering of electrons to fill the place of the missing captured electron). These types of decay involve the nuclear capture of electrons or emission of electrons or positrons, and thus acts to move a nucleus toward the ratio of neutrons to protons that has the least energy for a given total number of nucleons. This consequently produces a more stable (lower energy) nucleus.

(A theoretical process of positron capture, analogous to electron capture, is possible in antimatter atoms, but has not been observed, as complex antimatter atoms beyond antihelium are not experimentally available. Such a decay would require antimatter atoms at least as complex as beryllium-7, which is the lightest known isotope of normal matter to undergo decay by electron capture.)

Shortly after the discovery of the neutron in 1932, Enrico Fermi realized that certain rare beta-decay reactions immediately yield neutrons as a decay particle (neutron emission). Isolated proton emission was eventually observed in some elements. It was also found that some heavy elements may undergo spontaneous fission into products that vary in composition. In a phenomenon called cluster decay, specific combinations of neutrons and protons other than alpha particles (helium nuclei) were found to be spontaneously emitted from atoms.

Other types of radioactive decay were found to emit previously-seen particles, but via different mechanisms. An example is internal conversion, which results in an initial electron emission, and then often further characteristic X-rays and Auger electrons emissions, although the internal conversion process involves neither beta nor gamma decay. A neutrino is not emitted, and none of the electron(s) and photon(s) emitted originate in the nucleus, even though the energy to emit all of them does originate there. Internal conversion decay, like isomeric transition gamma decay and neutron emission, involves the release of energy by an excited nuclide, without the transmutation of one element into another.

Rare events that involve a combination of two beta-decay type events happening simultaneously are known. Any decay process that does not violate the conservation of energy or momentum laws (and perhaps other particle conservation laws) is permitted to happen, although not all have been detected. An interesting example discussed in a final section, is bound state beta decay of rhenium-187. In this process, beta electron-decay of the parent nuclide is not accompanied by beta electron emission, because the beta particle has been captured into the K-shell of the emitting atom. An antineutrino is emitted, as in all negative beta decays.

Radionuclides can undergo a number of different reactions. These are summarized in the following table. A nucleus with mass number A and atomic number Z is represented as (A, Z). The column "Daughter nucleus" indicates the difference between the new nucleus and the original nucleus. Thus, $(A - 1, Z)$ means that the mass number is one less than before, but the atomic number is the same as before.

If energy circumstances are favorable, a given radionuclide may undergo many competing types of decay, with some atoms decaying by one route, and others decaying by another. An example is copper-64, which has 29 protons, and 35 neutrons, which decays with a half-life of about 12.7 hours. This isotope has one unpaired proton and one unpaired neutron, so either the proton or the neutron can decay to the opposite particle. This particular nuclide (though not all nuclides in this situation) is almost equally likely to decay through positron emission (18%), or through electron capture (43%), as it does through electron emission (39%). The excited energy states resulting from these decays which fail to end in a ground energy state, also produce later internal conversion and gamma decay in almost 0.5% of the time.

More common in heavy nuclides is competition between alpha and beta decay. The daughter nuclides will then normally decay through beta or alpha, respectively, to end up in the same place.

Mode of decay	Participating particles	Daughter nucleus
Decays with emission of nucleons:		
Alpha decay	An alpha particle ($A = 4$, $Z = 2$) emitted from nucleus	$(A - 4, Z - 2)$
Proton emission	A proton ejected from nucleus	$(A - 1, Z - 1)$
Neutron emission	A neutron ejected from nucleus	$(A - 1, Z)$

Double proton emission	Two protons ejected from nucleus simultaneously	$(A - 2, Z - 2)$
Spontaneous fission	Nucleus disintegrates into two or more smaller nuclei and other particles	—
Cluster decay	Nucleus emits a specific type of smaller nucleus (A_1, Z_1) which is larger than an alpha particle	$(A - A_1, Z - Z_1) + (A_1, Z_1)$
Different modes of beta decay:		
β^- decay	A nucleus emits an electron and an electron antineutrino	$(A, Z + 1)$
Positron emission (β^+ decay)	A nucleus emits a positron and an electron neutrino	$(A, Z - 1)$
Electron capture	A nucleus captures an orbiting electron and emits a neutrino; the daughter nucleus is left in an excited unstable state	$(A, Z - 1)$
Bound state beta decay	A free neutron or nucleus beta decays to electron and antineutrino, but the electron is not emitted, as it is captured into an empty K-shell; the daughter nucleus is left in an excited and unstable state. This process is a minority of free neutron decays (0.0004%) due to the low energy of hydrogen ionization, and is suppressed except in ionized atoms that have K-shell vacancies.	$(A, Z + 1)$
Double beta decay	A nucleus emits two electrons and two antineutrinos	$(A, Z + 2)$
Double electron capture	A nucleus absorbs two orbital electrons and emits two neutrinos – the daughter nucleus is left in an excited and unstable state	$(A, Z - 2)$
Electron capture with positron emission	A nucleus absorbs one orbital electron, emits one positron and two neutrinos	$(A, Z - 2)$
Double positron emission	A nucleus emits two positrons and two neutrinos	$(A, Z - 2)$
Transitions between states of the same nucleus:		
Isomeric transition	Excited nucleus releases a high-energy photon (gamma ray)	(A, Z)
Internal conversion	Excited nucleus transfers energy to an orbital electron, which is subsequently ejected from the atom	(A, Z)

Radioactive decay results in a reduction of summed rest mass, once the released energy (the *disintegration energy*) has escaped in some way. Although decay energy is sometimes defined as associated with the difference between the mass of the parent nuclide products and the mass of the decay products, this is true only of rest mass measurements, where some energy has been removed from the product system. This is true because the decay energy must always carry mass with it, wherever it appears according to the formula $E = mc^2$. The decay energy is initially released as the energy of emitted photons plus the kinetic energy of massive emitted particles (that is, particles that have rest mass). If these particles come to thermal equilibrium with their surroundings and photons are absorbed, then the decay energy is transformed to thermal energy, which retains its mass.

Decay energy therefore remains associated with a certain measure of mass of the decay system, called invariant mass, which does not change during the decay, even though the energy of decay is distributed among decay particles. The energy of photons, the kinetic energy of emitted particles, and, later, the thermal energy of the surrounding matter, all contribute to the invariant mass of the system. Thus, while the sum of the rest masses of the particles is not conserved in radioactive decay, the *system* mass and system invariant mass (and also the system total energy) is conserved throughout any decay process. This is a restatement of the equivalent laws of conservation of energy and conservation of mass.

Radioactive Decay Rates

The *decay rate*, or *activity*, of a radioactive substance is characterized by:

Constant Quantities:

- The *half-life—$t_{1/2}$*, is the time taken for the activity of a given amount of a radioactive sub-stance to decay to half of its initial value.

- The *decay constant— λ*, "lambda" the inverse of the mean lifetime, sometimes referred to as simply *decay rate*.

- The *mean lifetime— τ*, "tau" the average lifetime (1/e life) of a radioactive particle before decay.

Although these are constants, they are associated with the statistical behavior of populations of atoms. In consequence, predictions using these constants are less accurate for minuscule samples of atoms.

In principle a half-life, a third-life, or even a $(1/\sqrt{2})$-life, can be used in exactly the same way as half-life; but the mean life and half-life $t_{1/2}$ have been adopted as standard times associated with exponential decay.

Time-variable Quantities:

- *Total activity— A*, is the number of decays per unit time of a radioactive sample.

- *Number of particles—N*, is the total number of particles in the sample.

- *Specific activity—S_A*, number of decays per unit time per amount of substance of the sample at time set to zero ($t = 0$). "Amount of substance" can be the mass, volume or moles of the initial sample.

These are related as follows:

$$t_{1/2} = \frac{\ln(2)}{\lambda} = \tau \ln(2)$$

$$A = -\frac{dN}{dt} = \lambda N$$

$$S_A a_0 = -\frac{dN}{dt}\Big|_{t=0} = \lambda N_0$$

where N_0 is the initial amount of active substance — substance that has the same percentage of unstable particles as when the substance was formed.

Mathematics of Radioactive Decay

Universal Law of Radioactive Decay

Radioactivity is one very frequently given example of exponential decay. The law describes the statistical behaviour of a large number of nuclides, rather than individual atoms. In the following formalism, the number of nuclides or the nuclide population N, is of course a discrete variable (a natural number)—but for any physical sample N is so large that it can be treated as a continuous variable. Differential calculus is needed to set up differential equations for the modelling the behaviour of the nuclear decay.

The mathematics of radioactive decay depend on a key assumption that a nucleus of a radionuclide has no "memory" or way of translating its history into its present behavior. A nucleus does not "age" with the passage of time. Thus, the probability of its breaking down does not increase with time, but stays constant no matter how long the nucleus has existed. This constant probability may vary greatly between different types of nuclei, leading to the many different observed decay rates. However, whatever the probability is, it does not change. This is in marked contrast to complex objects which do show aging, such as automobiles and humans. These systems do have a chance of breakdown per unit of time, that increases from the moment they begin their existence.

One-decay Process

Consider the case of a nuclide A that decays into another B by some process $A \rightarrow B$ (emission of ther particles, like electron neutrinos ve and electrons e⁻ as in beta decay, are irrelevant in what follows). The decay of an unstable nucleus is entirely random and it is impossible to predict when a particular atom will decay. However, it is equally likely to decay at any instant in time. Therefore, given a sample of a particular radioisotope, the number of decay events $-dN$ expected to occur in a small interval of time dt is proportional to the number of atoms present N, that is

$$-\frac{dN}{dt} \propto N.$$

Particular radionuclides decay at different rates, so each has its own decay constant λ. The expected decay $-dN/N$ is proportional to an increment of time, dt:

$$-\frac{dN}{N} = \lambda dt$$

The negative sign indicates that N decreases as time increases, as the decay events follow one after another. The solution to this first-order differential equation is the function:

$$N(t) = N_0 e^{-\lambda t} = N_0 e^{-t/\tau},$$

where N_0 is the value of N at time $t = 0$.

We have for all time t:

$$N_A + N_B = N_{\text{total}} = N_{A0},$$

where N_{total} is the constant number of particles throughout the decay process, which is equal to the initial number of A nuclides since this is the initial substance.

If the number of non-decayed A nuclei is:

$$N_A = N_{A0} e^{-\lambda t}$$

then the number of nuclei of B, i.e. the number of decayed A nuclei, is

$$N_B = N_{A0} - N_A = N_{A0} - N_{A0} e^{-\lambda t} = N_{A0} \left(1 - e^{-\lambda t}\right).$$

The number of decays observed over a given interval obeys Poisson statistics. If the average number of decays is $<N>$, the probability of a given number of decays N is

$$P(N) = \frac{N^N \exp(-N)}{N!}.$$

Chain-decay Processes

Chain of Two Decays

Now consider the case of a chain of two decays: one nuclide A decaying into another B by one process, then B decaying into another C by a second process, i.e. $A \rightarrow B \rightarrow C$. The previous equation cannot be applied to the decay chain, but can be generalized as follows. Since A decays into B, *then* B decays into C, the activity of A adds to the total number of B nuclides in the present sample, *before* those B nuclides decay and reduce the number of nuclides leading to the later sample. In other words, the number of second generation nuclei B increases as a result of the first generation nuclei decay of A, and decreases as a result of its own decay into the third generation nuclei C. The sum of these two terms gives the law for a decay chain for two nuclides:

$$\frac{dN_B}{dt} = -\lambda_B N_B + \lambda_A N_A.$$

The rate of change of N_B, that is dN_B/dt, is related to the changes in the amounts of A and B, N_B can increase as B is produced from A and decrease as B produces C.

Re-writing using the previous results:

$$\boxed{\frac{dN_B}{dt} = -\lambda_B N_B + \lambda_A N_{A0} e^{-\lambda_A t}}$$

The subscripts simply refer to the respective nuclides, i.e. N_A is the number of nuclides of type A, N_{A0} is the initial number of nuclides of type A, λ_A is the decay constant for A - and similarly for nuclide B. Solving this equation for N_B gives:

$$N_B = \frac{N_{A0} \lambda_A}{\lambda_B - \lambda_A} \left(e^{-\lambda_A t} - e^{-\lambda_B t}\right).$$

In the case where B is a stable nuclide ($\lambda_B = 0$), this equation reduces to the previous solution:

$$\lim_{\lambda_B \to 0} \left[\frac{N_{A0}\lambda_A}{\lambda_B - \lambda_A} \left(e^{-\lambda_A t} - e^{-\lambda_B t} \right) \right] = \frac{N_{A0}\lambda_A}{0 - \lambda_A} \left(e^{-\lambda_A t} - 1 \right) = N_{A0} \left(1 - e^{-\lambda_A t} \right),$$

as shown above for one decay. The solution can be found by the integration factor method, where the integrating factor is $e^{\lambda_B t}$. This case is perhaps the most useful, since it can derive both the one-decay equation (above) and the equation for multi-decay chains (below) more directly.

Chain of Any Number of Decays

For the general case of any number of consecutive decays in a decay chain, i.e. $A_1 \to A_2 \cdots \to A_i \cdots \to A_D$, where D is the number of decays and i is a dummy index ($i =$ 1, 2, 3, ...D), each nuclide population can be found in terms of the previous population. In this case $N_2 = 0$, $N_3 = 0,..., N_D = 0$. Using the above result in a recursive form:

$$\frac{dN_j}{dt} = -\lambda_j N_j + \lambda_{j-1} N_{(j-1)0} e^{-\lambda_{j-1} t}.$$

The general solution to the recursive problem is given by *Bateman's equations*:

Bateman's equations

$$N_D = \frac{N_1(0)}{\lambda_D} \sum_{i=1}^{D} \lambda_i c_i e^{-\lambda_i t}$$

$$c_i = \prod_{j=1, i \neq j}^{D} \frac{\lambda_j}{\lambda_j - \lambda_i}$$

Alternative Decay Modes

In all of the above examples, the initial nuclide decays into only one product. Consider the case of one initial nuclide that can decay into either of two products, that is $A \to B$ and $A \to C$ in parallel. For example, in a sample of potassium-40, 89.3% of the nuclei decay to calcium-40 and 10.7% to argon-40. We have for all time t:

$$N = N_A + N_B + N_C$$

which is constant, since the total number of nuclides remains constant. Differentiating with respect to time:

$$\frac{dN_A}{dt} = -\left(\frac{dN_B}{dt} + \frac{dN_C}{dt} \right)$$

$$-\lambda N_A = -N_A (\lambda_B + \lambda_C)$$

defining the *total decay constant* λ in terms of the sum of *partial decay constants* λ_B and λ_C:

$$\lambda = \lambda_B + \lambda_C.$$

Notice that

$$\frac{dN_A}{dt}\left\langle 0, \frac{dN_B}{dt}\right\rangle 0, \frac{dN_C}{dt} > 0.$$

Solving this equation for N_A:

$$N_A = N_{A0}e^{-\lambda t}.$$

where N_{A0} is the initial number of nuclide A. When measuring the production of one nuclide, one can only observe the total decay constant λ. The decay constants λ_B and λ_C determine the probability for the decay to result in products B or C as follows:

$$N_B = \frac{\lambda_B}{\lambda} N_{A0}\left(1-e^{-\lambda t}\right),$$

$$N_C = \frac{\lambda_C}{\lambda} N_{A0}\left(1-e^{-\lambda t}\right).$$

because the fraction λ_B/λ of nuclei decay into B while the fraction λ_C/λ of nuclei decay into C.

Corollaries of The Decay Laws

The above equations can also be written using quantities related to the number of nuclide particles N in a sample;

- The activity: $A = \lambda N$.

- The amount of substance: $n = N/L$.

- The mass: $M = A_r n = A_r N/L$.

where $L = 6.022\times10^{23}$ is Avogadro's constant, A_r is the relative atomic mass number, and the amount of the substance is in moles.

Decay Timing: Definitions and Relations

Time Constant and Mean-life

For the one-decay solution $A \rightarrow B$:

$$N = N_0 e^{-\lambda t} = N_0 e^{-t/\tau},$$

the equation indicates that the decay constant λ has units of t^{-1}, and can thus also be represented as $1/\tau$, where τ is a characteristic time of the process called the *time constant*.

In a radioactive decay process, this time constant is also the mean lifetime for decaying atoms. Each atom "lives" for a finite amount of time before it decays, and it may be shown that this mean lifetime is the arithmetic mean of all the atoms' lifetimes, and that it is τ, which again is related to the decay constant as follows:

$$\tau = \frac{1}{\lambda}.$$

This form is also true for two-decay processes simultaneously $A \rightarrow B + C$, inserting the equivalent values of decay constants (as given above)

$$\lambda = \lambda_B + \lambda_C$$

into the decay solution leads to:

$$\frac{1}{\tau} = \lambda = \lambda_B + \lambda_C = \frac{1}{\tau_B} + \frac{1}{\tau_C}$$

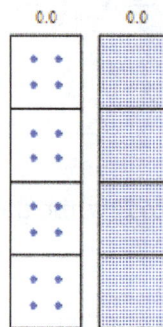

Simulation of many identical atoms undergoing radioactive decay, starting with either 4 atoms (left) or 400 (right). The number at the top indicates how many half-lives have elapsed.

Half-life

A more commonly used parameter is the half-life. Given a sample of a particular radionuclide, the half-life is the time taken for half the radionuclide's atoms to decay. For the case of one-decay nuclear reactions:

$$N = N_0 e^{-\lambda t} = N_0 e^{-t/\tau},$$

the half-life is related to the decay constant as follows: set $N = N_0/2$ and $t = T_{1/2}$ to obtain

$$t_{1/2} = \frac{\ln 2}{\lambda} = \tau \ln 2.$$

This relationship between the half-life and the decay constant shows that highly radioactive substances are quickly spent, while those that radiate weakly endure longer. Half-lives of known radionuclides vary widely, from more than 10^{19} years, such as for the very nearly stable nuclide ^{209}Bi, to 10^{-23} seconds for highly unstable ones.

The factor of ln(2) in the above relations results from the fact that concept of "half-life" is merely a way of selecting a different base other than the natural base e for the lifetime expression. The time constant τ is the e -1 -life, the time until only $1/e$ remains, about 36.8%, rather than the 50% in the half-life of a radionuclide. Thus, τ is longer than $t_{1/2}$. The following equation can be shown to be valid:

$$N(t) = N_0 e^{-t/\tau} = N_0 2^{-t/t_{1/2}}.$$

Since radioactive decay is exponential with a constant probability, each process could as easily be described with a different constant time period that (for example) gave its "(1/3)-life" (how long until only 1/3 is left) or "(1/10)-life" (a time period until only 10% is left), and so on. Thus, the choice of τ and $t_{1/2}$ for marker-times, are only for convenience, and from convention. They reflect a fundamental principle only in so much as they show that the *same proportion* of a given radioactive substance will decay, during any time-period that one chooses.

Mathematically, the nth life for the above situation would be found in the same way as above—by setting $N = N_0/n$, $t = T_{1/n}$ and substituting into the decay solution to obtain

$$t_{1/n} = \frac{\ln n}{\lambda} = \tau \ln n.$$

Example

A sample of ^{14}C has a half-life of 5,730 years and a decay rate of 14 disintegration per minute (dpm) per gram of natural carbon.

If an artifact is found to have radioactivity of 4 dpm per gram of its present C, we can find the approximate age of the object using the above equation:

$$N = N_0 e^{-t/\tau},$$

where: $\dfrac{N}{N_0} = 4/14 \approx 0.286,$

$$\tau = \frac{T_{1/2}}{\ln 2} \approx 8267 \text{ years},$$

$$t = -\tau \ln \frac{N}{N_0} \approx 10356 \text{ years}.$$

Changing Decay Rates

The radioactive decay modes of electron capture and internal conversion are known to be slightly sensitive to chemical and environmental effects that change the electronic structure of the atom, which in turn affects the presence of 1s and 2s electrons that participate in the decay process. A small number of mostly light nuclides are affected. For example, chemical bonds can affect the rate of electron capture to a small degree (in general, less than 1%) depending on the proximity of electrons to the nucleus. In ^{7}Be, a difference of 0.9% has been observed between half-lives in metallic and insulating environments. This relatively large effect is because beryllium is a small atom whose valence electrons are in 2s atomic orbitals, which are subject to electron capture in ^{7}Be because (like all s atomic orbitals in all atoms) they naturally penetrate into the nucleus.

In 1992, Jung et al. of the Darmstadt Heavy-Ion Research group observed an accelerated β decay of $^{163}Dy^{66+}$. Although neutral ^{163}Dy is a stable isotope, the fully ionized $^{163}Dy^{66+}$ undergoes β decay into the K and L shells with a half-life of 47 days.

Rhenium-187 is another spectacular example. ^{187}Re normally beta decays to ^{187}Os with a half-life of 41.6×10^9 years, but studies using fully ionised ^{187}Re atoms (bare nuclei) have found that this can decrease to only 33 years. This is attributed to "bound-state β^- decay" of the fully ionised atom – the electron is emitted into the "K-shell" (1s atomic orbital), which cannot occur for neutral atoms in which all low-lying bound states are occupied.

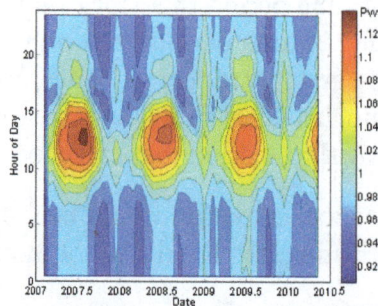

Decay rate of radon-222 as a function of date and time of day. The color-bar gives the power of the observed signal and represents ~4% seasonal decay rate variation.

A number of experiments have found that decay rates of other modes of artificial and naturally occurring radioisotopes are, to a high degree of precision, unaffected by external conditions such as temperature, pressure, the chemical environment, and electric, magnetic, or gravitational fields. Comparison of laboratory experiments over the last century, studies of the Oklo natural nuclear reactor (which exemplified the effects of thermal neutrons on nuclear decay), and astrophysical observations of the luminosity decays of distant supernovae (which occurred far away so the light has taken a great deal of time to reach us), for example, strongly indicate that unperturbed decay rates have been constant (at least to within the limitations of small experimental errors) as a function of time as well.

Recent results suggest the possibility that decay rates might have a weak dependence on environmental factors. It has been suggested that measurements of decay rates of silicon-32, manganese-54, and radium-226 exhibit small seasonal variations (of the order of 0.1%), while the decay of radon-222 exhibits large 4% peak-to-peak seasonal variations, proposed to be related to either solar flare activity or distance from the Sun. However, such measurements are highly susceptible to systematic errors, and a subsequent paper has found no evidence for such correlations in seven other isotopes (^{22}Na, ^{44}Ti, ^{108}Ag, ^{121}Sn, ^{133}Ba, ^{241}Am, ^{238}Pu), and sets upper limits on the size of any such effects.

Theoretical Basis of Decay Phenomena

The neutrons and protons that constitute nuclei, as well as other particles that approach close enough to them, are governed by several interactions. The strong nuclear force, not observed at the familiar macroscopic scale, is the most powerful force over subatomic distances. The electrostatic force is almost always significant, and, in the case of beta decay, the weak nuclear force is also involved.

The interplay of these forces produces a number of different phenomena in which energy may be released by rearrangement of particles in the nucleus, or else the change of one type of particle into others. These rearrangements and transformations may be hindered energetically, so that they do not occur immediately. In certain cases, random quantum vacuum fluctuations are theorized to promote relaxation to a lower energy state (the "decay") in a phenomenon known as quantum tunneling. Radioactive decay half-life of nuclides has been measured over timescales of 55 orders of magnitude, from 2.3×10^{-23} seconds (for hydrogen-7) to 6.9×10^{31} seconds (for tellurium-128). The limits of these timescales are set by the sensitivity of instrumentation only, and there are no known natural limits to how brief or long a decay half life for radioactive decay of a radionuclide may be.

The decay process, like all hindered energy transformations, may be analogized by a snowfield on a mountain. While friction between the ice crystals may be supporting the snow's weight, the system is inherently unstable with regard to a state of lower potential energy. A disturbance would thus facilitate the path to a state of greater entropy: The system will move towards the ground state, producing heat, and the total energy will be distributable over a larger number of quantum states. Thus, an avalanche results. The *total* energy does not change in this process, but, because of the second law of thermodynamics, avalanches have only been observed in one direction and that is toward the "ground state" — the state with the largest number of ways in which the available energy could be distributed.

Such a collapse (a *decay event*) requires a specific activation energy. For a snow avalanche, this energy comes as a disturbance from outside the system, although such disturbances can be arbitrarily small. In the case of an excited atomic nucleus, the arbitrarily small disturbance comes from quantum vacuum fluctuations. A radioactive nucleus (or any excited system in quantum mechanics) is unstable, and can, thus, *spontaneously* stabilize to a less-excited system. The resulting transformation alters the structure of the nucleus and results in the emission of either a photon or a high-velocity particle that has mass (such as an electron, alpha particle, or other type).

Occurrence and Applications

According to the Big Bang theory, stable isotopes of the lightest five elements (H, He, and traces of Li, Be, and B) were produced very shortly after the emergence of the universe, in a process called Big Bang nucleosynthesis. These lightest stable nuclides (including deuterium) survive to today, but any radioactive isotopes of the light elements produced in the Big Bang (such as tritium) have long since decayed. Isotopes of elements heavier than boron were not produced at all in the Big Bang, and these first five elements do not have any long-lived radioisotopes. Thus, all radioactive nuclei are, therefore, relatively young with respect to the birth of the universe, having formed later in various other types of nucleosynthesis in stars (in particular, supernovae), and also during ongoing interactions between stable isotopes and energetic particles. For example, carbon-14, a radioactive nuclide with a half-life of only 5,730 years, is constantly produced in Earth's upper atmosphere due to interactions between cosmic rays and nitrogen.

Nuclides that are produced by radioactive decay are called radiogenic nuclides, whether they themselves are stable or not. There exist stable radiogenic nuclides that were formed from short-lived extinct radionuclides in the early solar system. The extra presence of these stable radiogenic nuclides (such as Xe-129 from primordial I-129) against the background of primordial stable nuclides can be inferred by various means.

Radioactive decay has been put to use in the technique of radioisotopic labeling, which is used to track the passage of a chemical substance through a complex system (such as a living organism). A sample of the substance is synthesized with a high concentration of unstable atoms. The presence of the substance in one or another part of the system is determined by detecting the locations of decay events.

On the premise that radioactive decay is truly random (rather than merely chaotic), it has been used in hardware random-number generators. Because the process is not thought to vary significantly in mechanism over time, it is also a valuable tool in estimating the absolute ages of certain materials. For geological materials, the radioisotopes and some of their decay products become trapped when a rock solidifies, and can then later be used (subject to many well-known qualifications) to estimate the date of the solidification. These include checking the results of several simultaneous processes and their products against each other, within the same sample. In a similar fashion, and also subject to qualification, the rate of formation of carbon-14 in various eras, the date of formation of organic matter within a certain period related to the isotope's half-life may be estimated, because the carbon-14 becomes trapped when the organic matter grows and incorporates the new carbon-14 from the air. Thereafter, the amount of carbon-14 in organic matter decreases according to decay processes that may also be independently cross-checked by other means (such as checking the carbon-14 in individual tree rings, for example).

Szilard–Chalmers Effect

The *Szilard–Chalmers effect* is defined as the breaking of a chemical bond between an atom and the molecule which the atom is part of, as a result of a nuclear reaction of the atom. The effect can be used to separate isotopes by chemical means. The discovery of this effect is due to L. Szilárd and T.A. Chalmers.

Origins of Radioactive Nuclides

Radioactive primordial nuclides found in the Earth are residues from ancient supernova explosions which occurred before the formation of the solar system. They are the fraction of radionuclides that survived from that time, through the formation of the primordial solar nebula, through planet accretion, and up to the present time. The naturally occurring short-lived radiogenic radionuclides found in today's rocks, are the daughters of those radioactive primordial nuclides. Another minor source of naturally occurring radioactive nuclides are cosmogenic nuclides, that are formed by cosmic ray bombardment of material in the Earth's atmosphere or crust. The decay of the radionuclides in rocks of the Earth's mantle and crust contribute significantly to Earth's internal heat budget.

Decay Chains and Multiple Modes

The daughter nuclide of a decay event may also be unstable (radioactive). In this case, it too will decay, producing radiation. The resulting second daughter nuclide may also be radioactive. This can lead to a sequence of several decay events called a *decay chain*. Eventually, a stable nuclide is produced.

Gamma-ray energy spectrum of uranium ore (inset). Gamma-rays are emitted by decaying nuclides, and the gamma-ray energy can be used to characterize the decay (which nuclide is decaying to which). Here, using the gamma-ray spectrum, several nuclides that are typical of the decay chain of ^{238}U have been identified: ^{226}Ra, ^{214}Pb, ^{214}Bi.

An example is the natural decay chain of ^{238}U:

- Uranium-238 decays, through alpha-emission, with a half-life of 4.5 billion years to thorium-234

- which decays, through beta-emission, with a half-life of 24 days to protactinium-234

- which decays, through beta-emission, with a half-life of 1.2 minutes to uranium-234

- which decays, through alpha-emission, with a half-life of 240 thousand years to thorium-230

- which decays, through alpha-emission, with a half-life of 77 thousand years to radium-226

- which decays, through alpha-emission, with a half-life of 1.6 thousand years to radon-222

- which decays, through alpha-emission, with a half-life of 3.8 days to polonium-218

- which decays, through alpha-emission, with a half-life of 3.1 minutes to lead-214

- which decays, through beta-emission, with a half-life of 27 minutes to bismuth-214

- which decays, through beta-emission, with a half-life of 20 minutes to polonium-214

- which decays, through alpha-emission, with a half-life of 160 microseconds to lead-210

- which decays, through beta-emission, with a half-life of 22 years to bismuth-210

- which decays, through beta-emission, with a half-life of 5 days to polonium-210

- which decays, through alpha-emission, with a half-life of 140 days to lead-206, which is a stable nuclide.

Some radionuclides may have several different paths of decay. For example, approximately 36% of bismuth-212 decays, through alpha-emission, to thallium-208 while approximately 64% of bismuth-212 decays, through beta-emission, to polonium-212. Both thallium-208 and polonium-212 are radioactive daughter products of bismuth-212, and both decay directly to stable lead-208.

Nuclear Binding Energy

Nuclear binding energy is the energy that would be required to disassemble the nucleus of an atom into its component parts. These component parts are neutrons and protons, which are collectively called nucleons. The binding energy of nuclei is due to the attractive forces that hold these nucleons together and this is usually a positive number, since most nuclei would require the expenditure of energy to separate them into individual protons and neutrons. The mass of an atomic nucleus is usually less than the sum of the individual masses of the constituent protons and neutrons (according to Einstein's equation $E=mc^2$) and this 'missing mass' is known as the mass defect, and represents the energy that was released when the nucleus was formed.

The term nuclear binding energy may also refer to the energy balance in processes in which the nucleus splits into fragments composed of more than one nucleon. If new binding energy is available when light nuclei fuse, or when heavy nuclei split, either process can result in release of this binding energy. This energy may be made available as *nuclear energy* and can be used to produce electricity as in (nuclear power) or in a nuclear weapon. When a large nucleus splits into pieces, excess energy is emitted as photons (gamma rays) and as the kinetic energy of a number of different ejected particles (nuclear fission products).

The nuclear binding energies and forces are on the order of a million times greater than the electron binding energies of light atoms like hydrogen.

The mass defect of a nucleus represents the mass of the energy of binding of the nucleus, and is the difference between the mass of a nucleus and the sum of the masses of the nucleons of which it is composed.

Introduction

Nuclear binding energy is explained by the basic principles involved in nuclear physics.

Nuclear Energy

An absorption or release of nuclear energy occurs in nuclear reactions or radioactive decay; those that absorb energy are called endothermic reactions and those that release energy are exothermic reactions. Energy is consumed or liberated because of differences in the nuclear binding energy between the incoming and outgoing products of the nuclear transmutation.

The best-known classes of exothermic nuclear transmutations are fission and fusion. Nuclear energy may be liberated by atomic fission, when heavy atomic nuclei (like uranium and plutonium) are broken apart into lighter nuclei. The energy from fission is used to generate electric power in hundreds of locations worldwide. Nuclear energy is also released during atomic fusion, when light nuclei like hydrogen are combined to form heavier nuclei such as helium. The Sun and other stars use nuclear fusion to generate thermal energy which is later radiated from the surface, a type of stellar nucleosynthesis. In any exothermic nuclear process, nuclear mass might ultimately be converted to thermal energy, given off as heat.

In order to quantify the energy released or absorbed in any nuclear transmutation, one must know the nuclear binding energies of the nuclear components involved in the transmutation.

The Nuclear Force

Electrons and nuclei are kept together by electrostatic attraction (negative attracts positive). Furthermore, electrons are sometimes shared by neighboring atoms or transferred to them (by processes of quantum physics), and this link between atoms is referred to as a chemical bond, and is responsible for the formation of all chemical compounds.

The force of electric attraction does not hold nuclei together, because all protons carry a positive charge and repel each other. Thus, electric forces do not hold nuclei together, because they act in the opposite direction. It has been established that binding neutrons to nuclei clearly requires a non-electrical attraction.

Therefore, another force, called the *nuclear force* (or *residual strong force*) holds the nucleons of nuclei together. This force is a residuum of the strong interaction, which binds quarks into nucleons at an even smaller level of distance.

The nuclear force must be stronger than the electric repulsion at short distances, but weaker far away, or else different nuclei might tend to clump together. Therefore, it has short-range characteristics. An analogy to the nuclear force is the force between two small magnets: magnets are very difficult to separate when stuck together, but once pulled a short distance apart, the force between them drops almost to zero.

Unlike gravity or electrical forces, the nuclear force is effective only at very short distances. At greater distances, the electrostatic force dominates: the protons repel each other because they are positively charged, and like charges repel. For that reason, the protons forming the nuclei of ordinary hydrogen—for instance, in a balloon filled with hydrogen—do not combine to form helium (a process that also would require some protons to combine with electrons and become neutrons). They cannot get close enough for the nuclear force, which attracts them to each other, to become important. Only under conditions of extreme pressure and temperature (for example, within the core of a star), can such a process take place.

Physics of Nuclei

The nuclei of atoms are found in many different sizes. In hydrogen they contain just one proton, in deuterium or heavy hydrogen a proton and a neutron; in helium, two protons and two neutrons, and in carbon, nitrogen and oxygen - six, seven and eight of each particle, respectively. A helium nucleus weighs less than the sum of the weights of its components. The same phenomenon is found for carbon, nitrogen and oxygen. For example, the carbon nucleus is slightly lighter than three helium nuclei, which can combine to make a carbon nucleus. This illustrates the mass defect.

Mass Defect

Mass defect is the difference between the mass of a composite particle and the sum of the masses of its parts.

The "mass defect" can be explained using Albert Einstein's formula $E = m c^2$, expressing the equivalence of energy and mass. By this formula, adding energy also increases mass (both weight and inertia), whereas removing energy decreases mass.

If a combination of particles contains extra energy—for instance, in a molecule of the explosive TNT—weighing it reveals some extra mass, compared to its end products after an explosion. (The weighing must be done after the products have been stopped and cooled, however, as the extra mass must escape from the system as heat before its loss can be noticed, in theory.) On the other hand, if one must inject energy to separate a system of particles into its components, then the initial weight is less than that of the components after they are separated. In the latter case, the energy injected is "stored" as potential energy, which shows as the increased mass of the components that store it. This is an example of the fact that energy of all types is seen in systems as mass, since mass and energy are equivalent, and each is a "property" of the other.

The latter scenario is the case with nuclei such as helium: to break them up into protons and neutrons, one must inject energy. On the other hand, if a process existed going in the opposite direction, by which hydrogen atoms could be combined to form helium, then energy would be released. The energy can be computed using $E = \Delta m\, c^2$ for each nucleus, where Δm is the difference between the mass of the helium nucleus and the mass of four protons (plus two electrons, absorbed to create the neutrons of helium).

For elements heavier than oxygen, the energy that can be released by assembling them from lighter elements decreases, up to iron. For nuclei heavier than iron, one actually releases energy by breaking them up into 2 fragments. That is how energy is extracted by breaking up uranium nuclei in nuclear power reactors.

The reason the trend reverses after iron is the growing positive charge of the nuclei. The electric force may be weaker than the nuclear force, but its range is greater: in an iron nucleus, each proton repels the other 25 protons, while the nuclear force only binds close neighbors.

As nuclei grow bigger still, this disruptive effect becomes steadily more significant. By the time polonium is reached (84 protons), nuclei can no longer accommodate their large positive charge, but emit their excess protons quite rapidly in the process of alpha radioactivity—the emission of helium nuclei, each containing two protons and two neutrons. (Helium nuclei are an especially stable combination.) Because of this process, nuclei with more than 94 protons are not found naturally on Earth. The isotopes beyond uranium (atomic number 92) with the longest half-lives are plutonium-244 (80 million years) and curium-247 (16 million years).

Solar Binding Energy

The nuclear fusion process works as follows: five billion years ago, the new Sun formed when gravity pulled together a vast cloud of hydrogen and dust, from which the Earth and other planets also arose. The gravitational pull released energy and heated the early Sun, much in the way Helmholtz proposed.

Thermal energy appears as the motion of atoms and molecules: the higher the temperature of a collection of particles, the greater is their velocity and the more violent are their collisions. When the temperature at the center of the newly formed Sun became great enough for collisions between hydrogen nuclei to overcome their electric repulsion, and bring them into the short range of the attractive nuclear force, nuclei began to stick together. When this began to happen, protons combined into deuterium and then helium, with some protons changing in the process to neutrons (plus positrons, positive electrons, which combine with electrons and are both destroyed). This released nuclear energy now keeps up the high temperature of the Sun's core, and the heat also keeps the gas pressure high, keeping the Sun at its present size, and stopping gravity from compressing it any more. There is now a stable balance between gravity and pressure.

Different nuclear reactions may predominate at different stages of the Sun's existence, including the proton-proton reaction and the carbon-nitrogen cycle—which involves heavier nuclei, but whose final product is still the combination of protons to form helium.

A branch of physics, the study of controlled nuclear fusion, has tried since the 1950s to derive useful power from nuclear fusion reactions that combine small nuclei into bigger ones, typically to heat boilers, whose steam could turn turbines and produce electricity. Unfortunately, no earthly laboratory can match one feature of the solar powerhouse: the great mass of the Sun, whose weight keeps the hot plasma compressed and confines the nuclear furnace to the Sun's core. Instead, physicists use strong magnetic fields to confine the plasma, and for fuel they use heavy forms of hydrogen, which burn more easily. Magnetic traps can be rather unstable, and any plasma hot enough and dense enough to undergo nuclear fusion tends to slip out of them after a short time. Even with ingenious tricks, the confinement in most cases lasts only a small fraction of a second.

Combining Nuclei

Small nuclei that are larger than hydrogen can combine into bigger ones and release energy, but in combining such nuclei, the amount of energy released is much smaller compared to hydrogen fusion. The reason is that while the overall process releases energy from letting the nuclear attraction do its work, energy must first be injected to force together positively charged protons, which also repel each other with their electric charge.

For elements that weigh more than iron (a nucleus with 26 protons), the fusion process no longer releases energy. In even heavier nuclei energy is consumed, not released, by combining similar sized nuclei. With such large nuclei, overcoming the electric repulsion (which affects all protons in the nucleus) requires more energy than what is released by the nuclear attraction (which is effective mainly between close neighbors). Conversely, energy could actually be released by breaking apart nuclei heavier than iron.

With the nuclei of elements heavier than lead, the electric repulsion is so strong that some of them spontaneously eject positive fragments, usually nuclei of helium that form very stable combinations (alpha particles). This spontaneous break-up is one of the forms of radioactivity exhibited by some nuclei.

Nuclei heavier than lead (except for bismuth, thorium, uranium, and plutonium) spontaneously break up too quickly to appear in nature as primordial elements, though they can be produced artificially or as intermediates in the decay chains of lighter elements. Generally, the heavier the nuclei are, the faster they spontaneously decay.

Iron nuclei are the most stable nuclei (in particular iron-56), and the best sources of energy are therefore nuclei whose weights are as far removed from iron as possible. One can combine the lightest ones—nuclei of hydrogen (protons)—to form nuclei of helium, and that is how the Sun generates its energy. Or else one can break up the heaviest ones—nuclei of uranium or plutonium—into smaller fragments, and that is what nuclear power reactors do.

Nuclear Binding Energy

An example that illustrates nuclear binding energy is the nucleus of ^{12}C (carbon-12), which contains 6 protons and 6 neutrons. The protons are all positively charged and repel each other, but the nuclear force overcomes the repulsion and causes them to stick together. The nuclear force is a close-range force (it is strongly attractive at a distance of 1.0 fm and becomes extremely small beyond a distance of 2.5fm), and virtually no effect of this force is observed outside the nucleus. The nuclear force also pulls neutrons together, or neutrons and protons.

The energy of the nucleus is negative with regard to the energy of the particles pulled apart to infinite distance (just like the gravitational energy of planets of the solar system), because energy must be utilized to split a nucleus into its individual protons and neutrons. Mass spectrometers have measured the masses of nuclei, which are always less than the sum of the masses of protons and neutrons that form them, and the difference—by the formula $E = m\,c^2$—gives the binding energy of the nucleus.

Nuclear Fusion

The binding energy of helium is the energy source of the Sun and of most stars. The sun is composed of 74 percent hydrogen (measured by mass), an element whose nucleus is a single proton. Energy is released in the sun when 4 protons combine into a helium nucleus, a process in which two of them are also converted to neutrons.

The conversion of protons to neutrons is the result of another nuclear force, known as the weak (nuclear) force. The weak force, like the strong force, has a short range, but is much weaker than the strong force. The weak force tries to make the number of neutrons and protons into the most energetically stable configuration. For nuclei containing less than 40 particles, these numbers are usually about equal. Protons and neutrons are closely related and are sometimes collectively known as nucleons. As the number of particles increases toward a maximum of about 209, the number of neutrons to maintain stability begins to outstrip the number of protons, until the ratio of neutrons to protons is about three to two.

The protons of hydrogen combine to helium only if they have enough velocity to overcome each other's mutual repulsion sufficiently to get within range of the strong nuclear attraction. This means that fusion only occurs within a very hot gas. Hydrogen hot enough for combining to helium requires an enormous pressure to keep it confined, but suitable conditions exist in the central regions of the Sun, where such pressure is provided by the enormous weight of the layers above the core, pressed inwards by the Sun's strong gravity. The process of combining protons to form helium is an example of nuclear fusion.

The earth's oceans contain a large amount of hydrogen that could theoretically be used for fusion, and helium byproduct of fusion does not harm the environment, so some consider nuclear fusion a good alternative to supply humanity's energy needs. Experiments to generate electricity from fusion have so far have only partially succeeded. Sufficiently hot hydrogen must be ionized and confined. One technique is to use very strong magnetic fields, because charged particles (like those trapped in the Earth's radiation belt) are guided by magnetic field lines. Fusion experiments also rely on heavy hydrogen, which fuses more easily, and gas densities can be moderate. But even with these techniques far more net energy is consumed by the fusion experiments than is yielded by the process.

The Binding Energy Maximum and Ways to Approach It by Decay

In the main isotopes of light nuclei, such as carbon, nitrogen and oxygen, the most stable combination of neutrons and of protons are when the numbers are equal (this continues to element 20, calcium). However, in heavier nuclei, the disruptive energy of protons increases, since they are confined to a tiny volume and repel each other. The energy of the strong force holding the nucleus together also increases, but at a slower rate, as if inside the nucleus, only nucleons close to each other are tightly bound, not ones more widely separated.

The net binding energy of a nucleus is that of the nuclear attraction, minus the disruptive energy of the electric force. As nuclei get heavier than helium, their net binding energy per nucleon (deduced from the difference in mass between the nucleus and the sum of masses of component nucleons) grows more and more slowly, reaching its peak at iron. As nucleons are added, the total nuclear binding energy always increases—but the total disruptive energy of electric forces (positive protons repelling other protons) also increases, and past iron, the second increase outweighs the first. Iron-56 (^{56}Fe) is the most efficiently bound nucleus meaning that it has the least average mass per nucleon. However, nickel-62 is the most tightly bound nucleus in terms of energy of binding per nucleon. (Nickel-62's higher energy of binding does not translate to a larger mean mass loss than Fe-56, because Ni-62 has a slightly higher ratio of neutrons/protons than does iron-56, and the presence of the heavier neutrons increases nickel-62's average mass per nucleon).

To reduce the disruptive energy, the weak interaction allows the number of neutrons to exceed that of protons—for instance, the main isotope of iron has 26 protons and 30 neutrons. Isotopes also exist where the number of neutrons differs from the most stable number for that number of nucleons. If the ratio of protons to neutrons is too far from stability, nucleons may spontaneously change from proton to neutron, or neutron to proton.

The two methods for this conversion are mediated by the weak force, and involve types of beta decay. In the simplest beta decay, neutrons are converted to protons by emitting a negative electron and an antineutrino. This is always possible outside a nucleus because neutrons are more massive than protons by an equivalent of about 2.5 electrons. In the opposite process, which only happens within a nucleus, and not to free particles, a proton may become a neutron by ejecting a positron. This is permitted if enough energy is available between parent and daughter nuclides to do this (the required energy difference is equal to 1.022 MeV, which is the mass of 2 electrons). If the mass difference between parent and daughter is less than this, a proton-rich nucleus may still convert protons to neutrons by the process of electron capture, in which a proton simply electron captures one of the atom's K orbital electrons, emits a neutrino, and becomes a neutron.

Among the heaviest nuclei, starting with tellurium nuclei (element 52) containing 106 or more nucleons, electric forces may be so destabilizing that entire chunks of the nucleus may be ejected, usually as alpha particles, which consist of two protons and two neutrons (alpha particles are fast helium nuclei). (Beryllium-8 also decays, very quickly, into two alpha particles.) Alpha particles are extremely stable. This type of decay becomes more and more probable as elements rise in atomic weight past 106.

The curve of binding energy is a graph that plots the binding energy per nucleon against atomic mass. This curve has its main peak at iron and nickel and then slowly decreases again, and also a narrow isolated peak at helium, which as noted is very stable. The heaviest nuclei in nature, uranium ^{238}U, are unstable, but having a half-life of 4.5 billion years, close to the age of the Earth, they are still relatively abundant; they (and other nuclei heavier than iron) may have formed in a supernova explosion preceding the formation of the solar system. The most common isotope of thorium, ^{232}Th, also undergoes alpha particle emission, and its half-life (time over which half a number of atoms decays) is even longer, by several times. In each of these, radioactive decay produces daughter isotopes that are also unstable, starting a chain of decays that ends in some stable isotope of lead.

Determining Nuclear Binding Energy

Calculation can be employed to determine the nuclear binding energy of nuclei. The calculation involves determining the *mass defect*, converting it into energy, and expressing the result as energy per mole of atoms, or as energy per nucleon.

Conversion of Mass Defect Into Energy

Mass defect is defined as the difference between the mass of a nucleus, and the sum of the masses of the nucleons of which it is composed. The mass defect is determined by calculating three quantities. These are: the actual mass of the nucleus, the composition of the nucleus (number of protons and of neutrons), and the masses of a proton and of a neutron. This is then followed by converting the mass defect into energy. This quantity is the nuclear binding energy, however it must be expressed as energy per mole of atoms or as energy per nucleon.

Fission and Fusion

Nuclear energy is released by the splitting (fission) or merging (fusion) of the nuclei of atom(s). The conversion of nuclear mass-energy to a form of energy, which can remove some mass when the energy is removed, is consistent with the mass-energy equivalence formula:

$\Delta E = \Delta m\ c^2$,

in which,

ΔE = energy release,

Δm = mass defect,

and c = the speed of light in a vacuum (a physical constant).

Nuclear energy was first discovered by French physicist Henri Becquerel in 1896, when he found that photographic plates stored in the dark near uranium were blackened like X-ray plates (X-rays had recently been discovered in 1895).

Nuclear chemistry can be used as a form of alchemy to turn lead into gold or change any atom to any other atom (though this may require many intermediate steps). Radionuclide (radioisotope) production often involves irradiation of another isotope (or more precisely a nuclide), with alpha particles, beta particles, or gamma rays. Nickel-62 has the highest binding energy per nucleon of any isotope. If an atom of lower average binding energy is changed into two atoms of higher average binding energy, energy is given off. Also, if two atoms of lower average binding energy fuse into an atom of higher average binding energy, energy is given off. The chart shows that fusion of hydrogen, the combination to form heavier atoms, releases energy, as does fission of uranium, the breaking up of a larger nucleus into smaller parts. Stability varies between isotopes: the isotope U-235 is much less stable than the more common U-238.

Nuclear energy is released by three *exoenergetic* (or exothermic) processes:

- Radioactive decay, where a neutron or proton in the radioactive nucleus decays spontaneously by emitting either particles, electromagnetic radiation (gamma rays), or both. Note that for radioactive decay, it is not strictly necessary for the binding energy to increase. What is strictly necessary is that the mass decrease. If a neutron turns into a proton and the energy of the decay is less than 0.782343 MeV (such as rubidium-87 decaying to strontium-87), the average binding energy per nucleon will actually decrease.

- Fusion, two atomic nuclei fuse together to form a heavier nucleus

- Fission, the breaking of a heavy nucleus into two (or more rarely three) lighter nuclei

Binding Energy for Atoms

The binding energy of an atom (including its electrons) is not the same as the binding energy of the atom's nucleus. The measured mass deficits of isotopes are always listed as mass deficits of the neutral atoms of that isotope, and mostly in MeV. As a consequence, the listed mass deficits are not a measure for the stability or binding energy of isolated nuclei, but for the whole atoms. This has very practical reasons, because it is very hard to totally ionize heavy elements, i.e. strip them of all of their electrons.

This practice is useful for other reasons, too: stripping all the electrons from a heavy unstable nucleus (thus producing a bare nucleus) changes the lifetime of the nucleus, indicating that the nucleus cannot be treated independently (Experiments at the heavy ion accelerator GSI). This is also evident from phenomena like electron capture. Theoretically, in orbital models of heavy atoms, the electron orbits partially inside the nucleus (it doesn't *orbit* in a strict sense, but has a non-vanishing probability of being located inside the nucleus).

A nuclear decay happens to the nucleus, meaning that properties ascribed to the nucleus change in the event. In the field of physics the concept of "mass deficit" as a measure for "binding energy" means "mass deficit of the neutral atom" (not just the nucleus) and is a measure for stability of the whole atom.

Nuclear Binding Energy Curve

The region of increasing binding energy is followed by a region of relative stability (saturation) in the sequence from magnesium through xenon. In this region, the nucleus has become large enough that nuclear forces no longer completely extend efficiently across its width. Attractive nuclear forces in this region, as atomic mass increases, are nearly balanced by repellent electromagnetic forces between protons, as the atomic number increases.

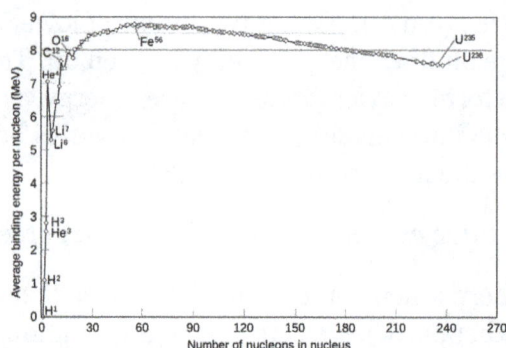

In the periodic table of elements, the series of light elements from hydrogen up to sodium is observed to exhibit generally increasing binding energy per nucleon as the atomic mass increases. This increase is generated by increasing forces per nucleon in the nucleus, as each additional nucleon is attracted by other nearby nucleons, and thus more tightly bound to the whole.

Finally, in elements heavier than xenon, there is a decrease in binding energy per nucleon as atomic number increases. In this region of nuclear size, electromagnetic repulsive forces are beginning to overcome the strong nuclear force attraction.

At the peak of binding energy, nickel-62 is the most tightly bound nucleus (per nucleon), followed by iron-58 and iron-56. This is the approximate basic reason why iron and nickel are very common metals in planetary cores, since they are produced profusely as end products in supernovae and in the final stages of silicon burning in stars. However, it is not binding energy per defined nucleon (as defined above), which controls which exact nuclei are made, because within stars, neutrons are free to convert to protons to release even more energy, per generic nucleon, if the result is a stable nucleus with a larger fraction of protons. In fact, it has been argued that photodisintegration of ^{62}Ni to form ^{56}Fe may be energetically possible in an extremely hot star core, due to this beta decay conversion of neutrons to protons. The conclusion is that at the pressure and temperature conditions in the cores of large stars, energy is released by converting all matter into ^{56}Fe nuclei (ionized atoms). (However, at high temperatures not all matter will be in the lowest energy state.) This energetic maximum should also hold for ambient conditions, say $T = 298$ K and $p = 1$ atm, for neutral condensed matter consisting of ^{56}Fe atoms—however, in these conditions nuclei of atoms are inhibited from fusing into the most stable and low energy state of matter.

It is generally believed that iron-56 is more common than nickel isotopes in the universe for mechanistic reasons, because its unstable progenitor nickel-56 is copiously made by staged build-up of 14 helium nuclei inside supernovas, where it has no time to decay to iron before being released into the interstellar medium in a matter of a few minutes, as the supernova explodes. However, nickel-56 then decays to cobalt-56 within a few weeks, then this radioisotope finally decays to iron-56 with a half life of about 77.3 days. The radioactive decay-powered light curve of such a process has been observed to happen in type II supernovae, such as SN 1987A. In a star, there are no good ways to create nickel-62 by alpha-addition processes, or else there would presumably be more of this highly stable nuclide in the universe.

Binding Energy and Nuclide Masses

The fact that the maximum binding energy is found in medium-sized nuclei is a consequence of the trade-off in the effects of two opposing forces that have different range characteristics. The attractive nuclear force (strong nuclear force), which binds protons and neutrons equally to each other, has a limited range due to a rapid exponential decrease in this force with distance. However, the repelling electromagnetic force, which acts between protons to force nuclei apart, falls off with distance much more slowly (as the inverse square of distance). For nuclei larger than about four nucleons in diameter, the additional repelling force of additional protons more than offsets any binding energy that results between further added nucleons as a result of additional strong force interactions. Such nuclei become increasingly less tightly bound as their size increases, though most of them are still stable. Finally, nuclei containing more than 209 nucleons (larger than about 6 nucleons in diameter) are all too large to be stable, and are subject to spontaneous decay to smaller nuclei.

Nuclear fusion produces energy by combining the very lightest elements into more tightly bound elements (such as hydrogen into helium), and nuclear fission produces energy by splitting the heaviest elements (such as uranium and plutonium) into more tightly bound elements (such as barium and krypton). Both processes produce energy, because middle-sized nuclei are the most tightly bound of all.

As seen above in the example of deuterium, nuclear binding energies are large enough that they may be easily measured as fractional mass deficits, according to the equivalence of mass and energy. The atomic binding energy is simply the amount of energy (and mass) released, when a collection of free nucleons are joined together to form a nucleus.

Nuclear binding energy can be computed from the difference in mass of a nucleus, and the sum of the masses of the number of free neutrons and protons that make up the nucleus. Once this mass difference, called the mass defect or mass deficiency, is known, Einstein's mass-energy equivalence formula $E = mc^2$ can be used to compute the binding energy of any nucleus. Early nuclear physicists used to refer to computing this value as a «packing fraction» calculation.

For example, the atomic mass unit (1 u) is defined as 1/12 of the mass of a ^{12}C atom—but the atomic mass of a ^{1}H atom (which is a proton plus electron) is 1.007825 u, so each nucleon in ^{12}C has lost, on average, about 0.8% of its mass in the form of binding energy.

Semiempirical Formula for Nuclear Binding Energy

For a nucleus with A nucleons, including Z protons and N neutrons, a semi-empirical formula for the binding energy (BE) per nucleon is:

$$\frac{\text{BE}}{A \cdot \text{MeV}} = a - \frac{b}{A^{1/3}} - \frac{cZ^2}{A^{4/3}} - \frac{d(N-Z)^2}{A^2} \pm \frac{e}{A^{7/4}}$$

where the coefficients are given by: $a = 14.0$; $b = 13.0$; $c = 0.585$; $d = 19.3$; $e = 33$.

The first term a is called the saturation contribution and ensures that the binding energy per nucleon is the same for all nuclei to a first approximation. The term $-b >$ is a surface tension effect and is proportional to the number of nucleons that are situated on the nuclear surface; it is largest for light nuclei. The term $-cZ^2 >$ is the Coulomb electrostatic repulsion; this becomes more important as Z increases. The symmetry correction term $-d(N-Z)^2 / A^2$ takes into account the fact that in the absence of other effects the most stable arrangement has equal numbers of protons and neutrons; this is because the n-p interaction in a nucleus is stronger than either the n-n or p-p interaction. The pairing term $\pm e \, / \, A^{7/4}$ is purely empirical; it is + for even-even nuclei and - for odd-odd nuclei.

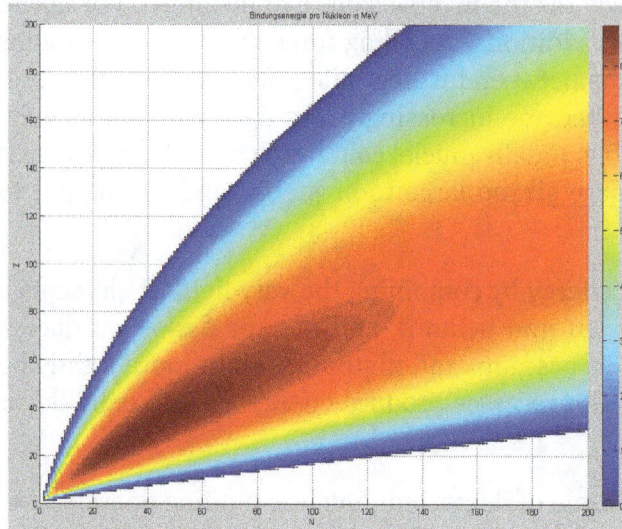

A graphical representation of the semi-empirical binding energy formula. The binding energy per nucleon in MeV (highest numbers in dark red, in excess of 8.5 MeV per nucleon) is plotted for various nuclides as a function of Z, the atomic number (y-axis), vs. N, the number of neutrons (x-axis). The highest numbers are seen for $Z = 26$ (iron).

Example Values Deduced from Experimentally Measured Atom Nuclide Masses

The following table lists some binding energies and mass defect values. Notice also that we use 1 u = (931.494028 ± 0.000023) MeV. To calculate the binding energy we use the formula $Z \, (m_p + m_e) + N \, m_n - m_{nuclide}$ where Z denotes the number of protons in the nuclides and N their number of neutrons. We take m_p = 938.2723 MeV, m_e = 0.5110 MeV and m_n = 939.5656 MeV. The letter A denotes the sum of Z and N (number of nucleons in the nuclide). If we assume the reference nucleon has the mass of a neutron (so that all "total" binding energies calculated are maximal) we could define the total binding energy as the difference from the mass of the nucleus, and the mass of a collection of A free neutrons. In other words, it would be $(Z + N) \, m_n - m_{nuclide}$. The "*total* binding energy per nucleon" would be this value divided by A.

Most strongly bound nuclides atoms									
nuclide	Z	N	mass excess	total mass	total mass / A	total binding energy / A	mass defect	binding energy	binding energy / A
^{56}Fe	26	30	−60.6054 MeV	55.934937 u	0.9988372 u	9.1538 MeV	0.528479 u	492.275 MeV	8.7906 MeV
^{58}Fe	26	32	−62.1534 MeV	57.932276 u	0.9988496 u	9.1432 MeV	0.547471 u	509.966 MeV	8.7925 MeV
^{60}Ni	28	32	−64.472 MeV	59.93079 u	0.9988464 u	9.1462 MeV	0.565612 u	526.864 MeV	8.7811 MeV
^{62}Ni	28	34	−66.7461 MeV	61.928345 u	0.9988443 u	9.1481 MeV	0.585383 u	545.281 MeV	8.7948 MeV

^{56}Fe has the lowest nucleon-specific mass of the four nuclides listed in this table, but this does not imply it is the strongest bound atom per hadron, unless the choice of beginning hadrons is completely free. Iron releases the largest energy if any 56 nucleons are allowed to build a nuclide—changing one to another if necessary, The highest binding energy per hadron, with the hadrons starting as the same number of protons Z and total nucleons A as in the bound nucleus, is ^{62}Ni. Thus, the true absolute value of the total binding energy of a nucleus depends on what we are allowed to construct the nucleus out of. If all nuclei of mass number A were to be allowed to be constructed of A neutrons, then ^{56}Fe would release the most energy per nucleon, since it has a larger fraction of protons than ^{62}Ni. However, if nucleons are required to be constructed of only the same number of protons and neutrons that they contain, then nickel-62 is the most tightly bound nucleus, per nucleon.

Some light nuclides resp. atoms								
nuclide	Z	N	mass excess	total mass	total mass $/A$	total binding energy $/A$	mass defect	binding energy	binding energy $/A$
n	0	1	8.0716 MeV	1.008665 u	1.008665 u	0.0000 MeV	0 u	0 MeV	0 MeV
^{1}H	1	0	7.2890 MeV	1.007825 u	1.007825 u	0.7826 MeV	0.0000000146 u	0.0000136 MeV	13.6 eV
^{2}H	1	1	13.13572 MeV	2.014102 u	1.007051 u	1.50346 MeV	0.002388 u	2.22452 MeV	1.11226 MeV
^{3}H	1	2	14.9498 MeV	3.016049 u	1.005350 u	3.08815 MeV	0.0091058 u	8.4820 MeV	2.8273 MeV
^{3}He	2	1	14.9312 MeV	3.016029 u	1.005343 u	3.09433 MeV	0.0082857 u	7.7181 MeV	2.5727 MeV

In the table above it can be seen that the decay of a neutron, as well as the transformation of tritium into helium-3, releases energy; hence, it manifests a stronger bound new state when measured against the mass of an equal number of neutrons (and also a lighter state per number of total hadrons). Such reactions are not driven by changes in binding energies as calculated from previously fixed N and Z numbers of neutrons and protons, but rather in decreases in the total mass of the nuclide/per nucleon, with the reaction. (Note that the Binding Energy given above for hydrogen-1 is the atomic binding energy, not the nuclear binding energy which would be zero.)

References

- M. Kikuchi, K. Lackner & M. Q. Tran (2012). Fusion Physics. International Atomic Energy Agency. p. 22. ISBN 9789201304100.

- Shultis, J.K. & Faw, R.E. (2002). Fundamentals of nuclear science and engineering. CRC Press. p. 151. ISBN 0-8247-0834-2.

- Atzeni, S. and Meyer-ter-Vehn, J. (2004). Chapter 1: "Nuclear fusion reactions" in The Physics of Inertial Fusion. University of Oxford Press. ISBN 978-0-19-856264-1

- Close, Frank E. (1992). Too Hot to Handle: The Race for Cold Fusion (2nd ed.). London: Penguin. pp. 32, 54. ISBN 0-14-015926-6

- Huizenga, John R. (1993). Cold Fusion: The Scientific Fiasco of the Century (2nd ed.). Oxford and New York: Oxford University Press. p. 112. ISBN 0-19-855817-1

- Negele, J. W.; Vogt, Erich (1998). Advances in nuclear physics (illustrated ed.). Springer. pp. 194–198. ISBN 9780306457579.

- Freidberg, Jeffrey P. (8 February 2007). Plasma Physics and Fusion Energy. Cambridge University Press. ISBN 978-0-521-85107-7.

- Atzeni, Stefano (3 June 2004). The Physics of Inertial Fusion: BeamPlasma Interaction, Hydrodynamics, Hot Dense Matter. OUP Oxford. pp. 12–13. ISBN 978-0-19-152405-9. |first2= missing |last2= in Authors list (help)

- McCracken, Garry; Stott, Peter (8 June 2012). Fusion: The Energy of the Universe. Academic Press. pp. 198–199. ISBN 978-0-12-384656-3. Retrieved 18 August 2012.

- Angelo, Joseph A. (30 November 2004). Nuclear Technology. Greenwood Publishing Group. p. 474. ISBN 978-1-57356-336-9. Retrieved 18 August 2012.

- Stabin, Michael G. (2007). "3". Radiation Protection and Dosimetry: An Introduction to Health Physics. Springer. doi:10.1007/978-0-387-49983-3. ISBN 978-0-387-49982-6.

- Best, Lara; Rodrigues, George; Velker, Vikram (2013). "1.3". Radiation Oncology Primer and Review. Demos Medical Publishing. ISBN 978-1-62070-004-4.

- Loveland, W.; Morrissey, D.; Seaborg, G.T. (2006). Modern Nuclear Chemistry. Wiley-Interscience. p. 57. ISBN 0-471-11532-0.

- Shultis, John K.; Faw, Richard E. (2007). Fundamentals of Nuclear Science and Engineering (2nd ed.). CRC Press. p. 175. ISBN 978-1-4398-9408-8.

- Mould, Richard F. (1995). A century of X-rays and radioactivity in medicine : with emphasis on photographic records of the early years (Reprint. with minor corr ed.). Bristol: Inst. of Physics Publ. p. 12. ISBN 978-0-7503-0224-1.

- L'Annunziata, Michael F. (2007). Radioactivity: Introduction and History. Amsterdam, Netherlands: Elsevier Science. p. 2. ISBN 9780080548883.

- Clayton, Donald D. (1983). Principles of Stellar Evolution and Nucleosynthesis (2nd ed.). University of Chicago Press. p. 75. ISBN 0-226-10953-4.

- Lilley, J.S. (2006). Nuclear physics : principles and applications (Repr. with corrections Jan. 2006. ed.). Chichester: J. Wiley. p. 7. ISBN 0-471-97936-8.

5

Nuclear Safety and Decommissioning

The radioactive isotopes used and produced in nuclear reactors make it imperative that there exist high standards of safety and security within the plant and in its operations. There have been several nuclear plant accidents that have led to stringent rules concerning the storage, usage and decommissioning of nuclear plant products and waste. This chapter highlights the efforts made in the direction of nuclear safety and security, nuclear reprocessing and nuclear decommissioning.

Nuclear Safety and Security

Nuclear safety is defined by the International Atomic Energy Agency (IAEA) as "The achievement of proper operating conditions, prevention of accidents or mitigation of accident consequences, resulting in protection of workers, the public and the environment from undue radiation hazards". The IAEA defines nuclear security as "The prevention and detection of and response to, theft, sabotage, unauthorized access, illegal transfer or other malicious acts involving nuclear material, other radioactive substances or their associated facilities".

A clean-up crew working to remove radioactive contamination after the Three Mile Island accident.

This covers nuclear power plants and all other nuclear facilities, the transportation of nuclear materials, and the use and storage of nuclear materials for medical, power, industry, and military uses.

The nuclear power industry has improved the safety and performance of reactors, and has proposed new and safer reactor designs. However, a perfect safety cannot be guaranteed. Potential sources of problems include human errors and external events that have a greater impact than anticipated: The designers of reactors at Fukushima in Japan did not anticipate that a tsunami generated by an earthquake would disable the backup systems that were supposed to stabilize the reactor after the earthquake. According to UBS AG, the Fukushima I nuclear accidents have cast doubt on whether even an advanced economy like Japan can master nuclear safety. Catastrophic scenarios involving terrorist attacks, insider sabotage, and cyberattacks are also conceivable.

In his book, *Normal accidents*, Charles Perrow says that multiple and unexpected failures are built into society's complex and tightly-coupled nuclear reactor systems. Such accidents are unavoidable and cannot be designed around. To date, there have been three serious accidents (core damage) in the world since 1970, involving five reactors (one at Three Mile Island in 1979; one at Chernobyl in 1986; and three at Fukushima-Daiichi in 2011), corresponding to the beginning of the operation of generation II reactors.

Nuclear weapon safety, as well as the safety of military research involving nuclear materials, is generally handled by agencies different from those that oversee civilian safety, for various reasons, including secrecy. There are ongoing concerns about terrorist groups acquiring nuclear bomb-making material.

Overview of Nuclear Processes and Safety Issues

As of 2011, nuclear safety considerations occur in a number of situations, including:

- Nuclear fission power used in nuclear power stations, and nuclear submarines and ships
- Nuclear weapons
- Fissionable fuels such as uranium and plutonium and their extraction, storage and use
- Radioactive materials used for medical, diagnostc, batteries for some space projects, and research purposes
- Nuclear waste, the radioactive waste residue of nuclear materials
- Nuclear fusion power, a technology under long-term development
- Unplanned entry of nuclear materials into the biosphere and food chain (living plants, animals and humans) if breathed or ingested.

With the exception of thermonuclear weapons and experimental fusion research, all safety issues specific to nuclear power stems from the need to limit the biological uptake of committed dose (ingestion or inhalation of radioactive materials), and external radiation dose due to radioactive contamination.

Nuclear safety therefore covers at minimum: -

- Extraction, transportation, storage, processing, and disposal of fissionable materials

- Safety of nuclear power generators

- Control and safe management of nuclear weapons, nuclear material capable of use as a weapon, and other radioactive materials

- Safe handling, accountability and use in industrial, medical and research contexts

- Disposal of nuclear waste

- Limitations on exposure to radiation

Responsible Agencies

International

Internationally the International Atomic Energy Agency "works with its Member States and multiple partners worldwide to promote safe, secure and peaceful nuclear technologies." Some scientists say that the 2011 Japanese nuclear accidents have revealed that the nuclear industry lacks sufficient oversight, leading to renewed calls to redefine the mandate of the IAEA so that it can better police nuclear power plants worldwide.

IAEA headquarters in Vienna, Austria

The IAEA Convention on Nuclear Safety was adopted in Vienna on 17 June 1994 and entered into force on 24 October 1996. The objectives of the Convention are to achieve and maintain a high level of nuclear safety worldwide, to establish and maintain effective defences in nuclear installations against potential radiological hazards, and to prevent accidents having radiological consequences.

The International Atomic Energy Agency was created in 1957 to encourage peaceful development of nuclear technology while providing international safeguards against nuclear proliferation.

The Convention was drawn up in the aftermath of the Three Mile Island and Chernobyl accidents at a series of expert level meetings from 1992 to 1994, and was the result of considerable work by States, including their national regulatory and nuclear safety authorities, and the International Atomic Energy Agency, which serves as the Secretariat for the Convention.

The obligations of the Contracting Parties are based to a large extent on the application of the safety principles for nuclear installations contained in the IAEA document Safety Fundamentals 'The Safety of Nuclear Installations' (IAEA Safety Series No. 110 published 1993). These obligations cover the legislative and regulatory framework, the regulatory body, and technical safety obligations related to, for instance, siting, design, construction, operation, the availability of adequate financial and human resources, the assessment and verification of safety, quality assurance and emergency preparedness.

The convention was amended in 2015 by the Vienna Declaration on Nuclear Safety This resulted in the following principles:

1. New nuclear power plants are to be designed, sited, and constructed, consistent with the objective of preventing accidents in the commissioning and operation and, should an accident occur, mitigating possible releases of radionuclides causing long-term off site contamination and avoiding early radioactive releases or radioactive releases large enough to require long-term protective measures and actions.

2. Comprehensive and systematic safety assessments are to be carried out periodically and regularly for existing installations throughout their lifetime in order to identify safety improvements that are oriented to meet the above objective. Reasonably practicable or achievable safety improvements are to be implemented in a timely manner.

3. National requirements and regulations for addressing this objective throughout the lifetime of nuclear power plants are to take into account the relevant IAEA Safety Standards and, as appropriate, other good practices as identified inter alia in the Review Meetings of the CNS.

There are several problems with the IAEA, says Najmedin Meshkati of University of Southern California, writing in 2011:

"It recommends safety standards, but member states are not required to comply; it promotes nuclear energy, but it also monitors nuclear use; it is the sole global organization overseeing the nuclear energy industry, yet it is also weighed down by checking compliance with the Nuclear Non-Proliferation Treaty (NPT)".

National

Many nations utilizing nuclear power have specialist institutions overseeing and regulating nuclear safety. Civilian nuclear safety in the U.S. is regulated by the Nuclear Regulatory Commission (NRC). However, critics of the nuclear industry complain that the regulatory bodies are too intertwined with the inustries themselves to be effective. The book *The Doomsday Machine* for example, offers a series of examples of national regulators, as they put it 'not regulating, just waving' (a pun on *waiving*) to argue that, in Japan, for example, "regulators and the regulated have long been friends, working together to offset the doubts of a public brought up on the horror of the nuclear bombs". Other examples offered include:

- in the United States, a dangerous custom whereby only supporters of the nuclear industry are allowed to supervise it and lobbyists have been allowed to have an effective veto over regulators.

- in China, where Kang Rixin, former general manager of the state-owned China National Nuclear Corporation, was sentenced to life in jail in 2010 for accepting bribes (and other abuses), a verdict raising questions about the quality of his work on the safety and trustworthiness of China's nuclear reactors.

- in India, where the nuclear regulator reports to the national Atomic Energy Commission, which champions the building of nuclear power plants there and the chairman of the Atomic Energy Regulatory Board, S. S. Bajaj, was previously a senior executive at the Nuclear Power Corporation of India, the company he is now helping to regulate.

- in Japan, where the regulator reports to the Ministry of Economy, Trade and Industry, which overtly seeks to promote the nuclear industry and ministry posts and top jobs in the nuclear business are passed among the same small circle of experts.

The book argues that nuclear safety is compromised by the suspicion that, as Eisaku Sato, formerly a governor of Fukushima province (with its infamous nuclear reactor complex), has put it of the regulators: "They're all birds of a feather".

The safety of nuclear plants and materials controlled by the U.S. government for research, weapons production, and those powering naval vessels is not governed by the NRC. In the UK nuclear safety is regulated by the Office for Nuclear Regulation (ONR) and the Defence Nuclear Safety Regulator (DNSR). The Australian Radiation Protection and Nuclear Safety Agency (ARPANSA) is the Federal Government body that monitors and identifies solar radiation and nuclear radiation risks in Australia. It is the main body dealing with ionizing and non-ionizing radiation and publishes material regarding radiation protection.

Other agencies include:

- Autorité de sûreté nucléaire
- Canadian Nuclear Safety Commission
- Radiological Protection Institute of Ireland
- Federal Atomic Energy Agency in Russia
- Kernfysische dienst, (NL)
- Pakistan Nuclear Regulatory Authority
- Bundesamt für Strahlenschutz, (DE)
- Atomic Energy Regulatory Board (India)

Nuclear Power Plant Safety and Security

Complexity

Nuclear power plants are some of the most sophisticated and complex energy systems ever designed. Any complex system, no matter how well it is designed and engineered, cannot be deemed failure-proof. Veteran journalist and author Stephanie Cooke has argued:

The reactors themselves were enormously complex machines with an incalculable number of things that could go wrong. When that happened at Three Mile Island in 1979, another fault line in the nuclear world was exposed. One malfunction led to another, and then to a series of others, until the core of the reactor itself began to melt, and even the world's most highly trained nuclear engineers did not know how to respond. The accident revealed serious deficiencies in a system that was meant to protect public health and safety.

The 1979 Three Mile Island accident inspired Perrow's book *Normal Accidents*, where a nuclear accident occurs, resulting from an unanticipated interaction of multiple failures in a complex system. TMI was an example of a normal accident because it was "unexpected, incomprehensible, uncontrollable and unavoidable".

Perrow concluded that the failure at Three Mile Island was a consequence of the system's immense complexity. Such modern high-risk systems, he realized, were prone to failures however well they were managed. It was inevitable that they would eventually suffer what he termed a 'normal accident'. Therefore, he suggested, we might do better to contemplate a radical redesign, or if that was not possible, to abandon such technology entirely.

A fundamental issue contributing to a nuclear power system's complexity is its extremely long lifetime. The timeframe from the start of construction of a commercial nuclear power station through the safe disposal of its last radioactive waste, may be 100 to 150 years.

Failure Modes of Nuclear Power Plants

There are concerns that a combination of human and mechanical error at a nuclear facility could result in significant harm to people and the environment:

Operating nuclear reactors contain large amounts of radioactive fission products which, if dispersed, can pose a direct radiation hazard, contaminate soil and vegetation, and be ingested by humans and animals. Human exposure at high enough levels can cause both short-term illness and death and longer-term death by cancer and other diseases.

It is impossible for a commercial nuclear reactor to explode like a nuclear bomb since the fuel is never sufficiently enriched for this to occur.

Nuclear reactors can fail in a variety of ways. Should the instability of the nuclear material generate unexpected behavior, it may result in an uncontrolled power excursion. Normally, the cooling system in a reactor is designed to be able to handle the excess heat this causes; however, should the reactor also experience a loss-of-coolant accident, then the fuel may melt or cause the vessel in which it is contained to overheat and melt. This event is called a nuclear meltdown.

After shutting down, for some time the reactor still needs external energy to power its cooling systems. Normally this energy is provided by the power grid to which that plant is connected, or by emergency diesel generators. Failure to provide power for the cooling systems, as happened in Fukushima I, can cause serious accidents.

Nuclear safety rules in the United States "do not adequately weigh the risk of a single event that would knock out electricity from the grid and from emergency generators, as a quake and tsunami recently did in Japan", Nuclear Regulatory Commission officials said in June 2011.

As a safeguard against mechanical failure, many nuclear plants are designed to shut down automatically after two days of continuous and unattended operation.

Vulnerability of Nuclear Plants to Attack

Nuclear reactors become preferred targets during military conflict and, over the past three decades, have been repeatedly attacked during military air strikes, occupations, invasions and campaigns:

- In September 1980, Iran bombed the Al Tuwaitha nuclear complex in Iraq in Operation Scorch Sword.

- In June 1981, an Israeli air strike completely destroyed Iraq's Osirak nuclear research facility in Operation Opera.

- Between 1984 and 1987, Iraq bombed Iran's Bushehr nuclear plant six times.

- On 8 January 1982, Umkhonto we Sizwe, the armed wing of the ANC, attacked South Africa's Koeberg nuclear power plant while it was still under construction.

- In 1991, the U.S. bombed three nuclear reactors and an enrichment pilot facility in Iraq.

- In 1991, Iraq launched Scud missiles at Israel's Dimona nuclear power plant

- In September 2007, Israel bombed a Syrian reactor under construction.

In the U.S., plants are surrounded by a double row of tall fences which are electronically monitored. The plant grounds are patrolled by a sizeable force of armed guards. The NRC's "Design Basis Threat" criterion for plants is a secret, and so what size of attacking force the plants are able to protect against is unknown. However, to scram (make an emergency shutdown) a plant takes fewer than 5 seconds while unimpeded restart takes hours, severely hampering a terrorist force in a goal to release radioactivity.

Attack from the air is an issue that has been highlighted since the September 11 attacks in the U.S. However, it was in 1972 when three hijackers took control of a domestic passenger flight along the east coast of the U.S. and threatened to crash the plane into a U.S. nuclear weapons plant in Oak Ridge, Tennessee. The plane got as close as 8,000 feet above the site before the hijackers' demands were met.

The most important barrier against the release of radioactivity in the event of an aircraft strike on a nuclear power plant is the containment building and its missile shield. Current NRC Chairman Dale Klein has said "Nuclear power plants are inherently robust structures that our studies show provide adequate protection in a hypothetical attack by an airplane. The NRC has also taken actions that require nuclear power plant operators to be able to manage large fires or explosions—no matter what has caused them."

In addition, supporters point to large studies carried out by the U.S. Electric Power Research Institute that tested the robustness of both reactor and waste fuel storage and found that they should be able to sustain a terrorist attack comparable to the September 11 terrorist attacks in the U.S. Spent fuel is usually housed inside the plant's "protected zone" or a spent nuclear fuel shipping cask; stealing it for use in a "dirty bomb" would be extremely difficult. Exposure to the intense radiation would almost certainly quickly incapacitate or kill anyone who attempts to do so.

Threat of Terrorist Attacks

Nuclear power plants are considered to be targets for terrorist attacks, these findings will be discussed on 11 September not only since the attacks. Even during the construction of the first nuclear power plants has been advised by security bodies on this issue. Even concrete threats of attack against nuclear power plants by terrorists or criminals are documented from several states. While older nuclear power plants were built without special protection against air accidents in Germany, the later nuclear power plants built with a massive concrete buildings are partially protected against air accidents. They are designed against the impact of combat aircraft at a speed of about 800 km / h. It was assumed as a basis of assessment of the impact of an aircraft of type Phantom II with a mass of 20 tonnes and speed of 215 m / s.

The dangers arising from a terrorist caused large aircraft crash on a nuclear power plant is currently being discussed. Such a terrorist attack could have catastrophic consequences. For example, the German government has confirmed that the nuclear power plant Biblis A not against the crash had secured a military aircraft. Following the terrorist attacks in Brussels in 2016 several nuclear power plants have been partially evacuated. At the same time it became known that the terrorists had spied on the nuclear power plants. Several employees access privileges has been withdrawn.v

Moreover, even "nuclear terrorism", for instance with a so-called "Dirty bomb" pose a considerable potantial hazard. For their production would come any radioactive waste or enriched for nuclear power plants uranium in question.

Plant Location

In many countries, plants are often located on the coast, in order to provide a ready source of cooling water for the essential service water system. As a consequence the design needs to take the risk of flooding and tsunamis into account. The World Energy Council (WEC) argues disaster risks are changing and increasing the likelihood of disasters such as earthquakes, cyclones, hurricanes, typhoons, flooding. High temperatures, low precipitation levels and severe droughts may lead to fresh water shortages. Failure to calculate the risk of flooding correctly lead to a Level 2 event on the International Nuclear Event Scale during the 1999 Blayais Nuclear Power Plant flood, while flooding caused by the 2011 Tōhoku earthquake and tsunami lead to the Fukushima I nuclear accidents.

earthquake map

Fort Calhoun Nuclear Generating Station surrounded by the 2011 Missouri River Floods on June 16, 2011

The design of plants located in seismically active zones also requires the risk of earthquakes and tsunamis to be taken into account. Japan, India, China and the USA are among the countries to have plants in earthquake-prone regions. Damage caused to Japan's Kashiwazaki-Kariwa Nuclear Power Plant during the 2007 Chūetsu offshore earthquake underlined concerns expressed by ex perts in Japan prior to the Fukushima accidents, who have warned of a *genpatsu-shinsai* (domi no-effect nuclear power plant earthquake disaster).

Angra Nuclear Power Plant in Rio de Janeiro state, Brazil

Multiple Reactors

The Fukushima nuclear disaster illustrated the dangers of building multiple nuclear reactor units close to one another. Because of the closeness of the reactors, Plant Director Masao Yoshida "was put in the position of trying to cope simultaneously with core meltdowns at three reactors and ex posed fuel pools at three units".

Nuclear Safety Systems

The three primary objectives of nuclear safety systems as defined by the Nuclear Regulatory Commission are to shut down the reactor, maintain it in a shutdown condition, and prevent the release of radioactive material during events and accidents. These objectives are accomplished using a variety of equipment, which is part of different systems, of which each performs specific functions.

Routine Emissions of Radioactive Materials

During everyday routine operations, emissions of radioactive materials from nuclear plants are released to the outside of the plants although they are quite slight amounts. The daily emissions go into the air, water and soil.

NRC says, "nuclear power plants sometimes release radioactive gases and liquids into the environment under controlled, monitored conditions to ensure that they pose no danger to the public or the environment", and "routine emissions during normal operation of a nuclear power plant are never lethal".

According to the United Nations (UNSCEAR), regular nuclear power plant operation including the nuclear fuel cycle amounts to 0.0002 millisieverts (mSv) annually in average public radiation exposure; the legacy of the Chernobyl disaster is 0.002 mSv/a as a global average as of a 2008 report; and natural radiation exposure averages 2.4 mSv annually although frequently varying depending on an individual's location from 1 to 13 mSv.

Japanese Public Perception of Nuclear Power Safety

In March 2012, Prime Minister Yoshihiko Noda said that the Japanese government shared the blame for the Fukushima disaster, saying that officials had been blinded by an image of the country's technological infallibility and were "all too steeped in a safety myth."

Japan has been accused by authors such as journalist Yoichi Funabashi of having an "aversion to facing the potential threat of nuclear emergencies." According to him, a national program to develop robots for use in nuclear emergencies was terminated in midstream because it "smacked too much of underlying danger." Though Japan is a major power in robotics, it had none to send in to Fukushima during the disaster. He mentions that Japan's Nuclear Safety Commission stipulated in its safety guidelines for light-water nuclear facilities that "the potential for extended loss of power need not be considered." However, this kind of extended loss of power to the cooling pumps caused the Fukushima meltdown. In other countries such as the UK, nuclear plants have not been claimed to be absolutely safe. It is instead claimed that a major accident has a likelihood of occurrence lower than (for example) 0.0001/year.

Incidents such as the Fukushima Daiichi nuclear disaster could have been avoided with stricter regulations over nuclear power. In 2002, TEPCO, the company that operated the Fukushima plant, admitted to falsifying reports on over 200 occasions between 1997 and 2002. TEPCO faced no fines for this. Instead, they fired four of their top executives. Three of these four later went on to take jobs at companies that do business with TEPCO.

Hazards of Nuclear Material

There is currently a total of 47,000 tonnes of high-level nuclear waste stored in the USA. Nuclear waste is approximately 94% Uranium, 1.3% Plutonium, 0.14% other Actinides, and 5.2% fission products. About 1.0% of this waste consists of long-lived isotopes ^{79}Se, ^{93}Zr, ^{99}Te, ^{107}Pd, ^{126}Sn, ^{129}I and ^{135}Cs. Shorter lived isotopes including ^{89}Sr, ^{90}Sr, ^{106}Ru, ^{125}Sn, ^{134}Cs, ^{137}Cs, and ^{147}Pm constitute 0.9% at one year, decreasing to 0.1% at 100 years. The remaining 3.3-4.1% consists of non-radioactive isotopes. There are technical challenges, as it is preferable to lock away the long-lived fission products, but the challenge should not be exaggerated. One tonne of waste, as described above, has measurable radioactivity of approximately 600 TBq equal to the natural radioactivity in one km³ of the Earth's crust, which if buried, would add only 25 parts per trillion to the total radioactivity.

Spent nuclear fuel stored underwater and uncapped at the Hanford site in Washington, USA.

The difference between short-lived high-level nuclear waste and long-lived low-level waste can be illustrated by the following example. As stated above, one mole of both ^{131}I and ^{129}I release 3×10^{23} decays in a period equal to one half-life. ^{131}I decays with the release of 970 keV whilst ^{129}I decays with the release of 194 keV of energy. 131gm of ^{131}I would therefore release 45 Gigajoules over eight days beginning at an initial rate of 600 EBq releasing 90 Kilowatts with the last radioactive decay occurring inside two years. In contrast, 129gm of ^{129}I would therefore release 9 Gigajoules over 15.7 million years beginning at an initial rate of 850 MBq releasing 25 microwatts with the radioactivity decreasing by less than 1% in 100,000 years.

One tonne of nuclear waste also reduces CO_2 emission by 25 million tonnes.

Radionuclides such as ^{129}I or ^{131}I, may be highly radioactive, or very long-lived, but they cannot be both. One mole of ^{129}I (129 grams) undergoes the same number of decays (3×10^{23}) in 15.7 million years, as does one mole of ^{131}I (131 grams) in 8 days. ^{131}I is therefore highly radioactive, but disappears very quickly, whilst ^{129}I releases a very low level of radiation for a very long time. Two long-lived fission products, Technetium-99 (half-life 220,000 years) and Iodine-129 (half-life 15.7 million years), are of somewhat greater concern because of a greater chance of entering the biosphere. The transuranic elements in spent fuel are Neptunium-237 (half-life two million years) and Plutonium-239 (half-life 24,000 years). will also remain in the environment for long periods of time. A more complete solution to both the problem of both Actinides and to the need for low-carbon energy may be the integral fast reactor. One tonne of nuclear waste after a complete burn in an IFR reactor will have prevented 500 million tonnes of CO_2 from entering the atmosphere. Otherwise, waste storage usually necessitates treatment, followed by a long-term management strategy involving permanent storage, disposal or transformation of the waste into a non-toxic form.

Governments around the world are considering a range of waste management and disposal options, usually involving deep-geologic placement, although there has been limited progress toward implementing long-term waste management solutions. This is partly because the timeframes in question when dealing with radioactive waste range from 10,000 to millions of years, according to studies based on the effect of estimated radiation doses.

Since the fraction of a radioisotope's atoms decaying per unit of time is inversely proportional to its half-life, the relative radioactivity of a quantity of buried human radioactive waste would diminish over time compared to natural radioisotopes (such as the decay chain of 120 trillion tons of thorium and 40 trillion tons of uranium which are at relatively trace concentrations of parts per million each over the crust's $3 * 10^{19}$ ton mass). For instance, over a timeframe of thousands of years, after the most active short half-life radioisotopes decayed, burying U.S. nuclear waste would increase the radioactivity in the top 2000 feet of rock and soil in the United States (10 million km²) by ≈ 1 part in 10 million over the cumulative amount of natural radioisotopes in such a volume, although the vicinity of the site would have a far higher concentration of artificial radioisotopes underground than such an average.

Safety Culture and Human Errors

One relatively prevalent notion in discussions of nuclear safety is that of safety culture. The International Nuclear Safety Advisory Group, defines the term as "the personal dedication and accountability of all individuals engaged in any activity which has a bearing on the safety of nuclear power plants". The goal is "to design systems that use human capabilities in appropriate ways, that protect systems from human frailties, and that protect humans from hazards associated with the system".

At the same time, there is some evidence that operational practices are not easy to change. Operators almost never follow instructions and written procedures exactly, and "the violation of rules appears to be quite rational, given the actual workload and timing constraints under which the operators must do their job". Many attempts to improve nuclear safety culture "were compensated by people adapting to the change in an unpredicted way".

According to Areva's Southeast Asia and Oceania director, Selena Ng, Japan's Fukushima nuclear disaster is "a huge wake-up call for a nuclear industry that hasn't always been sufficiently transparent about safety issues". She said "There was a sort of complacency before Fukushima and I don't think we can afford to have that complacency now".

An assessment conducted by the *Commissariat à l'Énergie Atomique* (CEA) in France concluded that no amount of technical innovation can eliminate the risk of human-induced errors associated with the operation of nuclear power plants. Two types of mistakes were deemed most serious: errors committed during field operations, such as maintenance and testing, that can cause an accident; and human errors made during small accidents that cascade to complete failure.

According to Mycle Schneider, reactor safety depends above all on a 'culture of security', including the quality of maintenance and training, the competence of the operator and the workforce, and the rigour of regulatory oversight. So a better-designed, newer reactor is not always a safer one, and older reactors are not necessarily more dangerous than newer ones. The 1979 Three Mile Island accident in the United States occurred in a reactor that had started operation only three months earlier, and the Chernobyl disaster occurred after only two years of operation. A serious loss of coolant occurred at the French Civaux-1 reactor in 1998, less than five months after start-up.

However safe a plant is designed to be, it is operated by humans who are prone to errors. Laurent Stricker, a nuclear engineer and chairman of the World Association of Nuclear Operators says that operators must guard against complacency and avoid overconfidence. Experts say that the "largest single internal factor determining the safety of a plant is the culture of security among regulators, operators and the workforce — and creating such a culture is not easy".

Risks

The routine health risks and greenhouse gas emissions from nuclear fission power are small relative to those associated with coal, but there are several "catastrophic risks":

The extreme danger of the radioactive material in power plants and of nuclear technology in and of itself is so well known that the US government was prompted (at the industry's urging) to enact provisions that protect the nuclear industry from bearing the full burden of such inherently risky nuclear operations. The Price-Anderson Act limits industry's liability in the case of accidents, and the 1982 Nuclear Waste Policy Act charges the federal government with responsibility for permanently storing nuclear waste.

Population density is one critical lens through which other risks have to be assessed, says Laurent Stricker, a nuclear engineer and chairman of the World Association of Nuclear Operators:

The KANUPP plant in Karachi, Pakistan, has the most people — 8.2 million — living within 30 kilometres of a nuclear plant, although it has just one relatively small reactor with an output of 125 megawatts. Next in the league, however, are much larger plants — Taiwan's 1,933-megawatt Kuosheng plant with 5.5 million people within a 30-kilometre radius and the 1,208-megawatt Chin Shan plant with 4.7 million; both zones include the capital city of Taipei.

172,000 people living within a 30 kilometre radius of the Fukushima Daiichi nuclear power plant, have been forced or advised to evacuate the area. More generally, a 2011 analysis by *Nature* and Columbia University, New York, shows that some 21 nuclear plants have populations larger than 1 million within a 30-km radius, and six plants have populations larger than 3 million within that radius.

Black Swan events are highly unlikely occurrences that have big repercussions. Despite planning, nuclear power will always be vulnerable to black swan events:

A rare event — especially one that has never occurred — is difficult to foresee, expensive to plan for and easy to discount with statistics. Just because something is only supposed to happen every 10,000 years does not mean that it will not happen tomorrow. Over the typical 40-year life of a plant, assumptions can also change, as they did on September 11, 2001, in August 2005 when Hurricane Katrina struck, and in March, 2011, after Fukushima.

The list of potential black swan events is "damningly diverse":

Nuclear reactors and their spent-fuel pools could be targets for terrorists piloting hijacked planes. Reactors may be situated downstream from dams that, should they ever burst, could unleash massive floods. Some reactors are located close to earthquake faults or shorelines, a dangerous scenario like that which emerged at Three Mile Island and Fukushima — a catastrophic coolant failure, the overheating and melting of the radioactive fuel rods, and a release of radioactive material.

- International Nuclear Events Scale

- Comparative Risk Assessment

- Statistical Risk Assessment

- Probabilistic risk assessment

  o *Severe Accident Risks: An Assessment for Five U.S. Nuclear Power Plants* NUREG-1150 1991

  o *Calculation of Reactor Accident Consequences* CRAC-II 1982

  o Rasmussen Report: *Reactor Safety Study* WASH-1400 1975

  o The Brookhaven Report: *Theoretical Possibilities and Consequences of Major Accidents in Large Nuclear Power Plants* WASH-740 1957

The AP1000 has a maximum core damage frequency of 5.09×10^{-7} per plant per year. The Evolutionary Power Reactor (EPR) has a maximum core damage frequency of 4×10^{-7} per plant per year. General Electric has recalculated maximum core damage frequencies per year per plant for its nuclear power plant designs:

BWR/4 -- 1×10^{-5}

BWR/6 -- 1×10^{-6}

ABWR -- 2×10^{-7}

ESBWR -- 3×10^{-8}

Beyond Design Basis Events

The Fukushima I nuclear accident was caused by a "beyond design basis event," the tsunami and associated earthquakes were more powerful than the plant was designed to accommodate, and the accident is directly due to the tsunami overflowing the too-low seawall. Since then, the possibility of unforeseen beyond design basis events has been a major concern for plant operators.

Transparency and Ethics

According to journalist Stephanie Cooke, it is difficult to know what really goes on inside nuclear power plants because the industry is shrouded in secrecy. Corporations and governments control what information is made available to the public. Cooke says "when information is made available, it is often couched in jargon and incomprehensible prose".

Kennette Benedict has said that nuclear technology and plant operations continue to lack transparency and to be relatively closed to public view:

Despite victories like the creation of the Atomic Energy Commission, and later the Nuclear Regular Commission, the secrecy that began with the Manhattan Project has tended to permeate the civilian nuclear program, as well as the military and defense programs.

In 1986, Soviet officials held off reporting the Chernobyl disaster for several days. The operators of the Fukushima plant, Tokyo Electric Power Co, were also criticised for not quickly disclosing information on releases of radioactivity from the plant. Russian President Dmitry Medvedev said there must be greater transparency in nuclear emergencies.

Historically many scientists and engineers have made decisions on behalf of potentially affected populations about whether a particular level of risk and uncertainty is acceptable for them. Many nuclear engineers and scientists that have made such decisions, even for good reasons relating to long term energy availability, now consider that doing so without informed consent is wrong, and that nuclear power safety and nuclear technologies should be based fundamentally on morality, rather than purely on technical, economic and business considerations.

Non-Nuclear Futures: The Case for an Ethical Energy Strategy is a 1975 book by Amory B. Lovins and John H. Price. The main theme of the book is that the most important parts of the nuclear power debate are not technical disputes but relate to personal values, and are the legitimate province of every citizen, whether technically trained or not.

Nuclear and Radiation Accidents

The nuclear industry has an excellent safety record and the deaths per megawatt hour are the lowest of all the major energy sources. According to Zia Mian and Alexander Glaser, the "past six decades have shown that nuclear technology does not tolerate error". Nuclear power is perhaps the primary example of what are called 'high-risk technologies' with 'catastrophic potential', because "no matter how effective conventional safety devices are, there is a form of accident that is inevitable, and such accidents are a 'normal' consequence of the system." In short, there is no escape from system failures.

Whatever position one takes in the nuclear power debate, the possibility of catastrophic accidents and consequent economic costs must be considered when nuclear policy and regulations are being framed.

Accident Liability Protection

Kristin Shrader-Frechette has said "if reactors were safe, nuclear industries would not demand government-guaranteed, accident-liability protection, as a condition for their generating electricity". No private insurance company or even consortium of insurance companies "would shoulder the fearsome liabilities arising from severe nuclear accidents".

Hanford Site

The Hanford site represents two-thirds of America's high-level radioactive waste by volume. Nuclear reactors line the riverbank at the Hanford Site along the Columbia River in January 1960.

The Hanford Site is a mostly decommissioned nuclear production complex on the Columbia River in the U.S. state of Washington, operated by the United States federal government. Plutonium manufactured at the site was used in the first nuclear bomb, tested at the Trinity site, and in Fat Man, the bomb detonated over Nagasaki, Japan. During the Cold War, the project was expanded to include nine nuclear reactors and five large plutonium processing complexes, which produced plutonium for most of the 60,000 weapons in the U.S. nuclear arsenal. Many of the early safety procedures and waste disposal practices were inadequate, and government documents have since confirmed that Hanford's operations released significant amounts of radioactive materials into the air and the Columbia River, which still threatens the health of residents and ecosystems. The weapons production reactors were decommissioned at the end of the Cold War, but the decades of manufacturing left behind 53 million US gallons (200,000 m³) of high-level radioactive waste, an additional 25 million cubic feet (710,000 m³) of solid radioactive waste, 200 square miles (520 km²) of contaminated groundwater beneath the site and occasional discoveries of undocumented contaminations that slow the pace and raise the cost of cleanup. The Hanford site represents two-thirds of the nation's high-level radioactive waste by volume. Today, Hanford is the most contaminated nuclear site in the United States and is the focus of the nation's largest environmental cleanup.

1986 Chernobyl Disaster

The Chernobyl disaster was a nuclear accident that occurred on 26 April 1986 at the Chernobyl Nuclear Power Plant in Ukraine. An explosion and fire released large quantities of radioactive contamination into the atmosphere, which spread over much of Western USSR and Europe. It is considered the worst nuclear power plant accident in history, and is one of only two classified as a level 7 event on the International Nuclear Event Scale (the other being the Fukushima Daiichi nuclear disaster). The battle to contain the contamination and avert a greater catastrophe ultimately involved over 500,000 workers and cost an estimated 18 billion rubles, crippling the Soviet economy. The accident raised concerns about the safety of the nuclear power industry, slowing its expansion for a number of years.

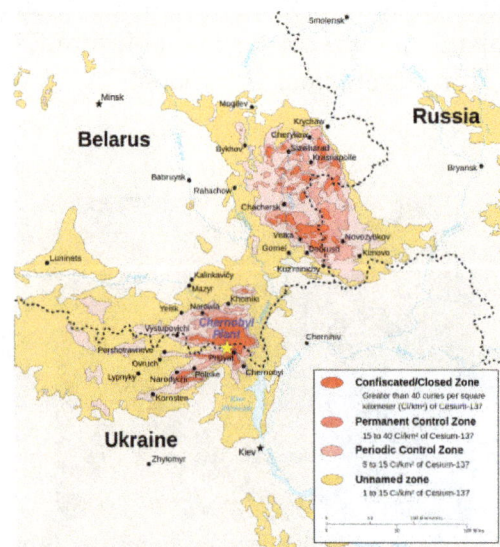

Map showing Caesium-137 contamination in Belarus, Russia, and Ukraine as of 1996.

UNSCEAR has conducted 20 years of detailed scientific and epidemiological research on the effects of the Chernobyl accident. Apart from the 57 direct deaths in the accident itself, UNSCEAR predicted in 2005 that up to 4,000 additional cancer deaths related to the accident would appear "among the 600 000 persons receiving more significant exposures (liquidators working in 1986–87, evacuees, and residents of the most contaminated areas)". Russia, Ukraine, and Belarus have been burdened with the continuing and substantial decontamination and health care costs of the Chernobyl disaster.

Eleven of Russia's reactors are of the RBMK 1000 type, similar to the one at Chernobyl Nuclear Power Plant. Some of these RBMK reactors were originally to be shut down but have instead been given life extensions and uprated in output by about 5%. Critics say that these reactors are of an "inherently unsafe design", which cannot be improved through upgrades and modernization, and some reactor parts are impossible to replace. Russian environmental groups say that the lifetime extensions "violate Russian law, because the projects have not undergone environmental assessments".

2011 Fukushima I Accidents

Despite all assurances, a major nuclear accident on the scale of the 1986 Chernobyl disaster happened again in 2011 in Japan, one of the world's most industrially advanced countries. Nuclear Safety Commission Chairman Haruki Madarame told a parliamentary inquiry in February 2012 that "Japan's atomic safety rules are inferior to global standards and left the country unprepared for the Fukushima nuclear disaster last March". There were flaws in, and lax enforcement of, the safety rules governing Japanese nuclear power companies, and this included insufficient protection against tsunamis.

Fukushima reactor control room.

Following the 2011 Japanese Fukushima nuclear disaster, authorities shut down the nation's 54 nuclear power plants. As of 2013, the Fukushima site remains highly radioactive, with some 160,000 evacuees still living in temporary housing, and some land will be unfarmable for centuries. The difficult cleanup job will take 40 or more years, and cost tens of billions of dollars.

A 2012 report in *The Economist* said: "The reactors at Fukushima were of an old design. The risks they faced had not been well analysed. The operating company was poorly regulated and did not know what was going on. The operators made mistakes. The representatives of the safety inspectorate fled. Some of the equipment failed. The establishment repeatedly played down the risks and suppressed information about the movement of the radioactive plume, so some people were evacuated from more lightly to more heavily contaminated places".

The designers of the Fukushima I Nuclear Power Plant reactors did not anticipate that a tsunami generated by an earthquake would disable the backup systems that were supposed to stabilize the reactor after the earthquake. Nuclear reactors are such "inherently complex, tightly coupled systems that, in rare, emergency situations, cascading interactions will unfold very rapidly in such a way that human operators will be unable to predict and master them".

Lacking electricity to pump water needed to cool the atomic core, engineers vented radioactive steam into the atmosphere to release pressure, leading to a series of explosions that blew out concrete walls around the reactors. Radiation readings spiked around Fukushima as the disaster widened, forcing the evacuation of 200,000 people. There was a rise in radiation levels on the outskirts of Tokyo, with a population of 30 million, 135 miles (210 kilometers) to the south.

Back-up diesel generators that might have averted the disaster were positioned in a basement, where they were quickly overwhelmed by waves. The cascade of events at Fukushima had been predicted in a report published in the U.S. several decades ago:

The 1990 report by the U.S. Nuclear Regulatory Commission, an independent agency responsible for safety at the country's power plants, identified earthquake-induced diesel generator failure and power outage leading to failure of cooling systems as one of the "most likely causes" of nuclear accidents from an external event.

The report was cited in a 2004 statement by Japan's Nuclear and Industrial Safety Agency, but it seems adequate measures to address the risk were not taken by TEPCO. Katsuhiko Ishibashi, a seismology professor at Kobe University, has said that Japan's history of nuclear accidents stems from an overconfidence in plant engineering. In 2006, he resigned from a government panel on nuclear reactor safety, because the review process was rigged and "unscientific".

According to the International Atomic Energy Agency, Japan "underestimated the danger of tsunamis and failed to prepare adequate backup systems at the Fukushima Daiichi nuclear plant". This repeated a widely held criticism in Japan that "collusive ties between regulators and industry led to weak oversight and a failure to ensure adequate safety levels at the plant". The IAEA also said that the Fukushima disaster exposed the lack of adequate backup systems at the plant. Once power was completely lost, critical functions like the cooling system shut down. Three of the reactors "quickly overheated, causing meltdowns that eventually led to explosions, which hurled large amounts of radioactive material into the air".

Louise Fréchette and Trevor Findlay have said that more effort is needed to ensure nuclear safety and improve responses to accidents:

The multiple reactor crises at Japan's Fukushima nuclear power plant reinforce the need for strengthening global instruments to ensure nuclear safety worldwide. The fact that a country that has been operating nuclear power reactors for decades should prove so alarmingly improvisational in its response and so unwilling to reveal the facts even to its own people, much less the International Atomic Energy Agency, is a reminder that nuclear safety is a constant work-in-progress.

David Lochbaum, chief nuclear safety officer with the Union of Concerned Scientists, has repeatedly questioned the safety of the Fukushima I Plant's General Electric Mark 1 reactor design, which is used in almost a quarter of the United States' nuclear fleet.

A report from the Japanese Government to the IAEA says the "nuclear fuel in three reactors probably melted through the inner containment vessels, not just the core". The report says the "inadequate" basic reactor design — the Mark-1 model developed by General Electric — included "the venting system for the containment vessels and the location of spent fuel cooling pools high in the buildings, which resulted in leaks of radioactive water that hampered repair work".

Following the Fukushima emergency, the European Union decided that reactors across all 27 member nations should undergo safety tests.

According to UBS AG, the Fukushima I nuclear accidents are likely to hurt the nuclear power industry's credibility more than the Chernobyl disaster in 1986:

The accident in the former Soviet Union 25 years ago 'affected one reactor in a totalitarian state with no safety culture,' UBS analysts including Per Lekander and Stephen Oldfield wrote in a report today. 'At Fukushima, four reactors have been out of control for weeks -- casting doubt on whether even an advanced economy can master nuclear safety.'

The Fukushima accident exposed some troubling nuclear safety issues:

Despite the resources poured into analyzing crustal movements and having expert committees determine earthquake risk, for instance, researchers never considered the possibility of a magnitude-9 earthquake followed by a massive tsunami. The failure of multiple safety features on nuclear power plants has raised questions about the nation's engineering prowess. Government flip-flopping on acceptable levels of radiation exposure confused the public, and health professionals provided little guidance. Facing a dearth of reliable information on radiation levels, citizens armed themselves with dosimeters, pooled data, and together produced radiological contamination maps far more detailed than anything the government or official scientific sources ever provided.

As of January 2012, questions also linger as to the extent of damage to the Fukushima plant caused by the earthquake even before the tsunami hit. Any evidence of serious quake damage at the plant would "cast new doubt on the safety of other reactors in quake-prone Japan".

Two government advisers have said that "Japan's safety review of nuclear reactors after the Fukushima disaster is based on faulty criteria and many people involved have conflicts of interest". Hiromitsu Ino, Professor Emeritus at the University of Tokyo, says "The whole process being undertaken is exactly the same as that used previous to the Fukushima Dai-Ichi accident, even though the accident showed all these guidelines and categories to be insufficient".

In March 2012, Prime Minister Yoshihiko Noda acknowledged that the Japanese government shared the blame for the Fukushima disaster, saying that officials had been blinded by a false belief in the country's "technological infallibility", and were all too steeped in a "safety myth".

Other Accidents

Serious nuclear and radiation accidents include the Chalk River accidents (1952, 1958 & 2008), Mayak disaster (1957), Windscale fire (1957), SL-1 accident (1961), Soviet submarine K-19 accident (1961), Three Mile Island accident (1979), Church Rock uranium mill spill (1979), Soviet submarine K-431 accident (1985), Goiânia accident (1987), Zaragoza radiotherapy accident (1990), Costa Rica radiotherapy accident (1996), Tokaimura nuclear accident (1999), Sellafield THORP leak (2005), and the Flerus IRE cobalt-60 spill (2006).

Health Impacts

Four hundred and thirty-seven nuclear power stations are presently in operation but, unfortunately, five major nuclear accidents have occurred in the past. These accidents occurred at Kyshtym (1957), Windscale (1957), Three Mile Island (1979), Chernobyl (1986), and Fukushima (2011). A report in *Lancet* says that the effects of these accidents on individuals and societies are diverse and enduring:

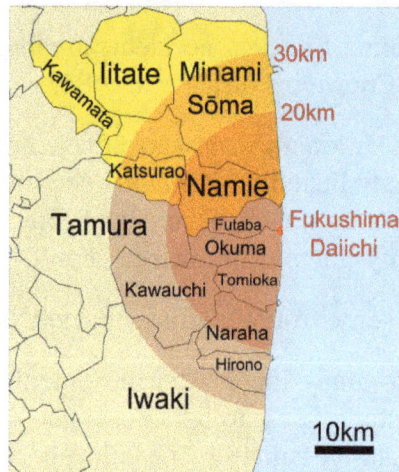

Japan towns, villages, and cities around the Fukushima Daiichi nuclear plant. The 20km and 30km areas had evacuation and sheltering orders, and additional administrative districts that had an evacuation order are highlighted.

> "Accumulated evidence about radiation health effects on atomic bomb survivors and other radiation-exposed people has formed the basis for national and international regulations about radiation protection. However, past experiences suggest that common issues were not necessarily physical health problems directly attributable to radiation exposure, but rather psychological and social effects. Additionally, evacuation and long-term displacement created severe health-care problems for the most vulnerable people, such as hospital inpatients and elderly people."

In spite of accidents like these, studies have shown that nuclear deaths are mostly in uranium mining and that nuclear energy has generated far fewer deaths than the high pollution levels that result from the use of conventional fossil fuels. However, the nuclear power industry relies on uranium mining, which itself is a hazardous industry, with many accidents and fatalities.

Journalist Stephanie Cooke says that it is not useful to make comparisons just in terms of number of deaths, as the way people live afterwards is also relevant, as in the case of the 2011 Japanese nuclear accidents:

"You have people in Japan right now that are facing either not returning to their homes forever, or if they do return to their homes, living in a contaminated area for basically ever... It affects millions of people, it affects our land, it affects our atmosphere ... it's affecting future generations ... I don't think any of these great big massive plants that spew pollution into the air are good. But I don't think it's really helpful to make these comparisons just in terms of number of deaths".

The Fukushima accident forced more than 80,000 residents to evacuate from neighborhoods around the plant.

A survey by the Iitate, Fukushima local government obtained responses from some 1,743 people who have evacuated from the village, which lies within the emergency evacuation zone around the crippled Fukushima Daiichi Plant. It shows that many residents are experiencing growing frustration and instability due to the nuclear crisis and an inability to return to the lives they were living before the disaster. Sixty percent of respondents stated that their health and the health of their families had deteriorated after evacuating, while 39.9 percent reported feeling more irritated compared to before the disaster.

"Summarizing all responses to questions related to evacuees' current family status, one-third of all surveyed families live apart from their children, while 50.1 percent live away from other family members (including elderly parents) with whom they lived before the disaster. The survey also showed that 34.7 percent of the evacuees have suffered salary cuts of 50 percent or more since the outbreak of the nuclear disaster. A total of 36.8 percent reported a lack of sleep, while 17.9 percent reported smoking or drinking more than before they evacuated."

Chemical components of the radioactive waste may lead to cancer. For example, Iodine 131 was released along with the radioactive waste when Chernobyl disaster and Three Mile Island accidents occurred. It was concentrated in leafy vegetation after absorption in the soil. It also stays in animals' milk if the animals eat the vegetation. When Iodine 131 enters the human body, it migrates to the thyroid gland in the neck and can cause thyroid cancer.

Other elements from nuclear waste can lead to cancer as well. For example, Strontium 90 causes breast cancer and leukemia, Plutonium 239 causes liver cancer.

Improvements to Nuclear Fission Technologies

Newer reactor designs intended to provide increased safety have been developed over time. These designs include those that incorporate passive safety and Small Modular Reactors. While these reactor designs "are intended to inspire trust, they may have an unintended effect: creating distrust of older reactors that lack the touted safety features".

The next nuclear plants to be built will likely be Generation III or III+ designs, and a few such are already in operation in Japan. Generation IV reactors would have even greater improvements in safety. These new designs are expected to be passively safe or nearly so, and perhaps even inherently safe (as in the PBMR designs).

Some improvements made (not all in all designs) are having three sets of emergency diesel generators and associated emergency core cooling systems rather than just one pair, having quench tanks (large coolant-filled tanks) above the core that open into it automatically, having a double containment (one containment building inside another), etc.

However, safety risks may be the greatest when nuclear systems are the newest, and operators have less experience with them. Nuclear engineer David Lochbaum explained that almost all serious nuclear accidents occurred with what was at the time the most recent technology. He argues that "the problem with new reactors and accidents is twofold: scenarios arise that are impossible to plan for in simulations; and humans make mistakes". As one director of a U.S. research laboratory put it, "fabrication, construction, operation, and maintenance of new reactors will face a steep learning curve: advanced technologies will have a heightened risk of accidents and mistakes. The technology may be proven, but people are not".

Developing Countries

There are concerns about developing countries "rushing to join the so-called nuclear renaissance without the necessary infrastructure, personnel, regulatory frameworks and safety culture". Some countries with nuclear aspirations, like Nigeria, Kenya, Bangladesh and Venezuela, have no significant industrial experience and will require at least a decade of preparation even before breaking ground at a reactor site.

The speed of the nuclear construction program in China has raised safety concerns. The challenge for the government and nuclear companies is to "keep an eye on a growing army of contractors and subcontractors who may be tempted to cut corners". China is advised to maintain nuclear safeguards in a business culture where quality and safety are sometimes sacrificed in favor of cost-cutting, profits, and corruption. China has asked for international assistance in training more nuclear power plant inspectors.

Nuclear Security and Terrorist Attacks

Nuclear power plants, civilian research reactors, certain naval fuel facilities, uranium enrichment plants, and fuel fabrication plants, are vulnerable to attacks which could lead to widespread radioactive contamination. The attack threat is of several general types: commando-like ground-based attacks on equipment which if disabled could lead to a reactor core meltdown or widespread dispersal of radioactivity; and external attacks such as an aircraft crash into a reactor complex, or cyber attacks.

The United States 9/11 Commission has said that nuclear power plants were potential targets originally considered for the September 11, 2001 attacks. If terrorist groups could sufficiently damage safety systems to cause a core meltdown at a nuclear power plant, and/or sufficiently damage spent fuel pools, such an attack could lead to widespread radioactive contamination. The Federation of American Scientists have said that if nuclear power use is to expand significantly, nuclear facilities will have to be made extremely safe from attacks that could release massive quantities of radioactivity into the community. New reactor designs have features of passive safety, which may help. In the United States, the NRC carries out "Force on Force" (FOF) exercises at all Nuclear Power Plant (NPP) sites at least once every three years.

Nuclear reactors become preferred targets during military conflict and, over the past three decades, have been repeatedly attacked during military air strikes, occupations, invasions and campaigns. Various acts of civil disobedience since 1980 by the peace group Plowshares have shown how nuclear weapons facilities can be penetrated, and the groups actions represent extraordinary breaches of security at nuclear weapons plants in the United States. The National Nuclear Security Administration has acknowledged the seriousness of the 2012 Plowshares action. Non-proliferation policy experts have questioned "the use of private contractors to provide security at facilities that manufacture and store the government's most dangerous military material". Nuclear weapons materials on the black market are a global concern, and there is concern about the possible detonation of a small, crude nuclear weapon by a militant group in a major city, with significant loss of life and property. *Stuxnet* is a computer worm discovered in June 2010 that is believed to have been created by the United States and Israel to attack Iran's nuclear facilities.

Nuclear Fusion Research

Nuclear fusion power is a developing technology still under research. It relies on fusing rather than fissioning (splitting) atomic nuclei, using very different processes compared to current nuclear power plants. Nuclear fusion reactions have the potential to be safer and generate less radioactive waste than fission. These reactions appear potentially viable, though technically quite difficult and have yet to be created on a scale that could be used in a functional power plant. Fusion power has been under theoretical and experimental investigation since the 1950s.

Construction of the International Thermonuclear Experimental Reactor facility began in 2007, but the project has run into many delays and budget overruns. The facility is now not expected to begin operations until the year 2027 – 11 years after initially anticipated. A follow on commercial nuclear fusion power station, DEMO, has been proposed. There is also suggestions for a power plant based upon a different fusion approach, that of a Inertial fusion power plant.

Fusion powered electricity generation was initially believed to be readily achievable, as fission power had been. However, the extreme requirements for continuous reactions and plasma containment led to projections being extended by several decades. In 2010, more than 60 years after the first attempts, commercial power production was still believed to be unlikely before 2050.

More Stringent Safety Standards

Matthew Bunn, the former US Office of Science and Technology Policy adviser, and Heinonen, the former Deputy Director General of the IAEA, have said that there is a need for more stringent nuclear safety standards, and propose six major areas for improvement:

- operators must plan for events beyond design bases;

- more stringent standards for protecting nuclear facilities against terrorist sabotage;

- a stronger international emergency response;

- international reviews of security and safety;

- binding international standards on safety and security; and

- international co-operation to ensure regulatory effectiveness.

Coastal nuclear sites must also be further protected against rising sea levels, storm surges, flooding, and possible eventual "nuclear site islanding".

Nuclear Decommissioning

Nuclear decommissioning is the process whereby a nuclear power plant site is dismantled to the point that it no longer requires measures for radiation protection. The presence of radioactive material necessitates processes that are occupationally dangerous, hazardous to the natural environment, expensive, and time-intensive.

The reactor pressure vessel being transported away from the site for burial.

Example of decommissioning work underway.

Decommissioning is an administrative and technical process. It includes clean-up of radioactive materials and progressive demolition of the plant. Once a facility is fully decommissioned, no radiologic danger should persist. The costs of decommissioning are spread over the lifetime of a facility and saved in a decommissioning fund. After a facility has been completely decommissioned, it is released from regulatory control and the plant licensee is no longer responsible for its nuclear safety. Decommissioning may proceed all the way to "greenfield" status.

Options

The International Atomic Energy Agency has defined three options for decommissioning:

- *Immediate Dismantling* (Early Site Release/DECON in the US): This option allows for the facility to be removed from regulatory control relatively soon after shutdown or termination of regulated activities. Final dismantling or decontamination activities begin within a few months or years, and depending on the facility, it could take five years or more. Following removal from regulatory control, the site becomes available for re-use.

- *Safe Enclosure* (or Safestor(e) SAFSTOR): This option postpones the final removal of controls for a longer period, usually on the order of 40 to 60 years. The facility is placed into a safe storage configuration until the eventual dismantling and decontamination activities occur.

- *Entombment/ENTOMB*: This option entails placing the facility into a condition that will allow the remaining radioactive material to remain on-site indefinitely. This option usually involves reducing the size of the area where the radioactive material is located and then encasing the facility in a long-lived material such as concrete, theoretically preventing a release of radioactive material.

Experience

A wide range of nuclear facilities have been decommissioned so far. This includes nuclear power plants (NPPs), research reactors, isotope production plants, particle accelerators, and uranium mines. The number of decommissioned power plants is small. Companies specialize in nuclear decommissioning; decommissioning has become a profitable business. More recently, construction and demolition companies in the UK have also begun to develop nuclear decommissioning services. The current estimate by the United Kingdom's Nuclear Decommissioning Authority is that it will cost at least £100 billion to decommission the 19 existing United Kingdom nuclear sites. Due to the radioactivity in the reactor structure, decommissioning takes place in stages. The plans of the Nuclear Decommissioning Authority for decommissioning reactors have an average 50 year time frame. The long time frame makes reliable cost estimates difficult. Cost overruns are common even for quick projects.

North America

The Pickering Nuclear Generating Station, viewed from the west. All eight reactors are visible; two units have been shut down.

Most nuclear plants currently operating in the United States were designed for a life of about 30–40 years and are licensed to operate for 40 years by the US Nuclear Regulatory Commission. The average age of these reactors is 32 years. Many are coming to the end of their licensing period. If their licenses are not renewed, the plants must go through a decontamination and decommissioning process.

Several nuclear reactors dismantled in North America, type, power, and decommissioning cost:

Dismantled nuclear reactors in Canada & USA					
Country:	**Location:**	**Reactor type:**	**Operative life:**	**Decommissioning phase:**	**Dismantling costs:**
Canada (Québec)	Gentilly-1	CANDU-BWR 250 MWe	180 days (between 1966 and 1973)	"Static state" since 1986	stage two: $25 million
Canada (Ontario)	Pickering NGS Units A2 and A3	CANDU-PWR 8 x 542 MWe	30 years (from 1974 to 2004)	Two units currently in "cold standby" Decommissioning to begin in 2020	(calculated: $270–430/ kWe ?)

United States	Fort St. Vrain	HTGR (helium-graphite) 380 MWe	12 years (1977–1989)	Immediate Decon	$195 million
USA	Rancho Seco	Multiunit: PWR 913 MWe	12 years (Closed after a referendum in 1989)	SAFSTOR: 5–10 years completion 2018	? ($200–500/ kWe)
USA	Three Mile Island 2	Multiunit: 913 MWe PWR	INCIDENT: core fusion (in 1979)	Post-Defuelling Phase 2 (1979)	$805 Million (estimated)
USA	Shippingport	(The first BWR) 60 MWe	25 years (closed in 1989)	Decon completed dismantled in 5 years (first small experimental reactor)	$98.4 million
USA	Piqua(Ohio)	OCM (Organically Cooled/ Moderated) reactor 46 MWe	2 years (closed in 1966)	ENTOMB (coolant design inadequate for neutron flux)	?
USA	Trojan	PWR 1,180 MWe	16 years (Closed in 1993 because nearby to seismic fault)	SAFSTOR (cooling tower demolished in 2006)	?
USA	Yankee Rowe	PWR 185 MWe	31 years (1960–1991)	Decon completed - Demolished (greenfield open to visitors)	$608 million with $8 million per year upkeep
USA	Maine Yankee	PWR 860 MWe	24 years (closed in 1996)	Decon completed - Demolished in 2004 (greenfield open to visitors)	$635 million
USA	Connecticut Yankee	PWR 590 MWe	28 years (closed in 1996)	Decon - Demolished in 2007 (greenfield open to visitors)	$820 million
USA	Exelon - Zion 1 & 2	PWR - Westinghouse 2 x 1040 MWe	25 years (1973–1998) (Incident in proceedings, abandoned because of the excessive cost of vaporizers substitution)	SAFSTOR- EnergySolutions (opening of the site to visitors for 2018)	$900–1,100 million (2007 dollars)

| USA | Pacific Gas & Electric - Humboldt Bay Nuclear Power Plant - Unit 3 | BWR 1 x 63 MWe | 13 years (1963–1976) (Shut down due to seismic retrofit) | On July 2, 1976, Humboldt Bay Power Plant (HBPP) Unit 3 was shut down for annual refueling and to conduct seismic modifications. In 1983, updated economic analyses indicated that restarting Unit 3 would probably not be cost-effective, and in June 1983, PG&E announced its intention to decommission the unit. On July 16, 1985, the U.S. Nuclear Regulatory Commission (NRC) issued Amendment No. 19 to the HBPP Unit 3 Operating License to change the status to possess-but-not-operate, and the plant was placed into a SAFSTOR status. | Unknown - Closure date: 12/31/2015 |

Asia

Reactors Not Located in Japan

Several nuclear reactors dismantled in Asia, type, power and decommissioning cost per kilowatt of electric power (source: World Nuclear Association)

Dismantled reactors in Asia					
Country:	Location:	Reactor type:	Operative life:	Decommissioning phase:	Dismantling cost:
China	Beijing (CIAE)	HWWR 10 MWe (multipurpose) (Heavy Water Experimental Reactor for the production of plutonium and tritium)	49 years (1958–2007)	SAFSTOR & Decon in 20 years (until 2027)	proposed: $6 Million for dismantling $5 Million for fuel remotion
North Korea	Yongbyon	Magnox-type (reactor for the production of nuclear weapons through PUREX treatment)	20 years (1985–2005) Deactivated after a treaty	SAFSTOR: cooling tower dismantled	?

India	Tarapur-1,2 (Maharashtra)	2x BWR 160 MWe	40 years ? (1969–2009?)	NOT deactivated	?
India	Rawatbhata Atomic Power Station-1,2 (Rajasthan)	1x PHWR 100 MWe 1x PHWR 200 MWe (similar to CANDU)	40 years ? (1970–2011?)	NOT deactivated	?
Iraq	Osiraq/Tammuz-1	BWR 40 MWe Nuclear reactor with weapons-grade plutonium production capability	(Destroyed by Israeli Air Force in 1981)	Not radioactive: never refurbished with uranium	?

Japan

The three damaged reactors at Fukushima Dai-ichi #1,#2,#3 are expected to be decommissioned as well as #4.

Heavily damaged reactors in Japan					
Nuclear Power Plant	Electric max. output (MW)	Type	Connection to electric grid	Situation	Decommissioning costs
Fukushima Dai-ichi NPP (Unit 1)	439	BWR	November 17, 1970	nuclear meltdown Hydrogen explosion (INES 7)	? Estimated at ¥10 trillion (US$ 100 billion) for decontaminating Fukushima and dismantling all reactors in Japan and considering long time damage to environment and economy, including agriculture, cattle breeding, fishery, water depuration, tourism (without considering further health care spending & reduction of life expectancy).
Fukushima Dai-ichi NPP (Unit 2)	760	BWR	December 24, 1973	partial nuclear meltdown (INES 6) (Risk of going into INES 7)	?
Fukushima Dai-ichi NPP (Unit 3)	760	BWR	October 26, 1974	nuclear meltdown Hydrogen explosion (INES 7)	?

Fukushima Dai-ichi NPP (Unit 4)	760	BWR	February 24, 1978	Reactor defuelled when tsunami hit Damage to spent fuel cooling-pool (INES 4) (Could worsen if pool collapses) (Other specialists disagree about this danger)	?
Fukushima Dai-ichi NPP (Unit 5)	760	BWR	September 22, 1977	SCRAM	?
Fukushima Daiichi NPP (Unit 6)	1067	BWR	May 4, 1979	SCRAM	?
Fukushima Daini NPP (Unit 1)	1067	BWR	July 31, 1981	SCRAM (leakage of coolant)	?
Fukushima Daini NPP (Units 2 - 4)	3 × 1067	BWR	June 23, 1983 December 14, 1984 December 17, 1986	3 x SCRAM	?
Tokai NPP (Reactor 2)	1100 MW	BWR/5	November 28, 1978	SCRAM Shutdown since March 2011 (anti-tsunami barrier stopped the waves) INES 1 (leakage of coolant)	?

Safely decommissioned reactors in Japan					
Nuclear Power Plant	Electric max. output (MW)	Type	Connection to electric grid	Situation	Decommissioning costs
Tokai NPP (Reactor 1)	160 MWe	Magnox (GCR)	(1966–1998)	Safstore: 10 years then decon until 2018	¥93 billion (€660 million of 2003)

Western Europe and Former Yugoslavia

Several nuclear reactors dismantled in Western Europe, type, power and decommissioning cost per kilowatt of power: European Union Website about Nuclear Decommissioning, World Nuclear Association (reactor building companies), United Kingdom.

Safely decommissioned reactors in Western Europe					
Country:	Location:	Reactor type:	Operative Life:	Decommissioning phase:	Dismantling cost:
Austria (Nuclear Free Country)	Zwentendorf NPP Google Maps	PWR 723 MWe	Never activated due to referendum in 1978	?	?

Belgium	SCK•CEN - BR3, located at Mol, Belgium	PWR (BR-3)	25 years (1962–1987)	Decon completed (2011) European pilot project (underwater cutting and remote operated tools)	?
France	Brennilis	HWGCR 70 MWe	12 years (1967–1979)	Phase 3	€480 million (20 times the forecasted amount)
France	Bugey-1	UNGG Gas cooled, graphite moderator	1972–1994	postponed	?
France	Chinon 1,2,3	Gas-graphite	(1973–1990)	postponed	?
France	Chooz-A	PWR 300 MW	24 years (1967–1991)	Fully decommissioned - Greenfield (Nuclear reactor was located inside a mountain cave)	?
France	Saint-Laurent Nuclear Power Plant	Gas-graphite	1969–1992 50 kg of Uranium in one of the reactors at the Saint-Laurent Nuclear Power Plant began to melt, an event classified at 'level 4' on the International Nuclear Event Scale (INES). As of March 2011, this is the most serious civil nuclear power accident in France.	postponed	?
France	Rapsodie at Cadarache	Experimental Fast breeder nuclear reactor (sodium-cooled) 40 MWe	15 years (1967–1983)	1983: Defuelling 1987: Remotion of neutron reflectors 1985-1989: Decontamination of sodium coolant Accident when cleaning residual sodium in vessel with ethyl carbitol (March 31° 1994)	The removed activity is estimated to around 4800 TBq. 600 TBq (^{60}Co) in 1990 still contained in 1ry vessel. The dose burden from 1987 to 1994 was 224 mSv. RAPSODIE reached IAEA level 2 of decomm in 2005 STAGE 3 is planned in 2020
France	Phénix at Marcoule	Experimental Fast breeder nuclear reactor (sodium-cooled) 233 MWe	36 years (1973–2009)	1) Defuelled	estimated for the future: $4000/kWe

Country	Plant	Type	Operation	Method	Cost
France	Superphénix at Creys-Malville	Fast breeder nuclear reactor (sodium-cooled)	11 years (1985–1996)	1) Defuelled 2) Extraction of Sodium Pipe cutting with a robot	estimated for the future: $4000/kWe
United Kingdom	Berkeley	Magnox (2 x 138 MWe)	27 years (1962–1989)	SAFSTOR: 30 years (internal demolition)	$2600/kWe
United Kingdom	Sellafield-Windscale	Windscale Advanced Gas Reactor WAGR (32 MWe)	18 years (1963–1981)	Remotion of reactor in 2009 pilot project (cutting with remote controlled robots, UV lasers)	More than $2600/kWe (WNI estimates) So far €117 million
United Kingdom	Dorset -Winfrith Operated in Dorset from 1958 to 1990. Had 9 reactors all mostly dismantled				
West Germany	Gundremmingen-A	BWR 250 MWe	11 years	Immediate dismantling pilot project (underwater cutting)	(~ $300–550/kWe)
Italy	Caorso NPP	BWR 840 MWe	3 years (1978 - Closed in 1987 after referendum in 1986)	SAFSTOR: 30 years (internal demolition)	€450 million (dismantling) + €300 million (fuel reprocessing)
Italy	Garigliano NPP (Caserta)	BWR 150 MWe	? (Closed on March 1, 1982)	SAFSTOR: 30 years (internal demolition)	?
Italy	Latina NPP (Foce Verde)	Magnox 210 MWe Gas-graphite	24 years (1962 - Closed in 1986 after referendum)	SAFSTOR: 30 years (internal demolition)	?
Italy	Trino Vercellese NPP	PWR Westinghouse, 270 MWe	? (Closed in 1986 after referendum)	SAFSTOR: 30 years (internal demolition)	?
Netherlands	Dodewaard NPP	BWR Westinghouse 58 MWe	28 years (1969–1997)	Defuelling completed SAFSTOR: 40 years	?
Spain	Vandellós NPP-1	UNGG 480 MWe (gas-graphite)	18 years Incident: fire in a turbo-generator (1989)	SAFSTOR: 30 years (internal demolition)	Phases 1 and 2: €93 million
Sweden	Barsebäck NPP 1 & 2	BWR 2 x 615 MW	Reactor 1: 24 years 1975 - 1999 Reactor 2: 28 years 1977 - 2005	SAFSTOR: demolition will begin in 2020	The Swedish Radiation Safety Authority has assessed that the costs for decommissioning and final disposal for the Swedish nuclear power industry may be underestimated by SKB by at least 11 billion Swedish crowns ($1.63 billion)

Switzerland	DIORIT	MWe Gas-graphite (experimental)	?	SAFSTOR (internal demolition)	?
Switzerland	LUCENS	8,3 MWe CO_2-heavy water (experimental)	(1962–1969) Incident: fire in 1969	Entombment for ? years SAFSTOR & Decon: 24 years (internal demolition)	?
Switzerland	SAPHIR	0,01-0,1 MWe (Light water pool)	39 years (1955–1994) (Experimental demonstrator)	In public display since inauguration	?

- Repository for radioactive waste Morsleben: €2.2 billion

Eastern Europe and Former Soviet Union

Several nuclear reactors dismantled in the nations born from the former Soviet Union: (Belarus, Russia, Ukraine and others) and reactors dismantled in countries formerly belonging to "Warsaw Pact" and/or to "Comecon", type, electric power and decommissioning cost per kilowatt of power: World Nuclear Association, OSTI (Russia & USA).

Decommissioned reactors in Eastern Europe					
Country:	Location:	Reactor typr:	Operative life:	Decommissioning phase:	Dismantling cost:
Bulgaria	Kozloduy NPP-1,2,3,4	PWR VVER-440 (4 x 408 MWe)	Reactors 1,2 closed in 2003, reactors 3,4 closed in 2006 (Closing forced by European Union)	De-fuelling	?
East Germany	Greifswald NPP-1,2,3,4,5,6	VVER-440 5 x 408 MWe	Reactors 1-5 closed in 1989/1990, reactor 6: finished but never operated	Immediate dismantling (underwater cutting)	~ $330/kWe
East Germany	Rheinsberg NPP-1	VVER-210 70–80 MWe	24 years (1966–1990)	In dismantling since 1996 Safstor (underwater cutting)	~ $330/kWe
East Germany	Stendal NPP-1,2,3,4	VVER-1000 (4 x 1000 MWe)	Never activated (1st reactor 85% completed)	Not radioactive (Cooling towers demolished with explosives)	? (Structure in exhibition inside an industrial park)
Russia	Mayak (Chelyabinsk-65)	PUREX plant for uranium enrichment	Several severe incidents (1946–1956)	?	?
Russia	Seversk (Tomsk-7)	Three plutonium reactors Plant for uranium enrichment	Two fast-breeder reactors closed (of three), after disarmaments agreements with USA in 2003.	?	?

Slovakia	Jaslovske Bohunice NPP-1,2 (180 km east from Vienna)	VVER 440/230 2 X 440 MWe	(1978–2006) (1980–2008)	?	?
Ukraine	Chernobyl NPP-4 (110 km from Kiev)	RBMK-1000 1000 MWe	hydrogen explosion, then graphite fire (1986) (INES 7)	ENTOMBMENT (armed concrete "sarcophagus")	Past: ? Future: riding sarcophagus in steel

Legal Aspects

The decommission of a nuclear reactor can only take place after the appropriate licence has been granted pursuant to the relevant legislation. As part of the licensing procedure, various documents, reports and expert opinions have to be written and delivered to the competent authority, e.g. safety report, technical documents and an environmental impact study (EIS).

In the European Union these documents are the basis for the environmental impact assessment (EIA) according to Council Directive 85/337/EEC. A precondition for granting such a licence is an opinion by the European Commission according to Article 37 of the Euratom Treaty. Article 37 obliges every Member State of the European Union to communicate certain data relating to the release of radioactive substances to the Commission. This information must reveal whether and if so what radiological impacts decommissioning – planned disposal and accidental release – will have on the environment, i.e. water, soil or airspace, of the EU Member States. On the basis of these general data, the Commission must be in a position to assess the exposure of reference groups of the population in the nearest neighbouring states.

Cost

In USA many utility estimates now average $325 million per reactor all-up (1998 $).

In France, decommissioning of Brennilis Nuclear Power Plant, a fairly small 70 MW power plant, already cost €480 million (20x the estimate costs) and is still pending after 20 years. Despite the huge investments in securing the dismantlement, radioactive elements such as plutonium, caesium-137 and cobalt-60 leaked out into the surrounding lake.

In the UK, decommissioning of the Windscale Advanced Gas Cooled Reactor (WAGR), a 32 MW prototype power plant, cost €117 million.

In Germany, decommissioning of Niederaichbach nuclear power plant, a 100 MW power plant, amounted to more than €143 million.

New methods for decommissioning have been developed in order to minimize the usual high decommissioning costs. One of these methods is in situ decommissioning (ISD), meaning that the reactor is entombed instead of dismantled. This method was implemented at the U.S. Department of Energy Savannah River Site in South Carolina for the closures of the P and R Reactors. With this tactic, the cost of decommissioning both reactors was $73 million. In comparison, the decommissioning of each reactor using traditional methods would have been an estimated $250 million. This results in a 71% decrease in cost by using ISD.

Decommissioning Funds

In Europe there is considerable concern over the funds necessary to finance final decommissioning. In many countries either the funds do not appear sufficient to cover decommissioning and in other countries decommissioning funds are used for other activities, putting decommissioning at risk, and distorting competition with parties who do not have such funds available.

In 2016 the European Commission assessed that European Union's nuclear decommissioning liabilities were seriously underfunded by about 118 billion euros, with only 150 billion euros of earmarked assets to cover 268 billion euros of expected decommissioning costs covering both dismantling of nuclear plants and storage of radioactive parts and waste. France had the largest shortfall with only 23 billion euros of earmarked assets to cover 74 billion euros of expected costs.

Similar concerns exist in the United States, where the U.S. Nuclear Regulatory Commission has located apparent decommissioning funding assurance shortfalls and requested 18 power plants to address that issue. The decommissioning cost of Small Modular Reactors is expected to be twice as much respect to Large Reactors.

International Collaboration

Organizations that promote the international sharing of information, knowledge, and experiences related to nuclear decommissioning include the International Atomic Energy Agency, the Organization for Economic Co-operation and Development's Nuclear Energy Agency and the European Atomic Energy Community. In addition, an online system called the Deactivation and Decommissioning Knowledge Management Information Tool was developed under the United States Department of Energy and made available to the international community to support the exchange of ideas and information. The goals of international collaboration in nuclear decommissioning are to reduce decommissioning costs and improve worker safety.

Ships, Mobile Reactors, Military Reactors

Many warships and a few civil ships have used nuclear reactors for propulsion. Former Soviet and American warships have been taken out of service and their power plants removed or scuttled. Dismantling of Russian submarines and ships and American submarines and ships is ongoing. Marine power plants are generally smaller than land-based electrical generating stations.

Reactor Protection System

A reactor protection system (RPS) is a set of nuclear safety components in a nuclear power plant designed to safely shut down the reactor and prevent the release of radioactive materials. The system can "trip" automatically (initiating a scram), or it can be tripped by the operators. Trips occurs when the parameters meet or exceed the limit setpoint. A trip of the RPS results in full insertion (by gravity in pressurized water reactors or high-speed injection in boiling water reactors) of all control rods and shutdown of the reactor.

Pressurized Water Reactors

Some of the measured parameters for US pressurized water plants would include:

- "High power", auctioneered between high nuclear power and high differential temperature (delta T) between the inlet and outlet of the reactor vessel (a measure of the thermal power for a given RCS flowrate).
- "High startup rate" (active below 10-4 percent power) at low power levels.
- "High pressurizer pressure"
- "Low reactor coolant flow"
- "Thermal margin / low pressure" (reactor power versus RCS pressure)
- "High containment pressure"
- "Low steam generator level"
- "Low steam generator pressure"
- "Loss of load" (main turbine trip)

Each parameter is measured by independent channels such that actuation of any two channels would result in an automatic SCRAM or reactor shutdown. The system also allows manual actuation by the operator.

Passive Nuclear Safety

Passive nuclear safety is a safety feature of a nuclear reactor that does not require operator actions or electronic feedback in order to shut down safely in the event of a particular type of emergency (usually overheating resulting from a loss of coolant or loss of coolant flow). Such reactors tend to rely more on the engineering of components such that their predicted behaviour according to known laws of physics would slow, rather than accelerate, the nuclear reaction in such circumstances. This is in contrast to older-yet-common reactor designs, where the natural tendency for the reaction was to accelerate rapidly from increased temperatures, such that either electronic feedback or operator triggered intervention was necessary to prevent damage to the reactor. The term "walk away" safety is also used to describe this feature.

Terminology

Terming a reactor 'passively safe' is more a description of the strategy used in maintaining a degree of safety, than it is a description of the level of safety. Whether a reactor employing passive safety systems is to be considered safe or dangerous will depend on the criteria used to evaluate the safety level. This said, modern reactor designs have focused on increasing the amount of passive safety, and thus most passively safe designs incorporate both active and passive safety systems, making them substantially safer than older installations. They can be said to be "relatively safe" compared to previous designs.

Reactor vendors like to call their new generation reactors 'passively safe' but this term is sometimes confused with 'inherently safe' in the public perception. It is very important to understand that there are no 'passively safe' reactors or 'passively safe' systems, only 'passively safe' *components* of safety systems exist. Safety systems are used to maintain control of the plant if it goes outside normal conditions in case of anticipated operational occurrences or accidents, while the control systems are used to operate the plant under normal conditions. Sometimes a system combines both features. Passive safety refers to safety system components, whereas inherent safety refers to control system process regardless of the presence or absence of safety specific subsystems.

As an example of a safety system with 'passively safe' components, let us consider the containment of a nuclear reactor. 'Passively safe' components are the concrete walls and the steel liner, but in order to fulfil its mission active systems have to operate, e.g. valves to ensure the closure of the piping leading outside the containment, feedback of reactor status to external instrumentation and control (I&C) both of which may require external power to function.

The International Atomic Energy Agency (IAEA) classifies the degree of "passive safety" of components from category A to D depending on what the system does not make use of:

1. no moving working fluid

2. no moving mechanical part

3. no signal inputs of 'intelligence'

4. no external power input or forces

In category A (1+2+3+4) is the fuel cladding, the protective and nonreactive outer layer of the fuel pellet, which uses none of the above features: It is always closed and keeps the fuel and the fission products inside and is not open before arriving at the reprocessing plant. In category B (2+3+4) is the surge line, which connects the hot leg with the pressurizer and helps to control the pressure in the primary loop of a PWR and uses a moving working fluid when fulfilling its mission. In category C (3+4) is the accumulator, which does not need signal input of 'intelligence' or external power. Once the pressure in the primary circuit drops below the set point of the spring-loaded accumulator valves, the valves open and water is injected into the primary circuit by compressed nitrogen. In category D (4 only) is the SCRAM which utilizes moving working fluids, moving mechanical parts and signal inputs of 'intelligence' but not external power or forces: the control rods drop driven by gravity once they have been released from their magnetic clamp. But nuclear safety engineering is never that simple: Once released the rod may not fulfil its mission: It may get stuck due to earthquake conditions or due to deformed core structures. This shows that though it is a passively safe system and has been properly actuated, it may not fulfil its mission. Nuclear engineers have taken this into consideration: Typically only a part of the rods dropped are necessary to shut down the reactor. Samples of safety systems with passive safety components can be found in almost all nuclear power stations: the containment, hydro-accumulators in PWRs or pressure suppression systems in BWRs.

In most texts on 'passively safe' components in next generation reactors, the key issue is that no pumps are needed to fulfil the mission of a safety system and that all active components (generally I&C and valves) of the systems work with the electric power from batteries.

IAEA explicitly uses the following caveat:

... passivity is not synonymous with reliability or availability, even less with assured adequacy of the safety feature, though several factors potentially adverse to performance can be more easily counteracted through passive design (public perception). On the other hand active designs employing variable controls permit much more precise accomplishment of safety functions; this may be particularly desirable under accident management conditions.

Nuclear reactor response properties such as Temperature coefficient of reactivity and Void coefficient of reactivity usually refer to the thermodynamic and phase-change response of the neutron moderator heat transfer *process* respectively. Reactors whose heat transfer process has the operational property of a negative void coefficient of reactivity are said to possess an *inherent safety* process feature. An operational failure mode could potentially alter the process to render such a reactor unsafe.

Reactors could be fitted with a hydraulic safety system component that increases the inflow pressure of coolant (esp. water) in response to increased outflow pressure of the moderator and coolant without control system intervention. Such reactors would be described as fitted with such a *passive safety* component that could - if so designed - render in a reactor a negative void coefficient of reactivity, regardless of the operational property of the reactor in which it is fitted. The feature would only work if it responded faster than an emerging (steam) void and the reactor components could sustain the increased coolant pressure. A reactor fitted with both safety features - if designed to constructively interact - is an example of a safety interlock. Rarer operational failure modes could render both such safety features useless and detract from the overall relative safety of the reactor.

Examples of Passive Safety in Operation

Traditional reactor safety systems are *active* in the sense that they involve electrical or mechanical operation on command systems (e.g., high-pressure water pumps). But some engineered reactor systems operate entirely passively, e.g., using pressure relief valves to manage overpressure. Parallel redundant systems are still required. Combined *inherent* and *passive* safety depends only on physical phenomena such as pressure differentials, convection, gravity or the *natural* response of materials to high temperatures to slow or shut down the reaction, not on the functioning of engineered components such as high-pressure water pumps.

Current pressurized water reactors and boiling water reactors are systems that have been designed with one kind of passive safety feature. In the event of an excessive-power condition, as the water in the nuclear reactor core boils, pockets of steam are formed. These steam voids moderate fewer neutrons, causing the power level inside the reactor to lower. The BORAX experiments and the SL-1 meltdown accident proved this principle.

A reactor design whose *inherently* safe process directly provides a *passive* safety component during a specific failure condition in *all* operational modes is typically described as relatively fail-safe to that failure condition. However most current water-cooled and -moderated reactors, when scrammed, can not remove residual production and decay heat without either process heat transfer or the active cooling system. In other words, whilst the inherently safe heat transfer process provides a passive safety component preventing excessive heat while the reactor is operating, the same inherently safe heat transfer process *does not* provide a passive safety component is shut down (SCRAMed). The Three Mile Island accident exposed this design deficiency: the reactor and steam generator were shut down but with loss of coolant it still suffered a partial meltdown.

Third generation designs improve on early designs by incorporating passive or inherent safety features which require *no* active controls or (human) operational intervention to avoid accidents in the event of malfunction, and may rely on pressure differentials, gravity, natural convection, or the natural response of materials to high temperatures.

In some designs the core of a fast breeder reactor is immersed into a pool of liquid metal. If the reactor overheats, thermal expansion of the metallic fuel and cladding causes more neutrons to escape the core, and the nuclear chain reaction can no longer be sustained. The large mass of liquid metal also acts as a heatsink capable of absorbing the decay heat from the core, even if the normal cooling systems would fail.

The pebble bed reactor is an example of a reactor exhibiting an inherently safe process that is also capable of providing a passive safety component for all operational modes. As the temperature of the *fuel* rises, Doppler broadening increases the probability that neutrons are captured by U-238 atoms. This reduces the chance that the neutrons are captured by U-235 atoms and initiate fission, thus reducing the reactor's power output and placing an inherent upper limit on the temperature of the fuel. The geometry and design of the fuel pebbles provides an important passive safety component.

Single fluid fluoride molten salt reactors feature fissile, fertile and actinide radioisotopes in molecular bonds with the fluoride coolant. The molecular bonds provide a passive safety feature in that a loss-of-coolant event corresponds with a loss-of-fuel event. The molten fluoride fuel can not itself reach criticality but only reaches criticality by the addition of a neutron reflector such as pyrolytic graphite. The higher density of the fuel along with additional lower density FLiBe fluoride coolant without fuel provides a flotation layer passive safety component in which lower density graphite that breaks off control rods or an immersion matrix during mechanical failure does not induce criticality. Gravity driven drainage of reactor liquids provides a passive safety component.

Low power pool-type reactors such as the SLOWPOKE and TRIGA have been licensed for *unattended* operation in research environments because as the temperature of the low-enriched (19.75% U-235) uranium alloy hydride fuel rises, the molecular bound hydrogen in the fuel cause the heat to be transferred to the fission neutrons as they are ejected. This Doppler shifting or spectrum hardening dissipates heat from the fuel more rapidly throughout the pool the higher the fuel temperature increases ensuring rapid cooling of fuel whilst maintaining a much lower water temperature than the fuel. Prompt, self-dispersing, high efficiency hydrogen-neutron heat transfer rather than inefficient radionuclide-water heat transfer ensures the fuel cannot melt through accident alone. In uranium-zirconium alloy hydride variants, the fuel itself is also chemically corrosion resistant ensuring a sustainable safety performance of the fuel molecules throughout their lifetime. A large expanse of water and the concrete surround provided by the pool for high energy neutrons to penetrate ensures the process has a high degree of intrinsic safety. The core is visible through the pool and verification measurements can be made directly on the core fuel elements facilitating total surveillance and providing nuclear non-proliferation safety. Both the fuel molecules themselves and the open expanse of the pool are passive safety components. Quality implementations of these designs are arguably the safest nuclear reactors.

Examples of Reactors Using Passive Safety Features

Three Mile Island Unit 2 was unable to contain about 480 PBq of radioactive noble gases from release into the environment and around 120 kL of radioactive contaminated cooling water from release beyond the containment into a neighbouring building. The pilot-operated relief valve at TMI-2 was designed to shut automatically after relieving excessive pressure inside the reactor into a quench tank. However the valve mechanically failed causing the PORV quench tank to fill, and the relief diaphragm to eventually rupture into the containment building. The containment building sump pumps automatically pumped the contaminated water outside the containment building. Both a working PORV with quench tank and separately the containment building with sump provided two layers of passive safety. An unreliable PORV negated its designed passive safety. The plant design featured only a single open/close indicator based on the status of its solenoid actuator, instead of a separate indicator of the PORV's actual position. This rendered the mechanical reliability of the PORV indeterminate directly, and therefore its passive safety status indeterminate. The automatic sump pumps and/or insufficient containment sump capacity negated the containment building designed passive safety.

The notorious RBMK graphite moderated, water-cooled reactors of Chernobyl Power Plant disaster were designed with a positive void coefficient with boron control rods on electromagnetic grapples for reaction speed control. To the degree that the control systems were reliable, this *design* did have a corresponding degree of *active* inherent safety. The reactor was unsafe at low power levels because erroneous control rod movement would have a counter-intuitively magnified effect. Chernobyl Reactor 4 was built instead with manual crane driven boron control rods that were tipped with the moderator substance, graphite, a neutron reflector. It was designed with an Emergency Core Cooling System (ECCS) that depended on either grid power or the backup Diesel generator to be operating. The ECCS safety component was decidedly not passive. The design featured a partial containment consisting of a concrete slab above and below the reactor - with pipes and rods penetrating, an inert gas filled metal vessel to keep oxygen away from the water-cooled hot graphite, a fire-proof roof, and the pipes below the vessel sealed in secondary water filled boxes. The roof, metal vessel, concrete slabs and water boxes are examples of passive safety components. The roof in the Chernobyl Power Plant complex was made of bitumen - against design - rendering it ignitable. Unlike the Three Mile Island accident, neither the concrete slabs nor the metal vessel could contain a steam, graphite and oxygen driven hydrogen explosion. The water boxes could not sustain high pressure failure of the pipes. The passive safety components as designed were inadequate to fulfill the safety requirements of the system.

The General Electric Company ESBWR (Economic Simplified Boiling Water Reactor, a BWR) is a design reported to use passive safety components. In the event of coolant loss, no operator action is required for three days.

The Westinghouse AP1000 ("AP" standing for "Advanced Passive") uses passive safety components. In the event of an accident, no operator action is required for 72 hours.[11] Recent version of the Russian VVER have added a passive heat removal system to the existing active systems, utilising a cooling system and water tanks built on top of the containment dome.[12]

The integral fast reactor was a fast breeder reactor run by the Argonne National Laboratory. It was a sodium cooled reactor capable of withstanding a loss of (coolant) flow without SCRAM and loss of heatsink without SCRAM. This was demonstrated throughout a series of safety tests in which the reactor successfully shut down without operator intervention. The project was canceled due to proliferation concerns before it could be copied elsewhere.

The Molten-Salt Reactor Experiment[13] (MSRE) was a molten salt reactor run by the Oak Ridge National Laboratory. It was nuclear graphite moderated and the coolant salt used was FLiBe, which also carried the uranium-233 fluoride fuel dissolved in it. The MSRE had a negative temperature coefficient of reactivity: as the FLiBe temperature increased, it expanded, along with the uranium ions it carried; this decreased density resulted in a reduction of fissile material in the core, which decreased the rate of fission. With less heat input, the net result was that the reactor would cool. Extending from the bottom of the reactor core was a pipe that lead to passively cooled drain tanks. The pipe had a "freeze valve" along its length, in which the molten salt was actively cooled to a solid plug by a fan blowing air over the pipe. If the reactor vessel developed excessive heat or lost electric power to the air cooling, the plug would melt; the FLiBe would be pulled out of the reactor core by gravity into dump tanks, and criticality would cease as the salt lost contact with the graphite moderator.

The General Atomics HTGR design features a fully passive and inherently safe decay heat removal system, termed the Reactor Cavity Cooling System (RCCS). In this design, an array of steel ducts line the concrete containment (and hence surround the reactor pressure vessel) which provide a flow path for air driven natural circulation from chimneys positioned above grade. Derivatives of this RCCS concept (with either air or water as the working fluid) has also been featured in other gas-cooled reactor designs, including the Japanese HTTR, the Chinese HTR-10, the South African PBMR, and the Russian GT-MHR. While none of these designs have been commercialized for power generation research in these areas is active, specifically in support of the Generation IV initiative and NGNP programs, with experimental facilities at Argonne National Laboratory (home to the Natural convection Shutdown heat removal Test Facility, a 1/2 scale air-cooled RCCS)[14] and the University of Wisconsin (home to separate 1/4 scale air and water-cooled RCCS),.[15][16]

References

- Ojovan, M. I.; Lee, W.E. (2005). An Introduction to Nuclear Waste Immobilisation. Amsterdam: Elsevier Science Publishers. p. 315. ISBN 0-08-044462-8.

- National Research Council (1995). Technical Bases for Yucca Mountain Standards. Washington, D.C.: National Academy Press. p. 91. ISBN 0-309-05289-0.

- Lovins, Amory B. and Price, John H. (1975). Non-nuclear Futures: The Case for an Ethical Energy Strategy (Cambridge, Mass.: Ballinger Publishing Company, 1975. xxxii + 223pp. ISBN 0-88410-602-0, ISBN 0-88410-603-9).

- Kagarlitsky, Boris (1989). "Perestroika: The Dialectic of Change". In Mary Kaldor; Gerald Holden; Richard A. Falk. The New Detente: Rethinking East-West Relations. United Nations University Press. ISBN 0-86091-962-5.

- Hallenbeck, William H (1994). Radiation Protection. CRC Press. p. 15. ISBN 0-87371-996-4. Reported thus far are 237 cases of acute radiation sickness and 31 deaths.

- "Paul Scherrer Institut (PSI) :: Severe Accidents in the Energy Sector (see pages 287,310,317)" (PDF). gabe. web.psi.ch. Retrieved 2015-02-07.

- House of Commons Committee of Public Accounts (4 February 2013). "Nuclear Decommissioning Authority: Managing risk at Sellafield" (PDF). London: The Stationery Office Limited. Retrieved 2 Dec 2013.

- "How old are U.S. nuclear power plants and when was the last one built? - FAQ - U.S. Energy Information Administration (EIA)". Eia.gov. Retrieved 2013-09-06.

- "Hanford Site: Hanford Overview". United States Department of Energy. Archived from the original on 2012-06-05. Retrieved 2012-02-13.

- Hiroko Tabuchi (March 3, 2012). "Japanese Prime Minister Says Government Shares Blame for Nuclear Disaster". The New York Times. Retrieved 2012-04-13.

- "Nuclear Power: During normal operations, do commercial nuclear power plants release radioactive material?". Radiation and Nuclear Power | Radiation Information and Answers. Radiation Answers. Retrieved March 12, 2012.

- "Radiation Dose". Factsheets & FAQs: Radiation in Everyday Life. International Atomic Energy Agency (IAEA). Retrieved March 12, 2012.

- "What happens to radiation produced by a plant?". NRC: Frequently Asked Questions (FAQ) About Radiation Protection. Nuclear Regulatory Commission. Retrieved March 12, 2012.

- "Is radiation exposure from a nuclear power plant always fatal?". NRC: Frequently Asked Questions (FAQ) About Radiation Protection. Nuclear Regulatory Commission. Retrieved March 12, 2012.

- "What you can do to protect yourself: Be Informed". Nuclear Power Plants | RadTown USA | US EPA. United States Environmental Protection Agency. Retrieved March 12, 2012.

Permissions

Index

www.ingramcontent.com/pod-product-compliance
Lightning Source LLC
Chambersburg PA
CBHW061319190326
41458CB00011B/3841